Aging
The Longevity Dividend

A subject collection from *Cold Spring Harbor Perspectives in Medicine*

Aging
The Longevity Dividend

A subject collection from *Cold Spring Harbor Perspectives in Medicine*

EDITED BY

S. Jay Olshansky
University of Illinois at Chicago

George M. Martin
University of Washington

James L. Kirkland
Mayo Clinic

COLD SPRING HARBOR LABORATORY PRESS
Cold Spring Harbor, New York • www.cshlpress.org

Aging: The Longevity Dividend

A subject collection from *Cold Spring Harbor Perspectives in Medicine*
Articles online at www.perspectivesinmedicine.org

Executive Editor	Richard Sever
Managing Editor	Maria Smit
Senior Project Manager	Barbara Acosta
Permissions Administrator	Carol Brown
Production Editor	Diane Schubach
Production Manager/Cover Designer	Denise Weiss
Publisher	John Inglis

Front cover artwork: This illustration exemplifies the mechanical nature of the human body; the fact that there is no aging or death clock, which means that aging is inherently modifiable; and the excitement for the future of aging science as many in the field now recognize that a therapeutic intervention to modulate aging may be in sight. (Artwork © 2002 J.W. Stewart, used with permission.)

Library of Congress Cataloging-in-Publication Data

Aging : the longevity dividend / edited by S. Jay Olshansky, University of Illinois at Chicago, George M. Martin, University of Washington, and James L. Kirkland, Mayo Clinic.
    pages cm
  Summary: "Aging affects us all and is characterized not only by increasing frailty but by increased susceptibility to conditions such as Alzheimer's, cardiovascular disease, and cancer. We are gaining an increasing understanding of the molecular mechanisms underlying aging, however, and uncovering clues to how life may be prolonged. This book examines the biological basis of aging and research into strategies that may extend lifespan"– Provided by publisher.
  ISBN 978-1-62182-080-2 (hardback); 978-1-62182-163-2 (paper)
  1. Aging–Physiological aspects. 2. Longevity–Physiological aspects. 3. Aging–Molecular aspects. I. Olshansky, Stuart Jay, 1954- II. Martin, George M., 1927- III. Kirkland, James (James L.)

QP86.A3757 2015
612.6'7--dc23
                                2015019164

All World Wide Web addresses are accurate to the best of our knowledge at the time of printing.

For a complete catalog of all Cold Spring Harbor Laboratory Press publications, visit our website at www.cshlpress.org.

Contents

Contents

Foreword

WITH THE GENERAL REALIZATION THAT THE population of our planet is rapidly becoming older, econ-omists, population health experts, epidemiologists, policy planners, physicians, scientists, and others have started considering implications of this "silver tsunami" for the society. At the level of physiological functioning and health maintenance in old age, it became apparent that this increase in longevity will be accompanied by multiple comorbidities in a significant proportion of the older population (Goldman et al. 2013). In many countries, particularly in the United States and others where the currently prevalent health-care delivery model is pay-for-service, rather than the more goal-oriented and holistic pay-for-outcome/health approach, this can be equated with a signifi-cantly higher future cost to care for an increasingly dependent, disabled, and unhealthy segment of the population. More importantly, in the current U.S. health-care delivery model, the care does not necessarily result in improved health posttreatment. The result is an inefficient and hugely expensive model of care for older adults.

Therefore, developing strategies to maintain optimal health in an increasingly aging population is becoming a global strategic imperative. This book represents the latest concerted and broad effort to shine a light on the potential of biology of aging research to implement what could be a revolutionary change in improving the health span of older adults. It is a culmination of more than 25 years of effort by the editors of this volume (Olshansky et al. 1990, 2007; Butler et al. 2008) to draw broader atten-tion to this issue. Specifically, beginning in 1990, Olshansky and colleagues, many of them co-authors of this volume, started to make a powerful case for why and how research into life-span and health-span extension, as part of the larger field of biology of aging, may have a unique potential to provide broad and far-reaching benefits to the aging human population. The effort, aptly named the Longevity Dividend Initiative, has already made substantial headway in the larger community of researchers and is beginning to extend to policy makers and society overall. This book is part of a groundswell of recent activities (reviewed in the chapter by Sierra and the chapters by Olshansky; also, see Nikolich-Žugich et al. 2015) to help scientists and public advocates of science reach the tipping point and bring about a coherent, conceptually innovative, scientifically based, and publicly as well as industry-supported and sponsored strategy to deal with health issues central to older adults from a revolutionary standpoint.

The argument for the Longevity Dividend is that the payback to society and individuals from extending health span via fundamental interventions based on knowledge of biology of aging will be considerable and broad. This case is made in great detail throughout the volume, but summarized best in the introductory article by Felipe Sierra and the two articles by S. Jay Olshansky. This argu-ment, in its entirety, seems intuitively appealing to the point of being a "no-brainer": The current approaches to treating age-related diseases that produce the highest morbidity and mortality in the older adult population are only incrementally effective at increasing life span and minimally effective in increasing health span, defined as the fraction of life span spent in good health and pros-perity. In fact, curing all cancers, for example, although desirable, merely replaces cancer with other chronic morbidities such as Alzheimer's, cardiovascular diseases, metabolic diseases, and so on (Olshanky et al. 1990; Miller 2002). By contrast, in numerous laboratory animal models, including some studies in nonhuman primates and humans, interventions based on manipulations of nutrient sensing and cellular metabolism have shown not only longevity extension but also significant post-ponement of multiple age-related diseases (including cancer, Alzheimer's, cardiovascular, and meta-bolic diseases). This, therefore, is close to, or achieves, health-span extension.

The promise of translating these interventions to human subjects, then, starkly contrasts with current, disease-specific research and treatment approach. Simply put, the choice would come down to the two extremes: (1) the current health-care approach, with most individuals enjoying a relatively long life span but reduced health span with multiple comorbidities and increased, ballooning health-care costs; or (2) the biology-of-aging-based health-span extension, which, if successfully translated to humans, would provide increased health span at a fraction of today's health-care cost, with a vigorous and engaged older adult population and even a potentially productive older workforce. One of the key strengths of this book is its further exploration of demographic and economic consequences of biological interventions aimed at modulating the aging process (see the chapters by Goldman and Beltrán-Sánchez et al.). The data provided in these chapters represent further, powerful arguments for biological modulation of the aging process. Other important chapters in the book discuss basic, fundamental physiological levers one could manipulate to modulate aging and increase health span (Part I of the book).

At present, two key issues stand in the way of broad application of health-span extension to humans. First, we are still not at the point of having applications that are distribution-ready. In that regard, the book provides welcome insights into possible translation of robust findings from model organism into clinical practice (see the chapter by Kirkland) and the review of the exciting, burgeoning literature on pharmacological intervention to extend life span and health span (see the chapters by Novelle et al. and Milman and Barzilai).

Second, serious additional roadblocks exist to implementation, including the omnipresent lack of funding for research and, even more so, advanced-stage clinical testing. There also remain ingrained views in society that aging is immutable and/or that intervening in the aging process will produce deleterious and unwanted consequences such as further overpopulation and shortages of resources (reviewed by Miller 2002). Various chapters in this volume go a long way toward debunking many of these myths and making a strong case for the longevity/health span dividend. I applaud the authors on an excellent job of reviewing the known facts and discussing future needs and actions and the editors for their vision to organize and compile an extremely timely set of contributions from the prominent, respected thought leaders in the field.

Janko Nikolich-Žugich, M.D., Ph.D.
Chairman of the Board and CEO, American Aging Association
Bowman Professor and Head, Department of Immunobiology
Co-Director of the Arizona Center on Aging
University of Arizona College of Medicine

REFERENCES

Butler RN, Miller RA, Perry D, Carnes BA, Williams TF, Cassel C, Brody J, Bernard MA, Partridge L, Kirkwood T, et al. 2008. New model of health promotion and disease prevention for the 21st century. *BMJ* 337: a399.

Goldman DP, Cutler D, Rowe JW, Michaud PC, Sullivan J, Peneva D, and Olshansky SJ. 2013. Substantial health and economic returns from delayed aging may warrant a new focus for medical research. *Health Aff (Millwood)* 32: 1698–1705.

Miller RA. 2002. Extending life: Scientific prospects and political obstacles. *Millbank Q* 80: 155–174.

Nikolich-Žugich J, Goldman DP, Cohen PR, Cortese D, Fontana L, Kennedy BK, Mohler MJ, Olshansky SJ, Perls T, Perry D, et al. 2015. Preparing for an aging world: Engaging biogerontologists, geriatricians and the society. *J Gerontol A Biol Sci Med Sci* doi:10.1093/gerona/glv164.

Olshansky SJ, Carnes BA, and Cassel C. 1990. In search of Methuselah: Estimating the upper limits to human longevity. *Science* 250: 634–640.

Olshansky SJ, Perry D, Miller RA, and Butler RN. 2007. Pursuing the longevity dividend: Scientific goals for an aging world. *Ann NY Acad Sci* 1114: 11–13.

Preface

THE INCREASE IN HUMAN LONGEVITY during the last century was one of humanity's most remarkable medical and technological accomplishments. As valuable as oil, gold, diamonds, fresh water, and clean air may seem, life itself is likely to be our most precious commodity—and we managed to manufacture more of it during the last 150 years than during all of humanity's existence prior to the 19th century. At first blush it would therefore seem like a rhetorical question whether we should be trying to extend life even further. After all, it is easy to justify almost all facets of modern medicine and public health as desirable. The answer to this question is no longer simple. A growing faction of scientists are concerned that further life extension in long-lived populations may extend the period of frailty and disability later in life as the biological processes of aging emerge as the most important risk factor for what ails us as we grow older. Living longer sounds good at one level, until it becomes clear that for people who already are expected to live long lives, it is health extension we should be pursuing, not life extension.

This is not a new argument. In the late 1970s, Dr. Bernice Neugarten and Dr. Robert Havighurst from the University of Chicago organized a meeting sponsored by the National Science Foundation in which they set out to answer the question on whether governments should be in the business of making people live longer. Neugarten and her coauthors identified two ways in which life extension could be accomplished: through continuing efforts to conquer disease (referred to as "disease control"), and through an effort to identify the intrinsic biological processes that are thought to underlie aging and that proceed independently from disease processes—that is, to discover the genetic and biochemical secrets of aging, then to alter them (referred to as "rate control"). Amid overly optimistic views by some suggesting that dramatic increases in longevity were forthcoming, Dr. Nathan Shock and others made it clear that life extension is not a legitimate goal of aging science, and that instead we should focus on making the years that we have good years.

Until recently, the idea that aging could be modified was little more than wishful thinking, but enough evidence has emerged just within the last decade to justify the pursuit of "rate control" as a new method of attacking diseases. Modern versions of the rate control idea emerged beginning in 2006 under the banner of the Longevity Dividend Initiative, and more recently, this has been called Geroscience. Scientists are now routinely arguing that modifying aging or its consequences represents perhaps the best opportunity to achieve the primary prevention of both fatal and disabling diseases among long-lived populations.

So, how exactly will this be accomplished? What are the various pathways that scientists in the fields of aging are pursuing to bring about this vision of rate control, morbidity compression, disease reduction, and the extension of healthy life? That is what this book is all about. We set out to put into one place a description of most of the major projects now underway or about to be pursued to achieve this end. Although it must be acknowledged that there are enticing approaches to aging science that are not described in the pages of this book, we are confident that many of the most interesting opportunities are described here.

This movement of aging science in the direction of a major public health intervention has gained significant momentum in recent years, and we expect this book will advance the cause and explain both the rationale for doing so and the consequences of failure. The longer lives we enjoy are a wondrous accomplishment to be sure, but we have placed ourselves in a precarious position by allowing aging to rear its ugly head with increasing frequency and duration. We may very well be on the precipice of a new public health movement—the seeds of which are most likely to be sown by the authors of this book and their colleagues across the globe now working in this exciting field of aging science.

On behalf of the editors I would like to thank Barbara Acosta, Richard Sever, and their colleagues at Cold Spring Harbor Laboratory Press for their excellent work in helping us to organize this book and for their patience in herding together scientists from a broad range of disciplines—all of whom are busy making history.

S. Jay Olshansky

The Emergence of Geroscience as an Interdisciplinary Approach to the Enhancement of Health Span and Life Span

Felipe Sierra

Division of Aging Biology, National Institute on Aging, National Institutes of Health, Bethesda, Maryland 20892

Correspondence: sierraf@nia.nih.gov

Research on the biology of aging has accelerated rapidly in the last two decades. It is now at the point where translation of the findings into useful approaches to improve the health of the elderly population seems possible. In trying to fill that gap, a new field termed geroscience will be articulated here that attempts to identify the biological underpinnings for the age dependency of most chronic diseases. Herein, I will review the major conceptual issues leading to the formulation of geroscience as a field, as well as give examples of current areas of inquiry in which basic aging biology research could lead to therapeutic approaches to address age-related chronic diseases, not one at a time, but most of them in unison.

The field of aging biology has exploded in the last few decades, with the initial focus on descriptive work that catalogued the myriad changes that occur during aging, first leading to a highly mechanistic phase in which the major molecular and cellular determinants of the process were identified, and to the current stage in which, without neglecting the still very unfinished mechanistic and discovery work, some of the findings are poised for possible application in humans. An interesting outcome of the descriptive work was the realization that not all age-related changes are necessarily bad for the organism. Although some phenomena appear indeed to be at least partially responsible for increasing the risk for age-related disease (e.g., the decrease in proteostasis leading to neurodegenerative diseases characterized by accumulation of misfolded proteins), others are neutral (cosmetic changes like hair graying) and, in fact, some of the changes appear to be adaptive to other changes occurring with age and as such, they might be beneficial to the health of the organism (changes in some hormones, e.g., such as possibly testosterone or insulin-like growth factor [IGF]) (Rincon et al. 2005; Corona et al. 2013; Matsumoto 2013). Other changes are the result of pathology and are therefore independent of the aging process per se, yet they are difficult to separate in the case of highly prevalent diseases and conditions. In that sense, the definition by Harrison is appropriate: "aging is what occurs to all individuals of a given species, while disease occurs to only a proportion of them" (Flurkey et al. 2007).

Although some fields, such as caloric restriction, cell senescence, and the free radical hypothesis, were potent initial drivers of

Cite this article as *Cold Spring Harb Perspect Med* doi: 10.1101/cshperspect.a025163

research into the biology of aging, the main transformative research leading to the current status was the genetic work initially encouraged by the National Institute of Agings's (NIA) Longevity Assurance Genes Initiative (LAG). Although presently we know of several hundred genes that, when modified, can increase life span in animal models (Kenyon 2010; genomics .senescence.info/genes/stats.php), and some variants of these genes have been identified in long-lived humans (Pawlikowska et al. 2009; Slagboom et al. 2011; Wheeler and Kim 2011; Milman et al. 2014), in the late 1990s there was widespread skepticism that even a single gene would be found, despite common recognition of the partially inheritable nature of longevity. The finding of molecular drivers of the process brought aging biology research into the mainstream and has resulted in the current renaissance of the field. The historical events leading to the current state of affairs have been reviewed previously and will not be repeated here (Warner 2005). Rather, I will focus on a discussion of the origins of geroscience and the importance of studying aging at the most basic biological level. I will finish with some reference to the main current areas of research, as identified during a recent summit organized by the trans-NIH (National Institutes of Health) GeroScience Interest Group (biomedgerontology.oxfordjournals.org/content/69/Suppl_1.toc).

GEROSCIENCE

The quest for eternal youth is as old as humanity itself. In fact, we have known for centuries that life span (and health span) can be extended, within limits, simply by adopting moderate changes in lifestyle, including diet and exercise. Unfortunately, this is easier said than done. Although public policy has shown that it is actually possible to change most people's behaviors when they understand it is in their own interest (seat belts, smoking, and laying babies in their backs represent successful recent examples), reversing behaviors that include both quantitatively and qualitatively unhealthful habits concerning diet and exercise is proving problematic for most people. For example, we know

that in many laboratory animals, substantially reducing caloric intake extends life span and improves health in old age. Yet, few people would subject themselves to the harshness of that regime, and the entire area of dietary restriction (DR) is more suitable for experimental investigations than as a practical approach to human health.

The need to address the issues posed by the graying of the world's population is urgent. The dramatic increase in the proportion of people aged 65 or older (and the even more dramatic increase in those 85 and older, including centenarians) poses challenges that as a species we are not yet equipped to handle. Furthermore, neither our health care systems, nor the economy or the societal system will be able to sustain this unprecedented increase in the proportion of elders in the human population (Bhattacharya et al. 2004). In addition to the obvious need for more properly trained geriatricians and social workers, there is also a need to better understand the biology driving the aging process. Epidemiological studies suggest that aging might be the major risk factor for most age-related chronic diseases (Niccoli and Partridge 2012). Experimental manipulation of the rate of aging in model organisms has borne that observation; increasing life span, which is assumed to either delay or slow down the process of aging, does indeed lead to a delay and softening of the diseases that normally accompany old age (Baur et al. 2006; Fernández and Fraga 2011; Wilkinson et al. 2012). Although these recent advances have made the idea of addressing the role of aging as the major risk factor for most chronic diseases closer to being feasible, the idea itself is not new. Indeed, the germ of the idea arose several decades ago, with the main concepts appearing as early as 1977 in a publication by Neugarten and Havighurst (1977). That idea was then further reformulated in 2006 in the form of the longevity dividend (Olshansky et al. 2006, 2007). Yet, recent advances have given the concept a more urgent and attainable form, exemplified by the current interest in the new field of geroscience, "an interdisciplinary field that aims to understand the relationship between aging and age-related

Cite this article as *Cold Spring Harb Perspect Med* doi: 10.1101/cshperspect.a025163

diseases and disabilities" (en.wikipedia.org/wiki/Geroscience).

The basic principles governing geroscience are simple: the ultimate goal of biomedical research is to improve the quality of life in humans, and because chronic diseases and conditions of the elderly represent the main hurdle toward reaching that goal, it follows that addressing these diseases should be—and is indeed—a priority. The twist comes with the concept that, because aging is malleable (at least in many animal models) and aging is also the main risk factor for those diseases and conditions, then addressing the basic biology of aging is likely to provide a better payoff than addressing diseases one at a time, as it is often done currently (Fig. 1). There are two major conceptual issues at play here. First is the complex nature of chronic diseases. Unlike infectious diseases or genetic disorders, chronic diseases of aging are multifactorial and complex. Although much research is devoted to fighting disease-specific risk factors (e.g., cholesterol and obesity for cardiovascular disease, amyloid β (Aβ) for Alzheimer's, glucose handling for diabetes), there is a growing realization that, for those disease-specific risk factors to lead to disease, additional elements also need to be present, including both environmental factors and what has been called a "receptive environment," more often than not provided by age itself (Krtolica and Campisi 2003). As an additional example, unless it is driven by a congenital abnormality, cancer usually does not affect humans until their sixth or seventh decade, yet it occurs frequently in most mice by the age of 2 years; that is, in both species, cancer strikes most frequently when individuals are at ∼2/3 of their expected life span. This difference could be driven, in part, by differences in repair rates (Kumar and Subramanian 2002; Gorbunova et al. 2014), and it has been shown that the rate of DNA repair changes with age (Gorbunova et al. 2007; Vaidya et al. 2014). These are still heavily debated topics (Promislow 1994; Jensen-Seaman et al. 2004). In addition, however, there is a striking commonality in that in both cases, cancer strikes when individuals are well past their midlife points. This suggests that for cancer to develop into a clinically recognizable disease, those mutated cells need to be in a "receptive environment."

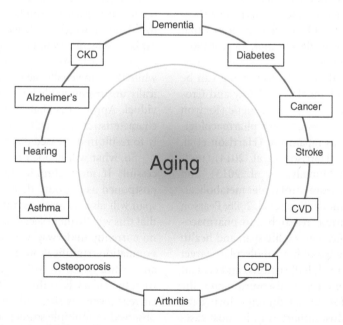

Figure 1. Aging is the major risk factor for most chronic diseases. CVD, Cardiovascular disease; COPD, chronic obstructive pulmonary disease; CKD, chronic kidney disease.

A separate issue that makes geroscience timely is the fact of comorbidities, and the inadequacy of the current model of addressing one disease at a time. Much current research in biomedicine aims at preventing, curing, or managing one disease at a time (hence, the NIH has a cancer institute, a diabetes institute, etc., and this model is emulated across the world). In addition, there is an entrenched operating procedure in clinical trials, whereby patients suffering from diseases other than the one under study are not included (and often, neither are older patients). Yet, the diseases we are trying to address seldom occur in isolation, and seldom in patients of the ages studied. By embracing the fact that aging biology facilitates multiple comorbidities, geroscience aims at reversing these trends and putting the emphasis on preventing, curing, or delaying not one, but all chronic diseases at once. In doing so, geroscience intends to shift the focus away from specific diseases and their role in mortality, toward an assessment of overall health in which the individual's physiology will determine his/her overall risk for chronic diseases, irrespective of which disease is more likely to affect the individual based on genetic or environmental factors.

As will be discussed later in this review, there is now plenty of evidence that indicates that biological aging is malleable, and, more importantly, when aging is delayed, so are age-related diseases and conditions. In fact, aging can be delayed by behavioral (DR) (McCay and Crowell 1934; McCay et al. 1935), genetic (Kenyon 2010), and, to an extent, by pharmacological means, such as rapamycin (Harrison et al. 2009), acarbose (Harrison et al. 2014), metformin (Martin-Montalvo et al. 2013), and others, including resveratrol under metabolically stressed conditions (Baur et al. 2006; Pearson et al. 2008). Current research into pharmacological means of extending life span and health span take advantage of both molecular (target of rapamycin [TOR], IGF, sirtuins) and cellular (senescence, stem cells) discoveries and the field is poised for further findings during the next 10 years. Most importantly, in most cases studied, these manipulations lead to significant improvements in physiology, including enhanced resistance to disease (Baur et al. 2006; Pearson et al. 2008; Baker et al. 2011; Fernández and Fraga 2011; Wilkinson et al. 2012; Martin-Montalvo et al. 2013).

It is common for people to question the goals of aging research, as most people—including the elderly—do not consider extending life span to be a worthy goal unless health span is improved and extended in parallel. Indeed, the goal of aging biology research, as well as geroscience, is not to increase life span (that is used in research as an easy-to-measure, binary surrogate), but, rather, to understand aging biology so as to improve health span and delay disease. The consequences of aging can be summarized as an increase in frailty and an attendant decrease in resilience, leading on the one hand to increased susceptibility to disease (frailty), and, on the other, to decreased ability to withstand the concomitant stress caused by disease (resilience). Together, these changes result in a decrease in the thresholds necessary for disease-specific insults to result in overt pathology. This explains, for example, why a given level of cholesterol, or burden of Aβ or oncogenic mutations, can be an asymptomatic risk factor in younger individuals, but disease causing in older counterparts. Similarly, the change in thresholds at which challenges can be overcome explains why, for example, a young person recuperates easily from a period being bedridden, while the same challenge can become a chronically unsurmountable block for an older individual. Approaches that target either of these characteristics of the aging process are thus likely to result in decreased disease burden.

So, what will happen if geroscience is successful? If most chronic diseases are indeed postponed as a group, then it follows that life span will also increase, and it has been argued that this will further exacerbate the silver tsunami currently underway, with dramatic shifts in population stratification by age (Mendelson and Schwartz 1993). Although that is certainly true, the fallacy is to think of these "elderly" in current terms, as sick and frail. As it has been observed in multiple studies in mice and other species, addressing aging (as opposed to addressing one disease at a time) leads to robust

Cite this article as *Cold Spring Harb Perspect Med* doi: 10.1101/cshperspect.a025163

elderly individuals, not sick ones (Baur et al. 2006; Pearson et al. 2008; Baker et al. 2011; Fernández and Fraga 2011; Wilkinson et al. 2012; Martin-Montalvo et al. 2013). In that scenario, the "new elderly" will not produce an undue burden on the health system or pensions, and in fact studies have shown that the opposite is true. Although curing cancer or cardiovascular disease (or both) would actually lead to a significant reduction in the number of people disabled from these often fatal conditions, the survivors would be disabled by other conditions later in life. The overall tradeoff would be negative (Miller 2002; Goldman et al. 2013). This occurs because curing one fatal disease allows the person to live longer but in the presence of other comorbid disabilities and conditions such as sarcopenia, osteoporosis, sensory loss, and others that, although not life threatening, considerably decrease the quality of life for so many elderly. If cured of only one fatal disease, individuals will keep on living with the other limitations until the next life-threatening disease (Alzheimer's, diabetes, cancer, etc.) does kill them.

In summary, geroscience aims at seeking innovative approaches to better identify the relationships between the biological processes of aging and the biological processes of age-related chronic diseases and disabilities, and, in so doing, we hope to understand why the former is the major risk factor for the latter. The underlying assumption is that these processes are likely to share much in common and to intersect and influence each other in manners that can be approached, experimentally first, and clinically in the not-too-distant future.

Based on these considerations, in 2011, we initiated a group internal to the NIH, dubbed the "GeroScience Interest Group" or GSIG. The purpose was to develop a collaborative framework that includes several NIH institutes with an interest in the biological mechanisms that drive the appearance of multiple diseases expressed in people of all ages, especially the elderly, with the aim of accelerating and coordinating efforts to promote further discoveries on the common risks and mechanisms behind such diseases. The goals of the GSIG are: (1) to promote discussion, sharing of ideas, and coordination of activities within the NIH, relating to the specific needs of the research community working on mechanisms underlying age-related changes, including those that could lead to increased disease susceptibility; (2) to raise awareness, both within and outside the NIH, of the relevant role played by aging biology in the development of age-related processes and chronic disease; (3) to develop potential public/private partnerships through interactions with scientific societies, industry, and other institutions with related interests; and (4) to develop trans-NIH funding initiatives that will encourage research on the basic biology of aging and its relationship to earlier life events, exposures, and diseases that will advance the goals and vision of the GSIG, and which complement and enhance the goals and vision of concerned institutes and centers (ICs).

The GSIG idea was received enthusiastically by several NIH directors and the group quickly grew to include official representatives from 20 different NIH institutes (for a complete list, please visit the GSIG website sigs.nih.gov/geroscience/Pages/default.aspx). In addition to activities designed to awaken interest in the topic within the NIH (seminars, forums, etc.), the GSIG organized a workshop in 2012 on "inflammation and age-related diseases," which resulted in a publication (Howcroft et al. 2013) and a funding announcement joined by eight NIH institutes (PAR-13-233, Chronic Inflammation and Age-related Disease). The group's next major activity was the organization of a summit in 2013, called "Advances in Geroscience: Impact on Healthspan and Chronic Disease." The meeting was held on the NIH campus in Bethesda and, after a series of keynote speeches, including the NIH Director Dr. Francis Collins, the main meeting focused on seven major areas of research encompassing mechanisms driving aging and most likely being involved in enabling chronic diseases. These areas overlap significantly with the topics identified by López-Otín et al. (2013) in a recent opinion piece titled "The Hallmarks of Aging," and a series of opinion pieces from the chairs of each session from the summit was published as a special issue of *The Journals of Gerontology—Biological Sciences* (Burch et al. 2014; biomedgerontology.oxfordjournals.org/content/69/Suppl_1.toc). A white paper summarizing the discussions from an ensuing executive session has been accepted for publication in the journal *Cell* (Kennedy et al. 2014). Some major aspects from these discussions are presented below.

THE MAIN PILLARS OF RESEARCH ON AGING BIOLOGY

It should be noted that the goal of geroscience (and the summit) is not to identify markers of aging, but rather, the goal has been to identify possible drivers of the process. Nevertheless, the search for biomarkers remains an under-appreciated area of research that deserves more attention. For years, biogerontologists have shied away from this line of research, under the assumption that such markers would be too elusive. Novel technical developments, such as multiple omics technologies, now open new possibilities that need to be explored, because in the absence of such markers, progress in the field remains hindered. On the other hand, it is expected that identification and further enhancement of our knowledge about the mechanistic drivers of the aging process, as well as their interactions, will lead to possible therapeutics to delay aging and with that, concomitantly delay the onset and/or severity of multiple chronic diseases and conditions that affect primarily the older population. Major areas currently considered as potential drivers, and discussed at the 2013 summit, include inflammation, responsiveness to stress, epigenetics, metabolism, macromolecular damage, proteostasis, and stem cells (Fig. 2). A brief overview of current thoughts about each of these topics follows.

Inflammation

The inflammatory response is crucial as a first level of defense of the organism against aggression by pathogens and recovery from tissue damage. Thus, it appears important that the acute inflammatory response be maintained even into old age. The molecular and cellular mechanisms involved in this response have been well studied in young organisms and a proper response is both swift and short-lived. Aged organisms appear quite capable of mounting a response to most challenges, although the response is not always "normal," with some aspects exacerbated and others blunted (Wu et al. 2007). However, in many instances, old organisms fail in the "shutting-off" phase of the response, leading to a lingering residue often called "sterile inflammation" (De Martinis

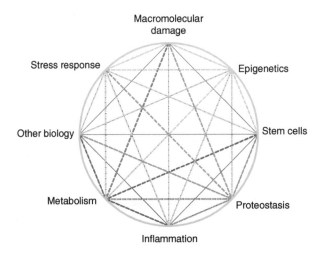

Figure 2. The main pillars of geroscience research. Aging physiology results from multiple interconnections among these pillars, including "other biology," to indicate our incomplete knowledge about the process.

et al. 2005). This is characterized by mild—but chronic—elevation in the serum levels of several cytokines and acute phase factors, the most studied being interleukin (IL)-6, tumor necrosis factor (TNF)-α and C-reactive protein (CRP) (Ferrucci et al. 1999; Harris et al. 1999; Bruunsgaard and Pedersen 2003; Bruunsgaard et al. 2003; Franceschi 2007). It is generally believed that this age-related low grade chronic inflammation (also termed "imflammaging" [Franceschi et al. 2007; Pawelec et al. 2014]) might be a significant contributing factor to several chronic diseases and conditions (Tchkonia et al. 2013; Franceschi and Campisi 2014). Efforts at curbing the inflammatory response are currently ongoing in the clinic, with 20 such trials currently listed in clinicaltrials.gov (clinicaltrials.gov/ct2/results?term=inflammation&search=search). However, considerable additional research is needed, including the identification of the source of cytokines (they do not appear to be derived from the classical source, macrophages, and both senescent cells and adipocytes are currently being investigated) (Howcroft et al. 2013; Franceschi and Campisi 2014). In addition, an important consideration is whether an intervention should be sought that dampens the inflammatory response altogether (e.g., anti-inflammatories) or whether a more appropriate approach might be represented by interventions that allow the response to proceed normally but improve the effectiveness of the shutting off mechanisms. In addition, whether sterile inflammation is really a maladaptive response needs to be clarified. Several lines of evidence indicate that reducing inflammation might be beneficial in terms of many diseases (Franceschi and Campisi 2014), but it is still entirely possible that this low level inflammation might be an adaptive response to age- or disease-induced tissue damage, and that reducing it might lead to unforeseen complications.

Adaptation to Stress

Stress is an inevitable part of life, for humans as well as other species. At the molecular level, cells are constantly bombarded with stressors including free radicals, environmental toxins, UV light, and others, although organisms are additionally exposed to cold, heat, and psychological stress in the form of predators, starvation, etc. Chronic, both physiological and psychological, stressors appear to accelerate the rate of aging (McEwen 2013; Epel and Lithgow 2014), and recent exciting work is beginning to show the interrelations between psychological stress and molecular responses, such as telomere shortening (Epel et al. 2004; Aydinonat et al. 2014; Zalli et al. 2014). It seems likely, but still unproven, that many—if not all—stresses might elicit a response from the organism via common mechanisms, and a further examination of these mechanisms is likely to yield insights into basic aging processes. If so, then the source of stress becomes less relevant and it is entirely possible that the focus might shift toward the mechanisms of response and coping. Because stress can be lowered but not eliminated, interventions aimed at increasing the ability of the organism to respond to stress might be more useful in our efforts to live to an old age in good health. Interestingly, both physiological and psychological stresses come in two very distinct flavors: mild stresses appear to be beneficial, probably through mechanisms related to hormesis (Calabrese et al. 2012). In contrast, both powerful acute or mild chronic stresses are detrimental (McEwen 2013; Epel and Lithgow 2014). A critical unanswered question is the identification of mechanisms that control that switch, and whether this switch can be manipulated so as to increase the positive and decrease the negative. This area of research is currently largely unexplored.

Epigenetics

The discovery of genes and pathways capable of increasing life span in a conserved fashion across many species was crucial in turning aging research from the descriptive to the mechanistic phase. Modern high throughput technologies have moved these discoveries into humans, in which some variants of genes initially described in lower organisms have been shown to be associated with extreme longevity (centenarians) (Wheeler and Kim 2011). More recent work in lower organisms has somewhat shifted toward the more malleable epigenome, in which significant changes have been described to occur as a function of age (Rando and Chang 2008; Wood et al. 2010; Greer et al. 2011; Kato et al. 2011). The origin of these changes and their downstream effects are currently the subject of intense study. Epigenetic changes have also been associated with a number of age-related diseases such as cancer, and the interrelationships between epigenetic changes caused by aging and those by disease are being explored (Brunet and Berger 2014). Because it is generally believed that epigenetic marks might integrate complex responses to the environment, another active area of research is the resolution between beneficial versus deleterious adaptations to stress, establishing to what extent these epigenetic changes can drive pathology, and to what extent those changes might be reversible or modifiable through pharmaceutical interventions.

Metabolism

Metabolic changes with age are widespread and in many cases they have been associated with age-related diseases, including diabetes, cancer, cardiovascular and neurodegenerative diseases, some of which are not classically considered "metabolic diseases." In fact, many of the genetic pathways affecting longevity have critical roles in the regulation of metabolism, including the insulin/IGF (Johnson 2008; Fontana et al. 2010; Barzilai et al. 2012) and mTOR (Harrison et al. 2009; Johnson et al. 2013; Miller et al. 2014) pathways, and the best characterized way of extending life span, caloric restriction, should be considered first and foremost a metabolic intervention. Another pathway that affects aging, sirtuins, has also been shown to have dramatic interactions with cellular metabolism, probably via regulation of NAD^+ levels, although this is still a subject of intense study (de Cabo et al. 2014; Imai and Guarente 2014; Rehan et al. 2014). It is telling that resveratrol, a molecule first studied because of its ability to activate sirtuins, extends life span in mice, but only if the animals are under severe metabolic stress (Baur et al. 2006). Additional metabolic

changes during aging are just now being identified, such as changes in circadian rhythms and changes in the microbiome, both of which might have dramatic metabolic effects. The roles of circadian clocks, changing microbiomes and changing intestinal leakiness in aging and age-related disease are likely to be extensively explored in the near future, and they are likely to have dramatic effects on metabolism, inflammation, and other aspects of aging. Because of their pivotal role in intracellular energy production, mitochondria have also been studied extensively in relationship to aging. Contrary to expectations, reducing the activity of the mitochondrial electron transport chain leads to increased longevity (Rea 2005; Munkácsy and Rea 2014), as do manipulations that slightly increase free radical production (Ristow and Schmeisser 2011).

Macromolecular Damage

The free radical theory of aging has been a cornerstone of aging biology research for more than half a century. In its simplest form, the theory posits that respiration-produced free radicals lead to macromolecular damage, and this damage results in the cellular and tissue loss-of-function observed during aging. Recent evidence coming from experiments in mice genetically manipulated to either increase or decrease free radical scavenging have put the theory into question, because most of these manipulations did indeed lead to the expected changes in macromolecular damage (decrease or increase, respectively) but did not affect either mean or maximal life span (Van Remmen et al. 2003; Pérez et al. 2009). A notable exception is the mCAT mouse model, in which expression of catalase in the mitochondria (but not other subcellular compartments) does indeed lead to increased longevity and a decrease in at least some age-related pathology, including cardiovascular disease (Schriner et al. 2005). Nevertheless, many studies have correlated free-radical damage with various age-related diseases, including cancer and cardiovascular diseases, the main killers in the Western world (Pala and Gürkan 2008). It remains to be seen whether free radical-driven damage can affect survival under conditions less pristine than IACUC (Institutional Animal Care and Use Committee)-approved mouse housing. Other molecular damage, such as DNA damage driven by mutations in either mitochondrial or nuclear DNA repair systems do lead to what some investigators have called "accelerated aging" (Trifunovic et al. 2004; Wallace 2005; Hoeijmakers 2009), and in independent research, it has been found that many human accelerated aging syndromes (Hutchinson–Gilford, Werner, Cockayne, and others) are characterized by mutations in genes involved in DNA repair or other DNA transactions, including structural integrity of the nuclear lamina (Rodríguez and Eriksson 2010; Worman 2012). As with response to stress, it remains to be seen whether the apparent acceleration of aging phenotypes in these instances is directly the result of DNA damage, or if it is rather the cell's response to that damage (for example, by apoptosis or senescence, leading to stem-cell depletion and/or inflammatory responses) that is directly responsible for the phenotype (Sierra 2006). An emerging area related to macromolecular damage is telomere integrity. Telomere shortening leads, in vitro at least, to cellular senescence (Bodnar et al. 1998), and telomere length has been associated with susceptibility to a variety of diseases (Cawthon et al. 2003; Epel et al. 2009). Whether sufficient to lead to that particular outcome or not, telomere shortening has been clearly associated with chronological aging and, perhaps more interestingly, it has been found that telomere shortening is accelerated by psychological stress (Herrera et al. 1999; Blasco 2007; Parks et al. 2009; Zalli et al. 2014). Whether causative or solely a biomarker, these findings are exciting and further research in this domain is likely to shed light on these relationships within the next few years.

Proteostasis

Just like the emphasis in the field of stress is shifting from stress itself toward the role of the response to stress, so is research on macromolecular damage shifting toward the mechanisms that control such damage. Proteostasis includes those mechanisms responsible for preserving the health of the proteome, including chaperones, autophagy, proteosomal degradations and others (Balch et al. 2008; Breusing and Grune 2008; Cuervo 2008; Morimoto and Cuervo 2014). A prominent role in aging has been ascribed to these mechanisms, including the unfolded protein response (UPR), both at the level of endoplasmic reticulum (ER) and mitochondria (for the moment; others such as nuclear UPR are likely to come to the fore in the near future). Changes in these processes are clearly related to age-related diseases including neurodegenerative and other diseases characterized by the accumulation of intracellular or extracellular aggregates, and it seems clear that modification of proteostasis represents an exciting possible therapeutic target. The possible role of proteostatic mechanisms in aging and age-related diseases is twofold, because not only is there a general decrease in the activity of the many protein quality control pathways with aging, but, in addition, there is an increased burden because of the accumulation of damaged proteins that need to be dealt with. So, aging organisms are confronted with a complex risk of losing protein quality control. On the positive side, this means that the problems of protein aggregation can be attacked on two fronts: decrease the damage or increase the defenses. Much recent effort has been placed in reducing the damage. Efforts at boosting the defenses are just starting. Exciting recent results indicate that the various protein quality control machineries can interact and supplement each other, even at a distance (Wong and Cuervo 2010; Durieux et al. 2011; Dillin et al. 2014). This gives hope that improving the entire system might not require a massive revamping of all the defective components, but partial alleviation in some crucial pathways and in relevant tissues might be sufficient to improve quality control even in tissues in which the system has not been specifically modified.

Stem Cells and Regeneration

Stem cells have generated much excitement both among scientists and the public alike, based on their promise as therapeutic agents for a wide range of diseases. In the case of chronic age-related diseases, that excitement needs to be tempered by an assessment of what happens with stem cells during aging. Elegant experiments using heterochronic parabiosis have shown that, at least in some tissues such as muscle and brain, the problem with aging is less with the stem cells themselves, but rather with their niche (Conboy et al. 2005). In other words, it appears that, in some cases, stem cells are still present in aged individuals, but their niche is incapable of activating them. Encouraging results from parabiosis experiments indicate that there are circulating factors capable of either activating or inhibiting stem cells (Villeda et al. 2011, 2014; Loffredo et al. 2013; Katsimpardi et al. 2014), and further detailing the relative importance of stem cells and their niches in different tissues and in different diseases will be crucial to better define useful therapies. Interestingly, it is likely that modifying the niche might be easier than injecting young stem cells, which in fact might not be effective in cases in which the niche is not receptive. A separate, exciting area of research involves induced pluripotent stem cells (iPSC) (Liu et al. 2012; Mahmoudi and Brunet 2012; Isobe et al. 2014). The availability of iPSC will most likely become an important tool in research aimed at identifying genetic determinants of age-related diseases, as well as possible therapeutic tools for at least a subset of complex chronic diseases. The field of stem-cell research is still in a rapid expansion phase and there are many areas still to be investigated, including the genetic basis for the decrease in their effectiveness with aging, their role in maintaining tissue function during aging in the absence of injury, and their impact on metabolism, inflammation, etc., and vice versa.

The areas discussed above are by no means all there is to learn about the basic biology of aging. Many additional leads and fields are continuously providing new information and there is no possibility of discussing them all in here. For example, no mention has been made of the findings coming from comparative biology, or the use of novel animal models. Similarly, no discussion has been included about the notable contributions of classical evolutionary biology or demography, fields that certainly shape the theoretical and conceptual contexts within which aging biology research is conducted. Comprehensive approaches, including systems biology, are also becoming an important aspect that will need to be developed to address important areas in the field. Finally, the importance of cross talk between basic researchers and clinicians working with elderly patients cannot be overemphasized.

CONCLUDING REMARKS

The fact that aging is the major risk factor for most chronic diseases and conditions has been known since the early days of civilization. However, because aging is usually understood as the chronological passing of time, rather than an integrated biological process, aging has traditionally been viewed—by physicians and the general public alike—as immutable. In contrast to chronological aging, however, recent research has shown that the rate of physiological aging can be manipulated by a variety of behavioral, genetic, and pharmacological means in many animal models, and there is reason to believe this can be accomplished in humans. Most importantly, when the rate of aging is decreased in animal models, there is often a delay (and decreased severity) of a number of age-associated diseases and conditions, suggesting that manipulations that either delay the onset or decrease the rate of aging could have a significant beneficial effect on the well-being of the elderly population.

In this review, I have described the currently recognized major pillars of aging, namely, inflammation, response to stress, epigenetics, metabolism, macromolecular damage, proteostasis, and stem cells. Recent discoveries in each of these domains, as well as novel approaches, should facilitate the elucidation of potential interventions in one or other of these pillars, so as to favorably alter the onset or progression of multiple chronic diseases affecting the elderly. Given the current demographic trends in human populations, there is an urgent need to develop these ideas into clinical practice.

REFERENCES

Aydinonat D, Penn DJ, Smith S, Moodley Y, Hoelzl F, Knauer F, Schwarzenberger F. 2014. Social isolation shortens telomeres in African grey parrots (*Psittacus erithacus erithacus*). *PLoS ONE* **9:** e93839.

Baker DJ, Wijshake T, Tchkonia T, LeBrasseur NK, Childs BG, van de Sluis B, Kirkland JL, van Deursen JM. 2011. Clearance of p16^{Ink4a}-positive senescent cells delays ageing-associated disorders. *Nature* **479:** 232–236.

Balch WE, Morimoto RI, Dillin A, Kelly JW. 2008. Adapting proteostasis for disease intervention. *Science* **319:** 916–919.

Barzilai N, Huffman DM, Muzumdar RH, Bartke A. 2012. The critical role of metabolic pathways in aging. *Diabetes* **61:** 1315–1322.

Baur JA, Pearson KJ, Price NL, Jamieson HA, Lerin C, Kalra A, Prabhu VV, Allard JS, Lopez-Lluch G, Lewis K, et al. 2006. Resveratrol improves health and survival of mice on a high-calorie diet. *Nature* **444:** 337–342.

Bhattacharya J, Cutler DM, Goldman DP, Hurd MD, Joyce GF, Lakdawalla DN, Panis CWA, Shang B. 2004. Disability forecasts and future Medicare costs. In *Frontiers in health policy research* (ed. Cutler DM, Garber AM), Vol 7, pp. 75–94. National Bureau of Economic Research, Cambridge, MA.

Blasco MA. 2007. Telomere length, stem cells and aging. *Nat Chem Biol* **3:** 640–649.

Bodnar AG, Ouellette M, Frolkis M, Holt SE, Chiu CP, Morin GB, Harley CB, Shay JW, Lichtsteiner S, Wright WE. 1998. Extension of life-span by introduction of telomerase into normal human cells. *Science* **279:** 349–352.

Breusing N, Grune T. 2008. Regulation of proteasome-mediated protein degradation during oxidative stress and aging. *Biol Chem* **389:** 203–209.

Brunet A, Berger SL. 2014. Epigenetics of aging and aging-related disease. *J Gerontol A Biol Sci Med Sci* **69:** S17–S20.

Bruunsgaard H, Pedersen BK. 2003. Age-related inflammatory cytokines and disease. *Immunol Allergy Clin North Am* **23:** 15–39.

Bruunsgaard H, Andersen-Ranberg K, Hjelmborg JB, Pedersen BK, Jeune B. 2003. Elevated levels of tumor necrosis factor alpha and mortality in centenarians. *Am J Med* **115:** 278–283.

Burch J, Augustine AD, Frieden LA, Hadley E, Howcroft TK, Johnson R, Khalsa PS, Kohanski RA, Li XL, Macchiarini F, et al. 2014. Advances in geroscience: Impact on healthspan and chronic disease. *J Gerontol A Biol Sci Med Sci* **69:** S1–S3.

Calabrese EJ, Iavicoli I, Calabrese V. 2012. Hormesis: Why it is important to biogerontologists. *Biogerontology* **13:** 215–235.

Cawthon RM, Smith KR, O'Brien E, Sivatchenko A, Kerber RA. 2003. Association between telomere length in blood and mortality in people aged 60 years or older. *Lancet* **361:** 393–395.

Conboy IM, Conboy MJ, Wagers AJ, Girma ER, Weissman IL, Rando TA. 2005. Rejuvenation of aged progenitor cells by exposure to a young systemic environment. *Nature* **433:** 760–764.

Corona G, Vignozzi L, Sforza A, Maggi M. 2013. Risks and benefits of late onset hypogonadism treatment: An expert opinion. *World J Mens Health* **31:** 103–125.

Cuervo AM. 2008. Autophagy and aging: Keeping that old broom working. *Trends Genet* **24:** 604–612.

De Cabo R, Carmona-Gutierrez D, Bernier M, Hall MN, Madeo F. 2014. The search for antiaging interventions: From elixirs to fasting regimens. *Cell* **157:** 1515–1526.

De Martinis M, Franceschi C, Monti D, Ginaldi L. 2005. Inflamm-ageing and lifelong antigenic load as major determinants of ageing rate and longevity. *FEBS Lett* **579:** 2035–2039.

Dillin A, Gottschling DE, Nyström T. 2014. The good and the bad of being connected: The integrons of aging. *Curr Opin Cell Biol* **26:** 107–112.

Durieux J, Wolff S, Dillin A. 2011. The cell-non-autonomous nature of electron transport chain-mediated longevity. *Cell* **114:** 79–91.

Epel ES, Lithgow GJ. 2014. Stress biology and aging mechanisms: Toward understanding the deep connection between adaptation to stress and longevity. *J Gerontol A Biol Sci Med Sci* **69:** S10–S16.

Epel ES, Blackburn EH, Lin J, Dhabhar FS, Adler NE, Morrow JD, Cawthon RM. 2004. Accelerated telomere shortening in response to life stress. *Proc Natl Acad Sci* **101:** 17312–17315.

Epel ES, Merkin SS, Cawthon R, Blackburn EH, Adler NE, Pletcher MJ, Seeman TE. 2009. The rate of leukocyte telomere shortening predicts mortality from cardiovascular disease in elderly men. *Aging (Albany)* **1:** 81–88.

Fernández AF, Fraga MF. 2011. The effects of the dietary polyphenol resveratrol on human healthy aging. *Epigenetics* **6:** 870–874.

Ferrucci L, Harris TB, Guralnik JM, Tracy RP, Corti MC, Cohen HJ, Penninx B, Pahor M, Wallace R, Havlik RJ. 1999. Serum IL-6 level and the development of disability in older persons. *J Am Geriatr Soc* **47:** 639–646.

Flurkey K, Currer JM, Harrison DE. 2007. The mouse in aging research. In *The mouse in biomedical research* (ed. Fox JG, et al.), 2nd ed., Vol. III, pp. 637–672. Academic, Burlington, MA.

Fontana L, Partridge L, Longo VD. 2010. Extending healthy life span—From yeast to humans. *Science* **328:** 321–326.

Franceschi C. 2007. Inflammaging as a major characteristic of old people: Can it be prevented or cured? *Nutr Rev* **65:** S173–S176.

Franceschi C, Campisi J. 2014. Chronic inflammation (inflammaging) and its potential contribution to age-associated diseases. *J Gerontol A Biol Sci Med Sci* **69:** S4–S9.

Franceschi C, Capri M, Monti D, Giunta S, Olivieri F, Sevini F, Panourgia MP, Invidia L, Celani L, Scurti M, et al. 2007. Inflammaging and anti-inflammaging: A systemic perspective on aging and longevity emerged from studies in humans. *Mech Ageing Dev* **128:** 92–105.

Goldman DP, Cutler D, Rowe JW, Michaud PC, Sullivan J, Peneva D, Olshansky SJ. 2013. Substantial health and economic returns from delayed aging may warrant a new focus for medical research. *Health Aff (Millwood)* **32:** 1698–1705.

Gorbunova V, Seluanov A, Mao Z, Hine C. 2007. Changes in DNA repair during aging. *Nucleic Acids Res* **35:** 7466–7474.

Gorbunova V, Seluanov A, Zhang Z, Gladyshev VN, Vijg J. 2014. Comparative genetics of longevity and cancer: Insights from long-lived rodents. *Nat Rev Genet* **15:** 531–540.

Greer EL, Maures TJ, Ucar D, Hauswirth AG, Mancini E, Lim JP, Benayoun BA, Shi Y, Brunet A. 2011. Transgenerational epigenetic inheritance of longevity in *Caenorhabditis elegans*. *Nature* **479:** 365–371.

Harris TB, Ferrucci L, Tracy RP, Corti MC, Wacholder S, Ettinger WH Jr, Heimovitz H, Cohen HJ, Wallace R. 1999. Associations of elevated interleukin-6 and C-reactive protein levels with mortality in the elderly. *Am J Med* **106:** 506–512.

Harrison DE, Strong R, Sharp ZD, Nelson JF, Astle CM, Flurkey K, Nadon NL, Wilkinson JE, Frenkel K, Carter CS, et al. 2009. Rapamycin fed late in life extends lifespan in genetically heterogeneous mice. *Nature* **460:** 392–395.

Harrison DE, Strong R, Allison DB, Ames BN, Astle CM, Atamna H, Fernandez E, Flurkey K, Javors MA, Nadon NL, et al. 2014. Acarbose, 17-α-estradiol, and nordihydroguaiaretic acid extend mouse lifespan preferentially in males. *Aging Cell* **13:** 273–282.

Herrera E, Samper E, Martín-Caballero J, Flores JM, Lee HW, Blasco MA. 1999. Disease states associated with telomerase deficiency appear earlier in mice with short telomeres. *EMBO J* **18:** 2950–2960.

Hoeijmakers JH. 2009. DNA damage, aging, and cancer. *New Engl J Med* **361:** 1475–1485.

Howcroft TK, Campisi J, Louis GB, Smith MT, Wise B, Wyss-Coray T, Augustine AD, McElhaney JE, Kohanski R, Sierra F. 2013. The role of inflammation in age-related disease. *Aging* **5:** 84–93.

Imai SI, Guarente L. 2014. NAD$^+$ and sirtuins in aging and disease. *Trends Cell Biol* **24:** 464–471.

Isobe KI, Cheng Z, Nishio N, Suganya T, Tanaka Y, Ito S. 2014. iPSCs, aging and age-related diseases. *N Biotechnol* **31:** 411–421.

Jensen-Seaman MI, Furey TS, Payseur BA, Lu Y, Roskin KM, Chen CF, Thomas MA, Haussler D, Jacob HJ. 2004. Comparative recombination rates in the rat, mouse, and human genomes. *Genome Res* **14:** 528–538.

Johnson TE. 2008. *Caenorhabditis elegans* 2007: The premier model for the study of aging. *Exp Gerontol* **43:** 1–4.

Johnson SC, Rabinovitch PS, Kaeberlein M. 2013. mTOR is a key modulator of ageing and age-related disease. *Nature* **493:** 338–345.

Kato M, Chen X, Inukai S, Zhao H, Slack FJ. 2011. Age-associated changes in expression of small, noncoding RNAs, including microRNAs, in *C. elegans*. *RNA* **17:** 1804–1820.

Katsimpardi L, Litterman NK, Schein PA, Miller CM, Loffredo FS, Wojtkiewicz GR, Chen JW, Lee RT, Wagers AJ, Rubin LL. 2014. Vascular and neurogenic rejuvenation of the aging mouse brain by young systemic factors. *Science* **344:** 630–634.

Kennedy BK, Berger SL, Brunet A, Campisi J, Cuervo AM, Epel ES, Franceschi C, Lithgow GJ, Morimoto RI, Pessin JE, et al. 2014. Geroscience: Linking aging to chronic disease. *Cell* **159:** 709–713.

Kenyon C. 2010. The genetics of aging. *Nature* **464:** 504–512.

Krtolica A, Campisi J. 2003. Integrating epithelial cancer, aging stroma and cellular senescence. *Adv Gerontol* **11:** 109–116.

Kumar S, Subramanian S. 2002. Mutation rates in mammalian genomes. *Proc Natl Acad Sci* **99:** 803–808.

Liu GH, Ding Z, Izpisua-Belmonte JC. 2012. iPSC technology to study human aging and aging-related disorders. *Curr Opin Cell Biol* **24:** 765–774.

Loffredo FS, Steinhauser ML, Jay SM, Gannon J, Pancoast JR, Yalamanchi P, Sinha M, Dall'Osso C, Khong D, Shadrach JL, et al. 2013. Growth differentiation factor 11 is a circulating factor that reverses age-related cardiac hypertrophy. *Cell* **153:** 828–839.

López-Otín C, Blasco MA, Partridge L, Serrano M, Kroemer G. 2013. The hallmarks of aging. *Cell* **153:** 1194–1217.

Mahmoudi S, Brunet A. 2012. Aging and reprogramming: A two-way street. *Curr Opin Cell Biol* **24:** 744–756.

Martin-Montalvo A, Mercken EM, Mitchell SJ, Palacios HH, Mote PL, Scheibye-Knudsen M, Gomes AP, Ward TM, Minor RK, Blouin MJ, et al. 2013. Metformin improves healthspan and lifespan in mice. *Nat Commun* **4:** 2192.

Matsumoto AM. 2013. Testosterone administration in older men. *Endocrinol Metab Clin North Am* **42:** 271–286.

McCay CM, Crowell MF. 1934. Prolonging the life span. *Sci Mon* **39:** 405–414.

McCay CM, Crowell MF, Maynard LA. 1935. The effect of retarded growth upon the length of life span and upon the ultimate body size. *J Nutr* **10:** 63–79.

McEwen BS. 2013. Brain on stress: How the social environment gets under the skin. *Proc Natl Acad Sci* **109:** 17180–17185.

Mendelson DN, Schwartz WB. 1993. The effects of aging and population growth on health care costs. *Health Aff* **12:** 119–125.

Miller RA. 2002. Extending life: Scientific prospects and political obstacles. *Milbank Q* **80:** 155–174.

Miller RA, Harrison DE, Astle CM, Fernandez E, Flurkey K, Han M, Javors MA, Li X, Nadon NL, Nelson JF, et al. 2014. Rapamycin-mediated lifespan increase in mice is dose and sex dependent and metabolically distinct from dietary restriction. *Aging Cell* **13:** 468–477.

Milman S, Atzmon G, Huffman DM, Wan J, Crandall JP, Cohen P, Barzilai N. 2014. Low insulin-like growth factor-1 level predicts survival in humans with exceptional longevity. *Aging Cell* **13:** 769–771.

Morimoto RI, Cuervo AM. 2014. Proteostasis and the aging proteome in health and disease. *J Gerontol A Biol Sci Med Sci* **69:** S33–S38.

Munkácsy E, Rea SL. 2014. The paradox of mitochondrial dysfunction and extended longevity. *Exp Gerontol* **56:** 221–233.

Neugarten BL, Havighurst RJ. 1977. Extending the human life span: Social policy and social ethics. *National Science Foundation*, Arlington, VA.

Niccoli T, Partridge L. 2012. Ageing as a risk factor for disease. *Curr Biol* **22:** R741–R752.

Olshansky SJ, Perry D, Miller RA, Butler RN. 2006. In pursuit of the longevity dividend. *The Scientist* **20:** 28–36.

Olshansky SJ, Perry D, Miller RA, Butler RN. 2007. Pursuing the longevity dividend. Scientific goals for an aging world. *Ann NY Acad Sci* **1114:** 11–13.

Pala FS, Gürkan H. 2008. The role of free radicals in ethiopathogenesis of diseases. *Adv Molec Biol* **1:** 1–9.

Parks CG, Miller DB, McCanlies EC, Cawthon RM, Andrew ME, DeRoo LA, Sandler DP. 2009. Telomere length, current perceived stress, and urinary stress hormones in women. *Cancer Epidemiol Biomarkers Prev* **18:** 551–560.

Pawelec G, Goldeck D, Derhovanessian E. 2014. Inflammation, ageing and chronic disease. *Curr Opin Immunol* **29:** 23–28.

Pawlikowska L, Hu D, Huntsman S, Sung A, Chu C, Chen J, Joyner AH, Schork NJ, Hsueh WC, Reiner AP, et al. 2009. Association of common genetic variation in the insulin/IGF-1 signaling pathway with human longevity. *Aging Cell* **8:** 460–472.

Pearson KJ, Baur JA, Lewis KN, Peshkin L, Price NL, Labinskyy N, Swindell WR, Kamara D, Minor RK, Perez E, et al. 2008. Resveratrol delays age-related deterioration and mimics transcriptional aspects of dietary restriction without extending life span. *Cell Metab* **8:** 157–168.

Pérez VI, Van Remmen H, Bokov A, Epstein CJ, Vijg J, Richardson A. 2009. The overexpression of major antioxidant enzymes does not extend the lifespan of mice. *Aging Cell* **8:** 73–75.

Promislow DEL. 1994. DNA repair and the evolution of longevity: A critical analysis. *J Theoret Biol* **170:** 291–300.

Rando TA, Chang HY. 2008. Aging, rejuvenation, and epigenetic reprogramming: Resetting the aging clock. *Ann Rev Biochem* **77:** 727–754.

Rea SL. 2005. Metabolism in the *Caenorhabditis elegans* Mit mutants. *Exp Gerontol* **40:** 841–849.

Rehan L, Laszki-Szczachor K, Sobieszczańska M, Polak-Jonkisz D. 2014. SIRT1 and NAD as regulators of ageing. *Life Sci* **105:** 1–6.

Rincon M, Rudin E, Barzilai N. 2005. The insulin/IGF-1 signaling in mammals and its relevance to human longevity. *Exp Gerontol* **40:** 873–877.

Ristow M, Schmeisser S. 2011. Extending life span by increasing oxidative stress. *Free Radic Biol Med* **51:** 327–336.

Rodríguez S, Eriksson M. 2010. Evidence for the involvement of lamins in aging. *Curr Aging Sci* **3:** 81–89.

Schriner SE, Linford NJ, Martin GM, Treuting P, Ogburn CE, Emond M, Coskun PE, Ladiges W, Wolf N, Van Remmen H, et al. 2005. Extension of murine life span by overexpression of catalase targeted to mitochondria. *Science* **308:** 1909–1911.

Sierra F. 2006. Is (your cellular response to) stress killing you? *J Gerontol A Biol Sci Med Sci* **61:** 557–561.

Slagboom PE, Beekman M, Passtoors WM, Deelen J, Vaarhorst AA, Boer JM, van den Akker EB, van Heemst D, de Craen AJ, Maier AB, et al. 2011. Genomics of human longevity. *Philos Trans R Soc Lond B Biol Sci* **366:** 35–42.

Tchkonia T, Zhu Y, van Deursen J, Campisi J, Kirkland J. 2013. Cellular senescence and the senescent secretory phenotype: Therapeutic opportunities. *J Clin Invest* **123:** 966–972.

Trifunovic A, Wredenberg A, Falkenberg M, Spelbrink JN, Rovio AT, Bruder CE, Bohlooly-Y M, Gidlöf S, Oldfors A, Wibom R, et al. 2004. Premature ageing in mice expressing defective mitochondrial DNA polymerase. *Nature* **429:** 417–423.

Vaidya A, Mao Z, Tian X, Spencer B, Seluanov A, Gorbunova V. 2014. Knock-in reporter mice demonstrate that DNA repair by non-homologous end joining declines with age. *PLoS Genet* **10:** e1004511.

Van Remmen H, Ikeno Y, Hamilton M, Pahlavani M, Wolf N, Thorpe SR, Alderson NL, Baynes JW, Epstein CJ, Huang TT, et al. 2003. Life-long reduction in MnSOD activity results in increased DNA damage and higher incidence of cancer but does not accelerate aging. *Physiol Genomics* **16:** 29–37.

Villeda SA, Luo J, Mosher KI, Zou B, Britschgi M, Bieri G, Stan TM, Fainberg N, Ding Z, Eggel A, et al. 2011. The ageing systemic milieu negatively regulates neurogenesis and cognitive function. *Nature* **477:** 90–94.

Villeda SA, Plambeck KE, Middeldorp J, Castellano JM, Mosher KI, Luo J, Smith LK, Bieri G, Lin K, Berdnik D, et al. 2014. Young blood reverses age-related impairments in cognitive function and synaptic plasticity in mice. *Nat Med* **20:** 659–663.

Wallace DC. 2005. A mitochondrial paradigm of metabolic and degenerative diseases, aging, and cancer: A dawn for evolutionary medicine. *Ann Rev Genet* **39:** 359–407.

Warner HR. 2005. Developing a research agenda in biogerontology: Basic mechanisms. *Sci Aging Knowl Environ* **44:** pe33.

Wheeler HE, Kim SK. 2011. Genetics and genomics of human aging. *Philos Trans R Soc Lond B Biol Sci* **366:** 43–50.

Wilkinson JE, Burmeister L, Brooks SV, Chan CC, Friedline S, Harrison DE, Hejtmancik JF, Nadon N, Strong R, Wood LK, et al. 2012. Rapamycin slows aging in mice. *Aging Cell* **11:** 675–682.

Wong E, Cuervo AM. 2010. Integration of clearance mechanisms: The proteasome and autophagy. *Cold Spring Harb Perspect Biol* **2:** a006734.

Wood JG, Hillenmeyer S, Lawrence C, Chang C, Hosier S, Lightfoot W, Mukherjee E, Jiang N, Schorl C, Brodsky AS, et al. 2010. Chromatin remodeling in the aging genome of Drosophila. *Aging Cell* **9:** 971–978.

Worman HJ. 2012. Nuclear lamins and laminopathies. *J Pathol* **226:** 316–325.

Wu D, Ren Z, Pae M, Guo W, Cui X, Merrill AH, Meydani SN. 2007. Aging up-regulates expression of inflammatory mediators in mouse adipose tissue. *J Immunol* **179:** 4829–4839.

Zalli A, Carvalho LA, Lin J, Hamer M, Erusalimsky JD, Blackburn EH, Steptoe A. 2014. Shorter telomeres with high telomerase activity are associated with raised allostatic load and impoverished psychosocial resources. *Proc Natl Acad Sci* **111:** 4519–4524.

The Role of the Microenvironmental Niche in Declining Stem-Cell Functions Associated with Biological Aging

Nathan A. DeCarolis[1], Elizabeth D. Kirby[2], Tony Wyss-Coray[2,3], and Theo D. Palmer[1,4]

[1]Institute for Stem Cell Biology and Regenerative Medicine, Stanford University School of Medicine, Stanford, California 94305

[2]Department of Neurology and Neurological Sciences, Stanford University School of Medicine, Stanford, California 94305

[3]Center for Tissue Regeneration, Repair, and Restoration, Veterans Administration, Palo Alto Health Care Systems, Palo Alto, California 94304

[4]Department of Neurosurgery, Stanford University School of Medicine, Stanford, California 94305

Correspondence: tpalmer@stanford.edu

Aging is strongly correlated with decreases in neurogenesis, the process by which neural stem and progenitor cells proliferate and differentiate into new neurons. In addition to stem-cell-intrinsic factors that change within the aging stem-cell pool, recent evidence emphasizes new roles for systemic and microenvironmental factors in modulating the neurogenic niche. This article focuses on new insights gained through the use of heterochronic parabiosis models, in which an old mouse and a young circulatory system are joined. By studying the brains of both young and old mice, researchers are beginning to uncover circulating pro-neurogenic "youthful" factors and "aging" factors that decrease stem-cell activity and neuro-genesis. Ultimately, the identification of factors that influence stem-cell aging may lead to strategies that slow or even reverse age-related decreases in neural-stem-cell (NSC) function and neurogenesis.

Aging is a process by which cells alter their biochemical and genetic functions through cell-intrinsic and cell-extrinsic (microenvironment and systemic) factors. Aging manifests in many ways including dysregulation of tissue homeostasis and the gradual loss of regenerative capacity (Lopez-Otin et al. 2013). One of the main goals of regenerative medicine and stem-cell biology is to overcome the deleterious cellular effects of aging and, ultimately, to reverse them. Stem cells play a two-pronged role in tissue maintenance through divisions: on one hand, stem cells divide asymmetrically to produce a daughter cell that can differentiate and maintain tissue homeostasis and repair tissue damage; on the other hand, stem cells must divide asymmetrically to maintain themselves ("self-renewal") and to provide a long-lasting source of cells with stem-like potential. To this end, one of the long-term effects associated with aging is the loss of cell "stemness" in aging tissue, either through stem cells dividing symmetrically into two new daughter cells and thus depleting the stem-cell pool, or by replicative senescence, whereby cells with stem-like potential exit the cell cycle and no longer contribute to tissue maintenance. In the either case, loss of stem cells can occur through cell-intrinsic effects or from loss of the microenvironmental niche that normally facilitates continued asymmetric divisions of stem cells and maintenance of homeostasis.

In the adult brain, stem cells persist in several discrete areas, contributing to adult neurogenesis. Neurogenesis is the process by which a proliferating cell exits the cell cycle and differentiates into a neuron, ultimately incorporating into the neuronal circuitry. Although it is widespread during embryogenesis, neurogenesis becomes increasingly restricted as the animal ages. Specifically in mice and humans, neurogenesis within the cortex of the brain is complete during the early postnatal period. However, there are at least two areas of the brain with well-established and substantial neurogenesis throughout the life of most mammals: the subventricular zone (SVZ) of the lateral ventricles and the subgranular zone (SGZ) of the hippocampal dentate gyrus. Despite ongoing research into the cellular origins of neurogenesis, debate continues as to the stem-like cell within each of these two regions (Carlen et al. 2009; Ma et al. 2009; Bonaguidi et al. 2011, 2012; Encinas et al. 2011; Goritz and Frisen 2012; DeCarolis et al. 2013). Although the identity of the stem cell remains controversial, one thing is clear: Cells with stem-like and neurogenic potential persist in the SVZ and SGZ and new neurons are born throughout the mammalian life, including in humans (Eriksson et al. 1998; Sanai et al. 2004, 2011; Curtis et al. 2007). In rodent models, SGZ stem-like populations give rise to new neurons that migrate a short distance in to the dentate gyrus granular layer and become new granule cells. In contrast, new neuroblasts derived from SVZ stem cells migrate a long way in what is known as the rostral migratory stream, from the SVZ to the olfactory bulb (OB), where they become new inhibitory neurons. In the adult hippocampus, new immature neurons are highly plastic and hypothesized to have crucial roles in memory function (Clelland et al. 2009; Sahay et al. 2011; Aimone et al. 2014; Rangel et al. 2014). New olfactory neurons may play a role in olfactory memory or discrimination (Lazarini and Lledo 2011).

Some aspects of these adult neurogenic systems appear to be conserved in humans. In the human, the dentate gyrus has decreased levels of neurogenesis with age, but recent work by Frisen and colleagues suggests that the age-related decline is much more gradual than previously thought (Spalding et al. 2013). Neurogenic precursor cells have been observed in the dentate gyrus of humans up to 100 years of age (Knoth et al. 2010). On the other hand, in the human OB, recent evidence suggests negligible amounts of adult neurogenesis (Sanai et al. 2011; Wang et al. 2011; Bergmann et al. 2012; Ernst et al. 2014), despite robust neurogenesis in mouse, rat, and nonhuman primates (Kornack and Rakic 1999; Pencea et al. 2001). Additional work is needed in humans to observe and characterize stem-cell function and neurogenesis.

To this end, we define neural-stem-like cells as cells, which can divide asymmetrically to produce a daughter cell with neurogenic potential while maintaining itself in an undifferentiated state. Our current understanding is that stem-like cells divide infrequently in vivo (Morshead et al. 1994). We define progenitor cells as rapidly dividing cells with neurogenic potential that cannot divide continuously and thus deplete after multiple rounds of successive division (Encinas et al. 2011).

Regarding the important intersections of aging and neurogenesis, two key features are well established. First, as an animal ages, there are fewer dividing cells and fewer neurogenic precursor cells in neurogenic regions in rodents and in humans (Kuhn et al. 1996; Rao et al. 2005, 2006; Ben Abdallah et al. 2010; Spalding et al. 2013). Second, as an animal ages, fewer cells maintain neurogenic potential as they differentiate (Ahlenius et al. 2009; Encinas et al. 2011; Villeda et al. 2011); in other words, as an animal gets older, fewer cells become neurons and more cells become astrocytes, the major cell fate alternative for neural precursors that are differentiating after cell cycle exit (although see Rao et al. 2005; Hattiangady and Shetty 2008). There are a number of possible explanations for the age-related decrease in neurogenesis, including the progressive loss of stem cells (as suggested by Encinas et al. 2011 and others), the potentially reversible loss of replicative activity in stem cells (Lugert et al. 2010), or the decrease in permissive microenvironment surrounding the stem cells (as suggested by Bernal and Peterson 2011; Villeda et al. 2011). It has also recently been found that stem and progenitor cells may also regulate their own microenvironment via secreted factors (Mosher et al. 2012; Butti et al. 2014; Kirby et al. 2015). For example, Wyss-Coray and colleagues recently showed that undifferentiated adult neural stem and progenitor cells secrete up to a third of the vascular endothelial growth factor (VEGF) in the young adult dentate gyrus, and that loss of VEGF from just the stem and progenitor population causes long-term depletion of the neurogenic pool (Kirby et al. 2015). VEGF is a neurogenic niche factor known to decrease with aging, and this recent work suggests the possibility that the loss of local growth factors that regulate neurogenesis may not be solely caused by local astrocytes. However, the secretome of stem and progenitor cells and its contribution to aging of the brain remain relatively unexplored.

In this review, we focus on changes associated with the neural-stem-cell (NSC) niches in the brain. Our current understanding is limited regarding the cellular and molecular mechanisms behind the diminished capacity for neurogenesis in the adult brain, but likely there is strong interplay between the stem-like cells in the brain and their permissive neurogenic niche. Recent studies have attributed the decline in neuron production with loss of NSCs in the hippocampus (Encinas et al. 2011) and in the SVZ (Maslov et al. 2004). However, other studies have argued that the number of stem-like cells (Hattiangady and Shetty 2008; Lugert et al. 2010) and the number of neurosphere-forming cells remains constant with aging (Ahlenius et al. 2009). These results suggest that the neu-

rogenic niche becomes less supportive (Luo et al. 2006; Ahlenius et al. 2009; Bouab et al. 2011) and/or that the NSCs shift into a quiescent state (Hattiangady and Shetty 2008; Lugert et al. 2010; Bouab et al. 2011). It is likely that a combination of factors contribute to the diminished neurogenic potential of the brain neurogenic niches, and we highlight what is known and emphasize that there is still much that we do not understand about aging and stem cells in the brain.

Although there has been a surge in research and reviews on aging and stem cells (Pollina and Brunet 2011; Artegiani and Calegari 2012; Conboy and Rando 2012; Lee et al. 2012; Lopez-Otin et al. 2013; Rolando and Taylor 2014), unfortunately, relatively little is known about aging NSCs, in part, because of technical challenges. Within the context of the adult mammalian brain, it has been difficult to identify stem cells from progenitor cells (DeCarolis et al. 2013; Knobloch et al. 2014). Further, protocols have been established to grow cells with stem-like potential in vitro as neurospheres or in monolayers (Reynolds and Weiss 1992; Morshead et al. 1994; Babu et al. 2007), but these protocols require removing cells from their in vivo microenvironmental niche. Given the complex interplay between neurogenic cells and their niches (Palmer et al. 2000; Shen et al. 2008; Tavazoie et al. 2008), new paradigms have emerged to explore age-related changes in neurogenesis within the complex stem-cell niche. The parabiosis model is one such method recently applied to the adult stem-cell niche and revealed numerous novel insights to how stem cells respond to an aging environment. In parabiosis, two mice are surgically connected and their circulatory systems become partially shared (Conboy et al. 2005). By pairing a young mouse with an older mouse in "heterochronic" pairings (Conboy and Rando 2012; Paul and Reddy 2014), researchers have explored age-associated changes in stem-cell function in muscle (Conboy et al. 2005; Brack et al. 2007), the heart (Loffredo et al. 2013), and, recently, the brain (Ruckh et al. 2012; Katsimpardi et al. 2014; Villeda et al. 2011, 2014). This work will explore briefly the cell-intrinsic changes associated with stem-cell decline in aging, then focus more broadly on the niche-specific effects that suggest that systemic circulation and the vasculature interposed within the neurogenic niche.

CELL-INTRINSIC CHANGES

Accumulating evidence suggests a battery of changes within stem cells that correlate with aging. For example, in nonneuronal systems like the hematopoietic system, stem cells reduce proliferation and differentiation capacity, accumulate marks of DNA damage, reduce activity of telomerase, change epigenetic marks, and alter transcription factor profiles (DeCarolis et al. 2008; Jaskelioff et al. 2011; Lopez-Otin et al. 2013). Importantly, although these age-related changes have been described in peripheral stem-cell pools, aging factors in the brain and NSC pools remain largely unexplored and poorly understood.

The best-characterized change related to neurogenesis is the significant decrease in proliferative capacity of the neurogenic regions with increasing age. The precipitous drop in proliferation of neurogenic cells was first characterized in the dentate gyrus of the rodent (Kuhn et al. 1996) and later in the OB (Molofsky et al. 2006), as noted above.

There is some limited evidence for cell-intrinsic aging of NSCs in adults. For example, recent studies suggest the accumulation of DNA damage (as indicated by genomic marks like γ-H2AX and 53BP1) in stem-cell populations like the hematopoietic system (Rossi et al. 2007) and in the dentate gyrus neurogenic niche (DeCarolis et al. 2014). Other reports have suggested the accumulation of mutations in DNA and genomic instability within the NSC pool with age (Mikheev et al. 2012; Dong et al. 2014) but additional research is needed. Other cell-intrinsic facets of aging have been characterized in peripheral stem-cell pools (Lopez-Otin et al. 2013), including decreased telomerase activity associated with shortened telomeres (Sahin and Depinho 2010), epigenetics changes (Webb et al. 2013; Brunet and Berger 2014), and asymmetric nonrandom chromosome segregation (Charville and Rando 2011).

However, similar changes in NSC pools are under investigation. Specifically, telomerase activity has been an active area of research, and promoting telomerase activity increases neurogenesis in vitro and in vivo (Caporaso et al. 2003; Ferron et al. 2009; Jaskelioff et al. 2011; Liu et al. 2012). These studies suggest that reversing the declines in telomerase activity can prevent aging-associated declines in stem-cell function and cognition.

SYSTEMIC FACTORS AND THE "NICHE"

The neurogenic niche that surrounds adult neural stem and progenitor cells is populated by a variety of cell types, including astrocytes, microglia, mature neurons, and endothelial cells. All of these cell types show age-related changes that could impact adult NSCs. Both astrocytes and microglia show increased activation with age (Conde and Streit 2006; Norden and Godbout 2013; Sierra et al. 2014; Rodriguez-Arellano et al. 2015) potentially as a part of increased immune activation with aging (termed "inflammaging"). This activation likely changes their secretory profile. For example, secretion of neurogenic growth factors commonly derived from glia, such as fibroblast growth factor 2 (FGF-2) and VEGF, decreases prominently with age (Shetty et al. 2005; Bernal and Peterson 2011). The vasculature also may deteriorate with age, occupying less volume and providing less blood flow to brain regions, such as the SVZ (Katsimpardi et al. 2014). If any of these changes are necessary or sufficient for inducing age-related neurogenic decline remains unclear. However, the potential involvement of the vasculature and immune responses strongly suggests a third player in inducing the aging phenotype in NSCs: the systemic environment.

SYSTEMIC CIRCULATION

The role of systemic circulation in aging of NSCs appears to be particularly potent. Like aging muscle, the aging brain has a persistent population of resident stem cells that lose proliferative activity with age. Using the heterochronic parabiosis model, Villeda, Wyss-Coray, and colleagues (Villeda et al. 2011) have recently shown that this proliferative capacity can be bidirectionally modulated by the "age" of the systemic circulation. Heterochronic young mice have significantly reduced numbers of new, doublecortin (DCX$^+$)-labeled neurons in the hippocampus, whereas their older counterpart shows partial restoration of new DCX$^+$ neuron number (Villeda et al. 2011). A similar increase in progenitor proliferation with parabiosis to a younger mouse has been recently found in the SVZ (Katsimpardi et al. 2014), suggesting that the rejuvenating effects of young blood extend to both neurogenic niches. However, it remains unknown which neurogenic populations are most affected by systemic factors. Are these stem cells or progenitors (or both) that are induced to proliferate? The answer to this question has relevance to the sustainability of young blood as a therapeutic intervention because stimulation of only progenitors could wane over repeated treatments if the stem-cell population does not replenish the progenitor pool.

Although parabiosis allows for sharing of both soluble blood-derived proteins and circulating cell populations, it appears that the operative components of the blood for aging of stem cells are soluble plasma proteins. Parabiotic pairing with a GFP$^+$ mouse reveals very little infiltration of GFP$^+$ cells in the brain of a wild-type parabionts (Villeda et al. 2011), suggesting that direct cellular contribution is unlikely. Moreover, the antineurogenic effects of heterochronic pairing with an old mouse can be recapitulated by injection with plasma isolated from aged mice (Villeda et al. 2011). Several recent studies have implicated growth differentiation factor 11 (GDF11), a transforming growth factor-β (TGF-β) family member, as a key blood-derived "youthful" factor that reverses aging of SVZ proliferation, vascular deterioration, skeletal muscle, and heart (Loffredo et al. 2013; Katsimpardi et al. 2014; Sinha et al. 2014). Recent work suggests that the role of systemic GDF11 in aging of other tissues may differ from the brain (Egerman et al. 2015). CCL11, a circulating immune cytokine, has also been shown to be associated with age-related decline in hippocampal neurogenesis (Villeda et al. 2011).

The question of which factors drive the rejuvenating properties of young blood remains open. Most likely, the rejuvenating cocktail is complex and possibly tissue-specific such that factors or mechanisms that rejuvenate myogenic progenitors may not be the same ones that rejuvenate neural progenitors. More broadly, the reliance on soluble factors may even differ for different tissues and cell types. Although muscle cell aging is rejuvenated by young serum-derived proteins much like CNS stem cells (Conboy et al. 2005; Brack et al. 2007), oligodendrocyte progenitor cells (OPCs) appear to be rejuvenated by the cellular component of young blood (Ruckh et al. 2012). Heterochronic parabiosis with a young mouse rescues age-related deficits in OPC-mediated remyelination after injury, but the rescue relies on recruitment of circulating young monocytes to clear myelin debris that then allows resident old OPCs to function better (Ruckh et al. 2012).

Part of determining what factors mediate systemic regulation of CNS stem cells is determining which cells respond directly to soluble factors. For example, although effects of CCL11 on isolated hippocampal NSCs suggest that these cells respond negatively to this "aging" factor (Villeda et al. 2011), numerous other aspects of hippocampal function are impacted by plasma-borne factors and aging, including hippocampal long-term potentiation (LTP) (Villeda et al. 2011, 2014), dendritic spine density (Villeda et al. 2014), immediate early gene expression (Villeda et al. 2014), and vascularization (Katsimpardi et al. 2014). Where do plasma proteins act first? The NSCs? The vasculature? The mature neurons and astrocytes? All these pieces of the neurogenic niche rely on each other extensively, and alterations in one cell population will likely echo through the others. The order of cause and effect is not yet clear. Teasing apart the varied players in maintaining a neurogenic niche on an aging systemic background remains a challenge.

THE VASCULAR NICHE

The reliance of the aging of adult NSCs on circulating plasma-borne proteins is perhaps not surprising given the highly vascularized nature of both the SVZ and SGZ neurogenic niches. Adult NSCs in both the SVZ and SGZ reside in complex, multicell niches in close association with local capillaries.

The SVZ is covered by a planar vascular plexus spanning its entire length, which is quite distinct from the more segmented capillary supply found in other, nonneurogenic brain regions, such as the cortex (Tavazoie et al. 2008). Proliferating SVZ progenitors and stem cells

are found in close association with this vascular supply, especially when repopulating after depletion by antimitotic treatment. The SVZ may also show unique permeability of the blood–brain barrier, allowing small molecules to diffuse through gaps between the astrocytic end-feet that typically insulate the brain from the circulating factors (Tavazoie et al. 2008).

In the SGZ, proliferating cells are similarly found in close apposition to blood vessels (Palmer et al. 2000). There is an intimate asso-

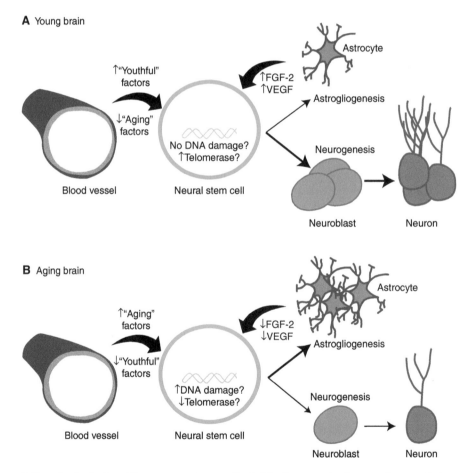

Figure 1. Hypothesized aging effects on neural stem cells. (*A*) In the young brain, neural stem cells (green) have low amounts of DNA damage but high amounts of telomerase activity. Astrocytes (purple) provide trophic support and circulating "youthful" factors from the blood (red) support a neurogenic environment, in which neural stem cells divide into neuroblasts (blue), which mature into neurons (magenta). (*B*) In contrast, in the aging brain, neural stem cells accumulate DNA damage and show decreased telomerase activity. Further, astrocytes provide less trophic support and "aging" factors increase in concentration in blood. In the aging brain, therefore, fewer neuroblasts are produced and neurogenesis is decreased; on the other hand, more astrocytes are produced. FGF, Fibroblast growth factor; VEGF, vascular endothelial growth factor.

ciation between the blood vasculature and neurogenic progenitors in the SGZ (Palmer et al. 2000; Pereira et al. 2007), which has been tied to the neurogenic response after exercise. Specifically, circulating increases in both VEGF and insulin-like growth factor 1 (IGF-1) have been suggested to drive exercise-induced increases in progenitor proliferation (Trejo et al. 2001; Fabel et al. 2003).

The vasculature may not simply be a means of conveying blood-borne factors, though, as it shows prominent age-related degradation that may be a driving factor in cognitive decline and neurodegenerative disease (Tarumi and Zhang 2014). Recent work by Katsimpardi, Rubin, and colleagues (2014) suggests that this decline in vascularization in the SVZ is closely linked with the decline in neurogenesis in this area and that both declines can be rescued by exposure to a young systemic environment. However, it remains unclear whether vascular endothelial cells and neural progenitors are both responding to systemic factors directly or to some indirect signal from each other or another niche cell type. The enhanced contact of adult neural stem and progenitor cells with the vasculature increases progenitor cell exposure both to the systemic circulation and to the vascular endothelium itself, which secretes a variety of proteins that impact NSC maintenance (Shen et al. 2004). The secretory profile of endothelium could change depending on the composition of the blood supply and thereby drive changes in closely associated stem and progenitor cells.

CONCLUSIONS

Taken together, current literature supports that a combination of factors contribute to changes in NSC function in aging, summarized in Figure 1. In the young brain, the local microenvironment combined with young stem cells produce an abundance of new neurons. Circulating "youthful" factors lead to the activation of cAMP-response element-binding protein (Creb), which promotes neurogenesis (Villeda et al. 2011). In a similar vein, Rubin and colleagues identified GDF11 as a circulating factor promoting vascularization and neurogenesis

(Katsimpardi et al. 2014). The niche astrocytes in the young brain also provide abundant trophic support to promote neurogenesis, including FGF-2, VEGF, and IGF (Shetty et al. 2005); as the brain ages, growth factor secretion decreases and, consequently, so does neurogenesis. Taken together, these results suggest that some (though likely not all) age-mediated decreases in neurogenesis can be rescued and perhaps even reverse the cognitive decline associated with aging.

REFERENCES

Ahlenius H, Visan V, Kokaia M, Lindvall O, Kokaia Z. 2009. Neural stem and progenitor cells retain their potential for proliferation and differentiation into functional neurons despite lower number in aged brain. *J Neurosci* **29:** 4408–4419.

Aimone JB, Li Y, Lee SW, Clemenson GD, Deng W, Gage FH. 2014. Regulation and function of adult neurogenesis: From genes to cognition. *Physiol Rev* **94:** 991–1026.

Artegiani B, Calegari F. 2012. Age-related cognitive decline: Can neural stem cells help us? *Aging* **4:** 176–186.

Babu H, Cheung G, Kettenmann H, Palmer TD, Kempermann G. 2007. Enriched monolayer precursor cell cultures from micro-dissected adult mouse dentate gyrus yield functional granule cell-like neurons. *PLoS ONE* **2:** e388.

Ben Abdallah NM, Slomianka L, Vyssotski AL, Lipp HP. 2010. Early age-related changes in adult hippocampal neurogenesis in C57 mice. *Neurobiol Aging* **31:** 151–161.

Bergmann O, Liebl J, Bernard S, Alkass K, Yeung MS, Steier P, Kutschera W, Johnson L, Landen M, Druid H, et al. 2012. The age of olfactory bulb neurons in humans. *Neuron* **74:** 634–639.

Bernal GM, Peterson DA. 2011. Phenotypic and gene expression modification with normal brain aging in GFAP-positive astrocytes and neural stem cells. *Aging Cell* **10:** 466–482.

Bonaguidi MA, Wheeler MA, Shapiro JS, Stadel RP, Sun GJ, Ming GL, Song H. 2011. In vivo clonal analysis reveals self-renewing and multipotent adult neural stem cell characteristics. *Cell* **145:** 1142–1155.

Bonaguidi MA, Song J, Ming GL, Song H. 2012. A unifying hypothesis on mammalian neural stem cell properties in the adult hippocampus. *Curr Opin Neurobiol* **22:** 754–761.

Bouab M, Paliouras GN, Aumont A, Forest-Berard K, Fernandes KJ. 2011. Aging of the subventricular zone neural stem cell niche: Evidence for quiescence-associated changes between early and mid-adulthood. *Neuroscience* **173:** 135–149.

Brack AS, Conboy MJ, Roy S, Lee M, Kuo CJ, Keller C, Rando TA. 2007. Increased Wnt signaling during aging alters muscle stem cell fate and increases fibrosis. *Science* **317:** 807–810.

Brunet A, Berger SL. 2014. Epigenetics of aging and aging-related disease. *J Gerontol A Biol Sci Med Sci* **69:** S17–20.

Butti E, Cusimano M, Bacigaluppi M, Martino G. 2014. Neurogenic and non-neurogenic functions of endogenous neural stem cells. *Front Neurosci* **8:** 92.

Caporaso GL, Lim DA, Alvarez-Buylla A, Chao MV. 2003. Telomerase activity in the subventricular zone of adult mice. *Mol Cell Neurosci* **23:** 693–702.

Carlen M, Meletis K, Goritz C, Darsalia V, Evergren E, Tanigaki K, Amendola M, Barnabe-Heider F, Yeung MS, Naldini L, et al. 2009. Forebrain ependymal cells are Notch-dependent and generate neuroblasts and astrocytes after stroke. *Nat Neurosci* **12:** 259–267.

Charville GW, Rando TA. 2011. Stem cell ageing and non-random chromosome segregation. *Philos Trans R Soc Lond B Biol Sci* **366:** 85–93.

Clelland CD, Choi M, Romberg C, Clemenson GD Jr, Fragniere A, Tyers P, Jessberger S, Saksida LM, Barker RA, Gage FH, et al. 2009. A functional role for adult hippocampal neurogenesis in spatial pattern separation. *Science* **325:** 210–213.

Conboy IM, Rando TA. 2012. Heterochronic parabiosis for the study of the effects of aging on stem cells and their niches. *Cell Cycle* **11:** 2260–2267.

Conboy IM, Conboy MJ, Wagers AJ, Girma ER, Weissman IL, Rando TA. 2005. Rejuvenation of aged progenitor cells by exposure to a young systemic environment. *Nature* **433:** 760–764.

Conde JR, Streit WJ. 2006. Microglia in the aging brain. *J Neuropathol Exp Neurol* **65:** 199–203.

Curtis MA, Kam M, Nannmark U, Anderson MF, Axell MZ, Wikkelso C, Holtas S, van Roon-Mom WM, Bjork-Eriksson T, Nordborg C, et al. 2007. Human neuroblasts migrate to the olfactory bulb via a lateral ventricular extension. *Science* **315:** 1243–1249.

DeCarolis NA, Wharton KA Jr, Eisch AJ. 2008. Which way does the Wnt blow? Exploring the duality of canonical Wnt signaling on cellular aging. *Bioessays* **30:** 102–106.

DeCarolis NA, Mechanic M, Petrik D, Carlton A, Ables JL, Malhotra S, Bachoo R, Gotz M, Lagace DC, Eisch AJ. 2013. In vivo contribution of nestin- and GLAST-lineage cells to adult hippocampal neurogenesis. *Hippocampus* **23:** 708–719.

DeCarolis NA, Rivera PD, Ahn F, Amaral WZ, LeBlanc JA, Malhotra S, Shih HY, Petrik D, Melvin N, Chen BP, et al. 2014. ^{56}Fe particle exposure results in a long-lasting increase in a cellular index of genomic instability and transiently suppresses adult hippocampal neurogenesis. *Life Sci Space Res (Amst)* **2:** 70–79.

Dong CM, Wang XL, Wang GM, Zhang WJ, Zhu L, Gao S, Yang DJ, Qin Y, Liang QJ, Chen YL, et al. 2014. A stress-induced cellular aging model with postnatal neural stem cells. *Cell Death Dis* **5:** e1116.

Egerman MA, Cadena SM, Gilbert JA, Meyer A, Nelson HN, Swalley SE, Mallozzi C, Jacobi C, Jennings LL, Clay I, et al. 2015. GDF11 increases with age and inhibits skeletal muscle regeneration. *Cell Metab* **22:** 164–174.

Encinas JM, Michurina TV, Peunova N, Park JH, Tordo J, Peterson DA, Fishell G, Koulakov A, Enikolopov G. 2011. Division-coupled astrocytic differentiation and age-related depletion of neural stem cells in the adult hippocampus. *Cell Stem Cell* **8:** 566–579.

Eriksson PS, Perfilieva E, Bjork-Eriksson T, Alborn AM, Nordborg C, Peterson DA, Gage FH. 1998. Neurogenesis in the adult human hippocampus. *Nat Med* **4:** 1313–1317.

Ernst A, Alkass K, Bernard S, Salehpour M, Perl S, Tisdale J, Possnert G, Druid H, Frisen J. 2014. Neurogenesis in the striatum of the adult human brain. *Cell* **156:** 1072–1083.

Fabel K, Fabel K, Tam B, Kaufer D, Baiker A, Simmons N, Kuo CJ, Palmer TD. 2003. VEGF is necessary for exercise-induced adult hippocampal neurogenesis. *Eur J Neurosci* **18:** 2803–2812.

Ferron SR, Marques-Torrejon MA, Mira H, Flores I, Taylor K, Blasco MA, Farinas I. 2009. Telomere shortening in neural stem cells disrupts neuronal differentiation and neuritogenesis. *J Neurosci* **29:** 14394–14407.

Goritz C, Frisen J. 2012. Neural stem cells and neurogenesis in the adult. *Cell Stem Cell* **10:** 657–659.

Hattiangady B, Shetty AK. 2008. Aging does not alter the number or phenotype of putative stem/progenitor cells in the neurogenic region of the hippocampus. *Neurobiol Aging* **29:** 129–147.

Jaskelioff M, Muller FL, Paik JH, Thomas E, Jiang S, Adams AC, Sahin E, Kost-Alimova M, Protopopov A, Cadinanos J, et al. 2011. Telomerase reactivation reverses tissue degeneration in aged telomerase-deficient mice. *Nature* **469:** 102–106.

Katsimpardi L, Litterman NK, Schein PA, Miller CM, Loffredo FS, Wojtkiewicz GR, Chen JW, Lee RT, Wagers AJ, Rubin LL. 2014. Vascular and neurogenic rejuvenation of the aging mouse brain by young systemic factors. *Science* **344:** 630–634.

Kirby ED, Kuwahara AA, Messer RL, Wyss-Coray T. 2015. Adult hippocampal neural stem and progenitor cells regulate the neurogenic niche by secreting VEGF. *Proc Natl Acad Sci* **112:** 4128–4133.

Knobloch M, von Schoultz C, Zurkirchen L, Braun SM, Vidmar M, Jessberger S. 2014. SPOT14-positive neural stem/progenitor cells in the hippocampus respond dynamically to neurogenic regulators. *Stem Cell Rep* **3:** 735–742.

Knoth R, Singec I, Ditter M, Pantazis G, Capetian P, Meyer RP, Horvat V, Volk B, Kempermann G. 2010. Murine features of neurogenesis in the human hippocampus across the lifespan from 0 to 100 years. *PLoS ONE* **5:** e8809.

Kornack DR, Rakic P. 1999. Continuation of neurogenesis in the hippocampus of the adult macaque monkey. *Proc Natl Acad Sci* **96:** 5768–5773.

Kuhn HG, Dickinson-Anson H, Gage FH. 1996. Neurogenesis in the dentate gyrus of the adult rat: Age-related decrease of neuronal progenitor proliferation. *J Neurosci* **16:** 2027–2033.

Lazarini F, Lledo PM. 2011. Is adult neurogenesis essential for olfaction? *Trends Neurosci* **34:** 20–30.

Lee SW, Clemenson GD, Gage FH. 2012. New neurons in an aged brain. *Behav Brain Res* **227:** 497–507.

Liu M, Hu Y, Zhu L, Chen C, Zhang Y, Sun W, Zhou Q. 2012. Overexpression of the *mTERT* gene by adenoviral vectors promotes the proliferation of neuronal stem cells in vitro and stimulates neurogenesis in the hippocampus of mice. *J Biomed Res* **26:** 381–388.

Loffredo FS, Steinhauser ML, Jay SM, Gannon J, Pancoast JR, Yalamanchi P, Sinha M, Dall'Osso C, Khong D, Shadrach JL, et al. 2013. Growth differentiation factor 11 is a circulating factor that reverses age-related cardiac hypertrophy. *Cell* **153:** 828–839.

Lopez-Otin C, Blasco MA, Partridge L, Serrano M, Kroemer G. 2013. The hallmarks of aging. *Cell* **153:** 1194–1217.

Lugert S, Basak O, Knuckles P, Haussler U, Fabel K, Gotz M, Haas CA, Kempermann G, Taylor V, Giachino C. 2010. Quiescent and active hippocampal neural stem cells with distinct morphologies respond selectively to physiological and pathological stimuli and aging. *Cell Stem Cell* **6:** 445–456.

Luo J, Daniels SB, Lennington JB, Notti RQ, Conover JC. 2006. The aging neurogenic subventricular zone. *Aging Cell* **5:** 139–152.

Ma DK, Bonaguidi MA, Ming GL, Song H. 2009. Adult neural stem cells in the mammalian central nervous system. *Cell Res* **19:** 672–682.

Maslov AY, Barone TA, Plunkett RJ, Pruitt SC. 2004. Neural stem cell detection, characterization, and age-related changes in the subventricular zone of mice. *J Neurosci* **24:** 1726–1733.

Mikheev AM, Ramakrishna R, Stoll EA, Mikheeva SA, Beyer RP, Plotnik DA, Schwartz JL, Rockhill JK, Silber JR, Born DE, et al. 2012. Increased age of transformed mouse neural progenitor/stem cells recapitulates age-dependent clinical features of human glioma malignancy. *Aging Cell* **11:** 1027–1035.

Molofsky AV, Slutsky SG, Joseph NM, He S, Pardal R, Krishnamurthy J, Sharpless NE, Morrison SJ. 2006. Increasing p16INK4a expression decreases forebrain progenitors and neurogenesis during ageing. *Nature* **443:** 448–452.

Morshead CM, Reynolds BA, Craig CG, McBurney MW, Staines WA, Morassutti D, Weiss S, van der Kooy D. 1994. Neural stem cells in the adult mammalian forebrain: A relatively quiescent subpopulation of subependymal cells. *Neuron* **13:** 1071–1082.

Mosher KI, Andres RH, Fukuhara T, Bieri G, Hasegawa-Moriyama M, He Y, Guzman R, Wyss-Coray T. 2012. Neural progenitor cells regulate microglia functions and activity. *Nat Neurosci* **15:** 1485–1487.

Norden DM, Godbout JP. 2013. Review: Microglia of the aged brain: Primed to be activated and resistant to regulation. *Neuropathol Appl Neurobiol* **39:** 19–34.

Palmer TD, Willhoite AR, Gage FH. 2000. Vascular niche for adult hippocampal neurogenesis. *J Comp Neurol* **425:** 479–494.

Paul SM, Reddy K. 2014. Young blood rejuvenates old brains. *Nat Med* **20:** 582–583.

Pencea V, Bingaman KD, Freedman LJ, Luskin MB. 2001. Neurogenesis in the subventricular zone and rostral migratory stream of the neonatal and adult primate forebrain. *Exp Neurol* **172:** 1–16.

Pereira AC, Huddleston DE, Brickman AM, Sosunov AA, Hen R, McKhann GM, Sloan R, Gage FH, Brown TR, Small SA. 2007. An in vivo correlate of exercise-induced neurogenesis in the adult dentate gyrus. *Proc Natl Acad Sci* **104:** 5638–5643.

Pollina EA, Brunet A. 2011. Epigenetic regulation of aging stem cells. *Oncogene* **30:** 3105–3126.

Rangel LM, Alexander AS, Aimone JB, Wiles J, Gage FH, Chiba AA, Quinn LK. 2014. Temporally selective contextual encoding in the dentate gyrus of the hippocampus. *Nat Commun* **5:** 3181.

Rao MS, Hattiangady B, Abdel-Rahman A, Stanley DP, Shetty AK. 2005. Newly born cells in the ageing dentate gyrus display normal migration, survival and neuronal fate choice but endure retarded early maturation. *Eur J Neurosci* **21:** 464–476.

Rao MS, Hattiangady B, Shetty AK. 2006. The window and mechanisms of major age-related decline in the production of new neurons within the dentate gyrus of the hippocampus. *Aging Cell* **5:** 545–558.

Reynolds BA, Weiss S. 1992. Generation of neurons and astrocytes from isolated cells of the adult mammalian central nervous system. *Science* **255:** 1707–1710.

Rodriguez-Arellano JJ, Parpura V, Zorec R, Verkhratsky A. 2015. Astrocytes in physiological aging and Alzheimer's disease. *Neuroscience* doi: 10.1016/j.neuroscience.2015.01.007.

Rolando C, Taylor V. 2014. Neural stem cell of the hippocampus: Development, physiology regulation, and dysfunction in disease. *Curr Top Dev Biol* **107:** 183–206.

Rossi DJ, Bryder D, Seita J, Nussenzweig A, Hoeijmakers J, Weissman IL. 2007. Deficiencies in DNA damage repair limit the function of haematopoietic stem cells with age. *Nature* **447:** 725–729.

Ruckh JM, Zhao JW, Shadrach JL, van Wijngaarden P, Rao TN, Wagers AJ, Franklin RJ. 2012. Rejuvenation of regeneration in the aging central nervous system. *Cell Stem Cell* **6:** 96–103.

Sahay A, Scobie KN, Hill AS, O'Carroll CM, Kheirbek MA, Burghardt NS, Fenton AA, Dranovsky A, Hen R. 2011. Increasing adult hippocampal neurogenesis is sufficient to improve pattern separation. *Nature* **472:** 466–470.

Sahin E, Depinho RA. 2010. Linking functional decline of telomeres, mitochondria and stem cells during ageing. *Nature* **464:** 520–528.

Sanai N, Tramontin AD, Quinones-Hinojosa A, Barbaro NM, Gupta N, Kunwar S, Lawton MT, McDermott MW, Parsa AT, Manuel-Garcia Verdugo J, et al. 2004. Unique astrocyte ribbon in adult human brain contains neural stem cells but lacks chain migration. *Nature* **427:** 740–744.

Sanai N, Nguyen T, Ihrie RA, Mirzadeh Z, Tsai HH, Wong M, Gupta N, Berger MS, Huang E, Garcia-Verdugo JM, et al. 2011. Corridors of migrating neurons in the human brain and their decline during infancy. *Nature* **478:** 382–386.

Shen Q, Goderie SK, Jin L, Karanth N, Sun Y, Abramova N, Vincent P, Pumiglia K, Temple S. 2004. Endothelial cells stimulate self-renewal and expand neurogenesis of neural stem cells. *Science* **304:** 1338–1340.

Shen Q, Wang Y, Kokovay E, Lin G, Chuang SM, Goderie SK, Roysam B, Temple S. 2008. Adult SVZ stem cells lie in a vascular niche: A quantitative analysis of niche cell–cell interactions. *Cell Stem Cell* **3:** 289–300.

Shetty AK, Hattiangady B, Shetty GA. 2005. Stem/progenitor cell proliferation factors FGF-2, IGF-1, and VEGF exhibit early decline during the course of aging in the hippocampus: Role of astrocytes. *Glia* **51:** 173–186.

Sierra A, Beccari S, Diaz-Aparicio I, Encinas JM, Comeau S, Tremblay ME. 2014. Surveillance, phagocytosis, and inflammation: How never-resting microglia influence adult hippocampal neurogenesis. *Neural Plast* **2014:** 610343.

Sinha M, Jang YC, Oh J, Khong D, Wu EY, Manohar R, Miller C, Regalado SG, Loffredo FS, Pancoast JR, et al. 2014. Restoring systemic GDF11 levels reverses age-related dysfunction in mouse skeletal muscle. *Science* **344:** 649–652.

Spalding KL, Bergmann O, Alkass K, Bernard S, Salehpour M, Huttner HB, Bostrom E, Westerlund I, Vial C, Buchholz BA, et al. 2013. Dynamics of hippocampal neurogenesis in adult humans. *Cell* **153:** 1219–1227.

Tarumi T, Zhang R. 2014. Cerebral hemodynamics of the aging brain: Risk of Alzheimer disease and benefit of aerobic exercise. *Front Physiol* **5:** 6.

Tavazoie M, Van der Veken L, Silva-Vargas V, Louissaint M, Colonna L, Zaidi B, Garcia-Verdugo JM, Doetsch F. 2008. A specialized vascular niche for adult neural stem cells. *Cell Stem Cell* **3:** 279–288.

Trejo JL, Carro E, Torres-Aleman I. 2001. Circulating insulin-like growth factor I mediates exercise-induced increases in the number of new neurons in the adult hippocampus. *J Neurosci* **21:** 1628–1634.

Villeda SA, Luo J, Mosher KI, Zou B, Britschgi M, Bieri G, Stan TM, Fainberg N, Ding Z, Eggel A, et al. 2011. The ageing systemic milieu negatively regulates neurogenesis and cognitive function. *Nature* **477:** 90–94.

Villeda SA, Plambeck KE, Middeldorp J, Castellano JM, Mosher KI, Luo J, Smith LK, Bieri G, Lin K, Berdnik D, et al. 2014. Young blood reverses age-related impairments in cognitive function and synaptic plasticity in mice. *Nat Med* **20:** 659–663.

Wang C, Liu F, Liu YY, Zhao CH, You Y, Wang L, Zhang J, Wei B, Ma T, Zhang Q, et al. 2011. Identification and characterization of neuroblasts in the subventricular zone and rostral migratory stream of the adult human brain. *Cell Res* **21:** 1534–1550.

Webb AE, Pollina EA, Vierbuchen T, Urban N, Ucar D, Leeman DS, Martynoga B, Sewak M, Rando TA, Guillemot F, et al. 2013. FOXO3 shares common targets with ASCL1 genome-wide and inhibits ASCL1-dependent neurogenesis. *Cell Rep* **4:** 477–491.

Biochemical Genetic Pathways that Modulate Aging in Multiple Species

Alessandro Bitto, Adrienne M. Wang, Christopher F. Bennett, and Matt Kaeberlein

Department of Pathology, University of Washington, Seattle, Washington 98195

Correspondence: kaeber@uw.edu

The mechanisms underlying biological aging have been extensively studied in the past 20 years with the avail of mainly four model organisms: the budding yeast *Saccharomyces cerevisiae*, the nematode *Caenorhabditis elegans*, the fruitfly *Drosophila melanogaster*, and the domestic mouse *Mus musculus*. Extensive research in these four model organisms has identified a few conserved genetic pathways that affect longevity as well as metabolism and development. Here, we review how the mechanistic target of rapamycin (mTOR), sirtuins, adenosine monophosphate-activated protein kinase (AMPK), growth hormone/insulin-like growth factor 1 (IGF-1), and mitochondrial stress-signaling pathways influence aging and life span in the aforementioned models and their possible implications for delaying aging in humans. We also draw some connections between these biochemical pathways and comment on what new developments aging research will likely bring in the near future.

The science of aging has rapidly advanced in the past two decades owing to the use of evolutionarily divergent model organisms in experimental studies. The most common of these include the budding yeast *Saccharomyces cerevisiae*, the nematode worm *Caenorhabditis elegans*, the fruit fly *Drosophila melanogaster*, the domestic mouse *Mus musculus*, and, to a lesser extent, the rat *Rattus norvegicus*. Studies of aging in nonhuman primates, such as rhesus monkeys and marmosets, have also had a large impact on the field in the last few years. Additionally, recent comparative biological approaches to aging have adopted less commonly used model organisms that feature exceptional longevity, such as the naked mole rat, short-lived fish, certain species of clams, or those with potent regenerative potential, such as hydra (Austad 2009; Valen-

zano et al. 2009; Ridgway et al. 2011; Petralia et al. 2014).

Historically, a major question in the field has been the degree to which studies of aging in nonhuman organisms will be informative about the mechanisms of human aging and the processes that contribute to age-related disease (Gershon and Gershon 2000; Warner 2003). Although it is still not possible to definitively answer this question, a large body of data exists supporting the idea that at least some basic principles of aging are broadly conserved, even in single-celled eukaryotes, such as budding yeast. In particular, several genetic and environmental factors that modulate longevity in two or more commonly used model organisms have been identified (Sutphin and Kaeberlein 2011). Manipulation of these conserved modifiers of longevity is sufficient to increase life span in evolutionarily divergent species, showing that they are likely to modulate the aging process itself. This, in turn, suggests that there are at least some fundamental aspects of aging that are shared between yeast, worms, flies, and mice. Given that the evolutionary distance between these species is much greater than the distance between mice and humans, it seems likely that at least a subset of the principles of aging gleaned from these model systems will also be applicable to people (Kaeberlein 2013).

A general principle of conserved modifiers of longevity is that they tend to regulate the relationship between growth and environmental cues, particularly nutrient status. The best studied of these is dietary restriction (DR), also referred to as caloric restriction or calorie restriction, which can be defined as a reduction in nutrient availability in the absence of malnutrition. DR has been shown to extend life span and improve health during aging in numerous species (Weindruch and Walford 1988). This may reflect a fundamental principle that rates of aging are linked to developmental pathways modulating growth and reproductive trajectories in response to variable environmental situations. For the remainder of this article, we will describe these conserved longevity pathways, their genetic and biochemical properties, and mechanisms by which they may act to modulate rates of aging and risk for age-related disease.

THE FOUR MOST COMMON MODEL ORGANISMS USED IN AGING RESEARCH

As mentioned above, the four major nonhuman model organisms used in aging-related research are budding yeast, nematode worms, fruit flies, and laboratory mice. Each of these species has its own strengths and weaknesses as a model for human aging, and it is important to consider the ways in which they are similar but also how they differ with respect to physiology, longevity, and aging traits. Notably, the shape of the survival curves in these organisms is superficially similar to each other and to human survival data as modeled by Gompertz–Makeham kinetics (Kaeberlein et al. 2001), although such similarity does not imply any conservation of aging mechanisms.

Yeast

Aging was first defined in budding yeast by Robert Mortimer and John Johnston more than 60 years ago, when it was discovered that yeast mother cells undergo a limited number of mitotic cell divisions before reaching a terminal replicative arrest (Mortimer and Johnston 1959). This form of aging in yeast has since been extensively studied and is referred to as replicative aging, with the corresponding longevity metric defined as replicative life span (RLS) (Steinkraus et al. 2008; Steffen et al. 2009; Kaeberlein 2010). A key feature of replicative aging in yeast is asymmetric division; mother cells retain and accumulate molecular damage during mitotic division, whereas daughter cells are generally spared from inheriting such damage and are born with a full RLS potential even when produced by aged mothers (Egilmez and Jazwinski 1989; Kennedy et al. 1994). There is good evidence for asymmetric inheritance of at least three different types of damage during yeast replicative aging: nuclear extrachromosomal ribosomal DNA (rDNA) circles (Sinclair and Guarente 1997; Defossez et al. 1998, 1999), oxidatively damaged or misfolded cytoplasmic proteins (Aguilaniu et al. 2003; Erjavec and Nystrom 2007; Erjavec et al. 2007), and dysfunctional mitochondria (Lai et al. 2002).

A second form of yeast aging has also been described, referred to as chronological aging. In contrast to replicative aging, chronological aging of yeast cells is accomplished by preventing mitotic cell division and maintaining cells in a nondividing state (Fabrizio and Longo 2007; Longo et al. 2012). The corresponding longevity metric, chronological life span (CLS), is defined as the length of time a yeast cell can survive in a nondividing state, with survival determined by the ability to reenter the cell cycle and resume vegetative growth on exposure to appropriate growth-promoting cues. Several different methods have been described for performing chronological aging experiments (Piper et al. 2006;

Cite this article as *Cold Spring Harb Perspect Med* doi: 10.1101/cshperspect.a025114

Murakami et al. 2008; Murakami and Kaeberlein 2009; Longo and Fabrizio 2012). Similar to replicative aging, there is evidence for accumulation of oxidatively damaged proteins and mitochondrial dysfunction during chronological aging.

The relationship between chronological and replicative aging and the relevance of these two distinct types of aging in yeast to aging in multicellular eukaryotes remains an area of active study. Large-scale studies have detected significant overlap between genetic control of longevity in *C. elegans* with genetic control of RLS (Smith et al. 2008a), but not CLS (Burtner et al. 2011). This could be related to the fact that acidification of the culture medium limits CLS under the most commonly used conditions for chronological aging experiments (Burtner et al. 2009; Murakami et al. 2011), which is not the case for RLS (Wasko et al. 2013). Nonetheless, key regulators of CLS also modulate RLS as well as aging in worms, flies, and mice (described further below). In addition, chronologically aged cells also show a reduction in subsequent RLS, suggesting that similar forms of age-associated damage may contribute to both mitotic (RLS) and postmitotic (CLS) aging in yeast cells (Ashrafi et al. 1999; Murakami et al. 2012; Delaney et al. 2013).

Worms

The nematode worm *C. elegans* is a facultative hermaphrodite that hatches from an egg and undergoes four larval stages (L1–L4) before reaching reproductive adulthood. Once adulthood has been reached, a typical hermaphrodite maintained at the standard temperature of 20°C will lay about 200 eggs over a period of 3–5 days before depletion of sperm, followed by an extended postreproductive period of 2–3 wk. With the exception of the germ line, adult *C. elegans* are thought to be completely postmitotic.

Life span in worms is typically defined as the length of time from hatching until death, which is determined manually by the absence of movement on gentle prodding. There are several features to consider when designing *C. elegans* life span studies, including the temperature (generally 15°C–25°C), the amount, strain, and metabolic state of bacterial food to provide, and whether to use the drug 5-fluorodeoxyuridine (FUDR) to prevent hatching of progeny (Sutphin and Kaeberlein 2009). Nearly all studies of aging in *C. elegans* use the standard wild-type N2 control strain. Depending on the conditions chosen, N2 life span can range from about 15 days (25°C, live food) to 35 days (15°C, growth-arrested food) (Leiser et al. 2011). In addition to life span, *C. elegans* affords the ability to monitor age-associated measures of health span, such as maintenance of muscle function, tissue atrophy, and accumulation of autofluorescent pigment (Herndon et al. 2002; Huang et al. 2004). As in yeast, there is good evidence that maintenance of protein homeostasis and mitochondrial function play a central role in *C. elegans* aging.

Flies

The fruit fly *D. melanogaster* was the first invertebrate organisms to be widely used in aging research, with life-span studies dating back to 1916 (Loeb and Northrop 1916). The fly life cycle consists of three easily distinguishable growth stages (embryo, larva, and pupae) occurring over a span of 10 days at 25°C, followed by reproductive adulthood. Flies are typically maintained in the laboratory in vials with an agar-based cornmeal–sugar–yeast or sugar–yeast food source. Unlike yeast and worms, methods for creating long-term frozen stocks are not available in *Drosophila*, but stock centers into which researchers deposit published strains as well as large centers that maintain libraries of transgenic, RNAi, and mutant fly lines are a valuable tool to fly researchers.

Life-span studies in *Drosophila* measure the length of time between eclosion (emergence of an adult fly from its pupal case) and death, with a median life span of wild-type strains averaging ~2–3 mo when maintained at 25°C. Because of genetic drift often found in individual laboratory strains and the differences in food composition between laboratories, genetic background and food type need to be tightly controlled in fly life-span and behavioral studies (Tatar et al. 2014). The adult fly shows many structures homologous to mammalian organs, such as heart, lung, kidney, gut, and reproductive tract. Importantly, *Drosophila* have a relatively simple nervous system that contains the same basic neural circuitry as mammals, modulating complex behavior and allowing for measures of health span and cognitive function often seen in human aging and neurodegenerative disease. As seen in yeast and worms, strong evidence exists linking protein homeostasis and mitochondrial function to organismal aging and life span (Cho et al. 2011; Rera et al. 2011; Bai et al. 2013).

Mice

The laboratory mouse *M. musculus* has become the premier mammalian model organism for aging research, in large part owing to the availability of several well-characterized inbred strains and the relatively early development of methods for knocking out and transgenically expressing different genes. Although several different inbred strains have been commonly used for aging-related studies, C57BL/6 has become the strain of choice for most mouse aging and longevity studies. Median life span can vary greatly among strains, and there is wide variation in reported life span even for C57BL/6, with median life spans ranging from ∼800 to >920 days (Coschigano et al. 2003; Perez et al. 2009). For many years, the National Institute on Aging has provided aged C57BL/6 mice to the research community, which has spurred numerous studies of aging-related traits in this background. In recent years, the importance of quantifying health span measures in addition to life span has become more widely recognized, and mice provide an opportunity to query numerous age-dependent measures of health that are shared with humans but not invertebrate species. The limitations of working in a single inbred strain background have also become better recognized, and the use of genetically heterogeneous mice for aging studies, such as the UM-HET3 four-way cross background, is becoming more common.

CONSERVED LONGEVITY PATHWAYS

Studies from model organisms have identified several orthologous genes that similarly modulate longevity across broad evolutionary distance (Kaeberlein 2007; Sutphin and Kaeberlein 2011). In some cases, the mechanisms by which the corresponding proteins affect aging are poorly understood, although in many cases they can be ascribed to one or more "conserved longevity pathways." The remainder of this review will discuss the best characterized of these pathways and the growing evidence for complex interrelationships among them.

Mechanistic Target of Rapamycin (mTOR)

mTOR is a serine/threonine protein kinase of the phosphoinositide-3-kinase-related family that is highly conserved among eukaryotes (Keith and Schreiber 1995; Stanfel et al. 2009). mTOR activity generally promotes cellular growth and cell division in response to nutrient and growth factor cues through a complicated network of interactions (Laplante and Sabatini 2012; Shimobayashi and Hall 2014; Johnson et al. 2015). The mTOR protein acts in at least two complexes, mTOR complex 1 (mTORC1) and mTOR complex 2 (mTORC2), each of which has distinct components, upstream regulators, and downstream substrates (Weber and Gutmann 2012; Takahara and Maeda 2013; Huang and Fingar 2014). The mTORC1 complex has been characterized extensively and functions as a central regulator of longevity. mTORC1 is known to promote global messenger RNA (mRNA) translation, repress autophagy, and modulate mitochondrial metabolism, with each of these downstream functions implicated in its role in aging (Johnson et al. 2015). The mTORC2 complex, in contrast, is less well characterized. Evidence from *D. melanogaster, Dictyostelium discoideum*, and human cells suggest that mTORC2 responds to insulin and Ras signaling (Lee et al. 2005; Sarbassov et al. 2005; Charest et al. 2010), but its precise mechanism of activation remains elusive. mTORC2 regulates the activity of several substrates involved in cytoskeleton reorganization and cell polarity, but also in cell growth and metabolism (Loewith et al. 2002; Sarbassov et al. 2004, 2005; García-Martínez and Alessi 2008; Ikenoue et al. 2008).

The first evidence that reduced mTORC1 signaling is sufficient to increase life span came from studies in yeast, in which mutation of the S6 kinase homolog Sch9 was shown to extend chronological life span (Fabrizio et al. 2001); although, at the time, the link between Sch9 and mTORC1 was not appreciated. A few years later, genetic inhibition of mTOR itself, as well as other components of the mTORC1 complex, was found to extend life span in worms (Vellai et al. 2003; Jia et al. 2004), flies (Kapahi et al. 2004), and, soon thereafter, replicative life span in yeast (Kaeberlein et al. 2005a).

The importance of mTORC1 in mammalian aging was first documented in studies performed by the National Institute on Aging's Interventions Testing Program (ITP). The ITP is a program designed to test the effect of interventions solicited from the scientific community on life span in the genetically heterogeneous mouse strain UM-HET3 background (Miller et al. 2007; Nadon et al. 2008). In 2009, the ITP reported that treating mice with the mTORC1 inhibitor rapamycin beginning at 600 days of age was sufficient to increase median and maximum life span in both male and female UM-HET3 mice (Harrison et al. 2009). As with genetic inhibition of mTOR, pharmacological inhibition of mTOR with rapamycin has also been found to increase life span in yeast (Powers et al. 2006; Medvedik et al. 2007), worms (Robida-Stubbs et al. 2012), and flies (Bjedov et al. 2010).

Since publication of the initial study showing life-span extension from rapamycin in UM-HET3 mice, the ITP has also reported life-span extension following rapamycin treatment initiated at 9 mo of age (Miller et al. 2011), as well as a partial dose–response trial testing three different doses of the drug, with the highest dose yielding the most robust increase in life span (Miller et al. 2014). Other groups have also reported life-span extension from rapamycin treatment in the C57BL/6 and 129/Sv mouse strain backgrounds (Anisimov et al. 2011; Neff et al. 2013; Zhang et al. 2014). In addition to extending life span, rapamycin also appears to improve numerous age-related health span parameters, including cancer risk, cardiac function, cognitive function, and immune function (Chen et al. 2009; Halloran et al. 2012; Majumder et al. 2012; Wilkinson et al. 2012; Flynn et al. 2013; Neff et al. 2013; Dai et al. 2014).

In addition to rapamycin treatment, several genetic models of reduced mTORC1 signaling have improved health span and enhanced longevity in mice. Knockout of the *S6k1* gene encoding S6 kinase increases life span in female mice (Selman et al. 2009) and confers enhanced resistance to diet-induced obesity (Um et al. 2004). Mice heterozygous for both mTOR and mLST8, two components of the mTORC1 complex, are also long-lived (Lamming et al. 2012), as are mice expressing hypomorphic alleles of mTOR, which also show preservation of function in many organ systems during aging (Wu et al. 2013).

The complexity of mTOR signaling presents a challenge in untangling the mechanistic basis for the positive effects of mTORC1 inhibition on life span and health span. At a molecular level, there is accumulating evidence that inhibition of mTORC1 increases life span through a combination of differential translation of target mRNAs (for more details, see Zid et al. 2009; Carson et al. 2012), induction of autophagy, and altered mitochondrial metabolism (Fig. 1) (Kaeberlein 2013; Johnson et al. 2015). It remains unclear which of these mechanisms is more important for life-span extension or which tissues and cell types are most critical for mediating their effects on longevity. In mice, a general reduction in inflammation and reduced cancer incidence are also important effects of mTOR inhibition, which contribute to improved longevity and health span. Regardless of the mechanisms of action, the large number of studies documenting life-span extension and improved health span from mTORC1 inhibition provide some expectation that this pathway similarly modulates human aging (discussed below).

Sirtuins

Sirtuins are a family of nicotinamide adenine dinucleotide (NAD^+)-dependent enzymes

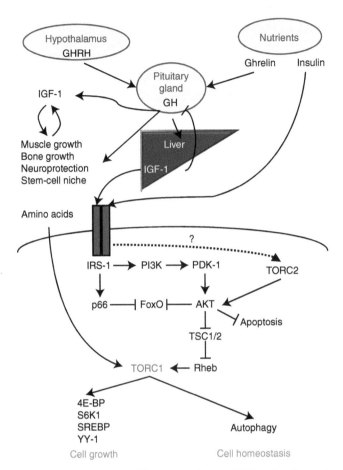

Figure 1. The growth hormone (GH)/insulin-like growth-factor (IGF)-1-signaling axis and its intracellular mediators. GH released from the pituitary gland stimulates IGF-1 production from the liver (systemic IGF-1) and peripheral tissues (local IGF-1). IGF-1 signals to several conserved players in aging research through the IGF-1 receptor, such as the transcription factor FoxO and the mechanistic target of rapamycin complex 1 (TORC1) kinase complex. These factors regulate cellular growth and homeostasis pathways, which can have a profound effect on longevity and aging phenotypes. GHRH, Growth-hormone-releasing hormone; IRS, insulin receptor substrate; PDK-1, phosphoinositide-dependent protein kinase 1; BP, binding protein; SREBP, sterol receptor element-binding protein.

that catalyze posttranslational modification of proteins—primarily deacetylation (Imai et al. 2000; Landry et al. 2000; Smith et al. 2000), but also adenosine diphosphate (ADP)-ribosylation (Haigis et al. 2006). In some cases, removal of succinyl, malonyl, glutaryl, and perhaps other moieties from lysine residues can also be catalyzed by sirtuins (Nakagawa et al. 2009; Du et al. 2011; Peng et al. 2011; Park et al. 2013; Rardin et al. 2013; Tan et al. 2014). The key biochemical feature that distinguishes sirtuins from other protein deacetylases is their

absolute requirement for NAD^+, which gets consumed during these reactions. The term "sirtuin" is derived from the budding yeast Sir2 protein, which was the first member of this family of enzymes to be extensively characterized, and was initially identified as a factor required for transcriptional silencing at the heterothallic (*HM*)-mating loci and subsequently at subtelomeric regions and ribosomal DNA in *S. cerevisiae* (Klar et al. 1979; Ivy et al. 1986; Rine and Herskowitz 1987; Aparicio et al. 1991; Fritze et al. 1997; Smith and Boeke 1997). Yeast and flies have five sirtuin proteins, *C. elegans* has four, and both mice and humans have seven sirtuins (SIRT1-7) (Frye 2000; Giblin et al. 2014).

The importance of sirtuins in aging was first defined in budding yeast, in which overexpression of Sir2 was shown to increase the RLS of mother cells (Kaeberlein et al. 1999). The primary mechanism by which Sir2 promotes replicative longevity in yeast appears to involve increased genomic stability within the rDNA (Gottlieb and Esposito 1989; Kaeberlein et al. 1999), which is composed of dozens of 9.1-kb DNA sequences arrayed in tandem (Petes and Botstein 1977; Philippsen et al. 1978). In addition to enhancing rDNA stability, yeast Sir2 has also been suggested to promote asymmetric retention of damaged proteins within the mother cell, which may also contribute to its prolongevity role (Aguilaniu et al. 2003).

Following the initial report that Sir2 overexpression extends life span in yeast, subsequent studies in both *C. elegans* and *D. melanogaster* reported a similar life-span extension from overexpression of the Sir2 orthologs in those organisms, *sir-2.1* and dSir2, respectively (Tissenbaum and Guarente 2001; Rogina and Helfand 2004). Thus far, no definitive mechanism of action has emerged for how worm and fly sirtuins enhance longevity, and there has been ample controversy over both of the initial reports (Burnett et al. 2011; Viswanathan and Guarente 2011). Although a comprehensive review of the literature on this topic is beyond the scope of this work, the emerging consensus appears to be that the importance of Sir2 orthologs in worm and fly aging is more modest than initially suggested, is likely dependent on specific experimental conditions and genetic background, and that the effects on life span from overexpression may be exquisitely sensitive to dosage (Lombard et al. 2011).

As mentioned above, there are seven mammalian sirtuins with distinct tissue and subcellular distributions, as well as differing biochemical activities (Finkel et al 2009; Satoh et al. 2011; Giblin et al. 2014). SIRT1 is the mammalian ortholog of yeast Sir2 and has been implicated (both as a protective and as risk factor) in a wide array of age-associated disorders, including

heart failure and cardiovascular disease (Hsu et al. 2008; Luo et al. 2014), different forms of cancer (Firestein et al. 2008; Lim et al. 2010; Srisuttee et al. 2012; Yuan et al. 2013), metabolic syndrome (Li 2013; Chang and Guarente 2014), and various neurodegenerative diseases (Herskovits and Guarente 2014). Whole-body overexpression of SIRT1 does not appear to extend life span in mice, even though it improves some metabolic parameters that may be associated with health span (Bordone et al. 2007). Several different promoters, including a variety of tissue- and cell-type-specific drivers, have also failed to show increased life span from SIRT1 overexpression in mice, although in some cases improved health measures were detected, similar to whole body overexpression (Alcendor et al. 2007; Herranz et al. 2010; Jeong et al. 2012). Recently, a mouse overexpressing SIRT1 specifically in the hypothalamus (BRASTO mice—brain-specific SIRT1 overexpressing) was reported to be both long-lived and have improved health span, suggesting that SIRT1 activity in the brain may be particularly important for healthy aging in mice (Satoh et al. 2013).

Although SIRT1 has been studied most extensively, there is a growing body of literature suggesting that the other sirtuin family members, SIRT2-7, may also modulate healthy aging in mammals. SIRT6, in particular, seems to be important for mouse longevity, as $Sirt6^{-/-}$ mice show phenotypes consistent with accelerated aging in some tissues (Mostoslavsky et al. 2006), and male but not female mice overexpressing SIRT6 have extended life span (Kanfi et al. 2012). Life-span extension has thus far not been reported for other sirtuins in wild-type mice; however, overexpression of SIRT2 in short-lived BUBR1 hypomorphic mice can improve cardiac function and partially suppress the life-span defect of male animals (North et al. 2014). SIRT3 is a mitochondrial sirtuin that has also been reported to influence age-related metabolic dysfunction and to promote improved mitochondrial function in a variety of contexts, including during DR (Shi et al. 2005; Nakagawa et al. 2009; Palacios et al. 2009; Schwer et al. 2009; Hirschey et al. 2010; Someya et al. 2010; Hallows et al. 2011).

A large body of work links pharmacological activation of sirtuins with improved metabolic and health span parameters. The first so-called sirtuin-activating compound (STAC) to be studied was resveratrol, which was reported to increase replicative life span in yeast (Howitz et al. 2003). Resveratrol has also been reported to increase life span in worms and flies (Wood et al. 2004) and to increase survival in short-lived mice fed a high-fat diet (Baur et al. 2006). This work has been clouded, however, by reports that resveratrol does not directly activate Sir2 or SIRT1 (Borra et al. 2005; Kaeberlein et al. 2005b), and by studies that have failed to reproduce the reported life span extension from resveratrol in yeast (Kaeberlein et al. 2005b), worms, or flies (Bass et al. 2007). The ITP has also reported that resveratrol has no effect on life span or health span of UM-HET3 mice fed a normal diet (Miller et al. 2011). Additional studies using second-generation STACs have provided further evidence that pharmacological activation of sirtuins may improve some measures of health span during aging in mice, particularly metabolic deficits associated with normative aging or diet-induced obesity, although with only modest increase in life span (Minor et al. 2011; Mercken et al. 2014; Mitchell et al. 2014).

Adenosine Monophosphate-Activated Kinase

The 5′-AMP-activated protein kinase (AMPK) is an important regulator of cellular energy homeostasis. AMPK coordinates the regulation of metabolic pathways, including glucose uptake and utilization as well as oxidation of fats, in response to changes in the relative abundance of AMP and ATP. AMPK is activated by AMP and inhibited by ATP, making it directly responsive to the AMP/ATP ratio, an indicator of energy availability in the cell (Hardie 2007). Several cellular stresses activate AMPK by affecting the AMP/ATP ratio, including starvation, impaired mitochondrial respiration, and hypoxia. Other stresses, such as DNA damage, inhibit AMPK without affecting the AMP/ATP ratio (Budanov and Karin 2008).

Most studies examining the role of AMPK in aging suggest that activation of AMPK is associated with increased longevity and improved health span. Overexpression of one of two worm orthologs of the AMPK-α subunit, AAK-2, is sufficient to extend life span, and deletion of *aak-2* prevents life-span extension in response to certain forms of DR (Apfeld et al. 2004; Schulz et al. 2007; Greer and Brunet 2009). More recently, it has been shown in flies that activation of AMPK in neurons or intestine increases life span and slows aging in a non-cell-autonomous manner (Ulgherait et al. 2014). In mice, chronic activation of AMPK protects against diet-induced obesity (Yang et al. 2008), and two related AMPK activators, phenformin and metformin, have also been reported to enhance longevity in mice (Dilman and Anisimov 1980; Anisimov et al. 2003; Martin-Montalvo et al. 2013) and worms (Onken and Driscoll 2010). Metformin is the most widely prescribed antidiabetic drug in the world. It clearly enhances life expectancy for diabetic patients, and there is some suggestive evidence that diabetics taking metformin may live longer than nondiabetics (Bannister et al. 2014).

One of the challenges associated with interpreting the effects of metformin is that, unlike rapamycin, it is a relatively "dirty drug." The mechanism by which metformin activates AMPK is indirect and poorly understood, and it is unclear whether the effects of metformin on life span and health span are directly mediated solely through AMPK or also through "off-target" effects. For example, metformin is also an inhibitor of complex I of the mitochondrial electron transport chain (El-Mir et al. 2000; Owen et al. 2000) and, in worms, has been suggested to increase life span through reducing folate production by the gut microbiota (Cabreiro et al. 2013) and through activation of the peroxiredoxin PRDX-2 via a mitochondrial stress response (mitohormesis, discussed below) (De Haes et al. 2014).

Insulin-Like Growth Factor 1 (IGF-1) Signaling

The insulin/IGF-1-like signaling axis (IIS) is perhaps the best-studied longevity pathway,

Cite this article as *Cold Spring Harb Perspect Med* doi: 10.1101/cshperspect.a025114

and is known to regulate longevity in worms, flies, mice, and perhaps humans. The genetic and biochemical features of this pathway were first worked out in *C. elegans*, and later studies showed a similar overall structure in flies and mammals (Fontana et al. 2010; Kenyon 2011). Insulin and insulin-like peptides are the main determinants of development, growth, and body size in animals, with highly conserved genes across the evolutionary spectrum. In invertebrate animals, the insulin receptor regulates the response to nutrients as well as growth and development (Fernandez et al. 1995; Kimura et al. 1997; Edgar 2006). In vertebrates, insulin generally regulates nutrient consumption and storage, and IGF-1 promotes growth and proliferation (Upton et al. 1997; Reinecke and Collet 1998). IGF-1 is also responsive to nutrients, although indirectly: nutrients stimulate production of the hypothalamic growth-hormone-releasing hormone (GHRH) (Gelato and Merriam 1986) and Ghrelin, a hormone produced by the digestive system (Kojima et al. 1999). These two hormones signal to the somatotropic cells of the anterior pituitary gland and promote the release of growth hormone (GH) (Forsyth and Wallis 2002), which, in turn, stimulates the production of IGF-1 in the liver and in peripheral tissues. Ghrelin, GHRH, and GH add a further level of complexity to growth, size, and aging regulation in vertebrates.

IGF-1 promotes cell growth, proliferation, and cell-cycle progression by engaging the IGF-1 receptor (IGF-1R). The IGF-1R activates the TOR pathway (Oldham and Hafen 2003), promotes cell survival via AKT (Song et al. 2005), curtails stress responses and cell-cycle arrest by inhibiting the transcription factor FoxO (Stitt et al. 2004), and can activate the mitogen-activated protein kinase (MAPK) pathway through p66[Shc] (Migliaccio et al. 1999; Giorgio et al. 2012).

The IIS was first implicated in aging with the discovery that loss-of-function mutations in the *C. elegans* phosphatidylinositol-3-kinase gene *age-1* and the insulin-like receptor gene *daf-2* could double life span, and this life-span extension was dependent on the FOXO-family transcription factor DAF-16 (Friedman and Johnson 1988; Kenyon et al. 1993; Dorman et al.

1995). Both *daf-2* and *daf-16* had been previously implicated in control of entry into the *C. elegans dauer*, a stress-resistant developmentally arrested alternative developmental state (Vowels and Thomas 1992). At the time their role in longevity control was discovered, the genes for *age-1*, *daf-2*, and *daf-16* had not yet been cloned, and it was not until a few years later that the homologies to the mammalian proteins were appreciated (Morris et al. 1996; Kimura et al. 1997; Ogg et al. 1997). Additionally, components of the pathway were subsequently identified in worms (Tissenbaum and Ruvkun 1998), and similar effects were reported for mutations that reduce IIS in flies (Clancy et al. 2001; Tatar et al. 2001; Tu et al. 2002).

In mice, longer life span and delayed aging have been recorded in Ames dwarf, Snell dwarf, GH-receptor knockout (GHRKO), GHRHKO, GHRH receptor mutants (GHRHR[lit]), and, to a lesser extent, in IGF-1R[+/−] mice (Brown-Borg et al. 1996; Coschigano et al. 2000; Flurkey et al. 2001; Holzenberger et al. 2003; Liou et al. 2013) and IGF-1-deficient mice (Lorenzini et al. 2013). Ames dwarf and Snell dwarf mice produce low amounts of GH, prolactin, and thyroid-stimulating hormone owing to mutations that impair the development of the anterior pituitary gland, and together with GHRKO mice have extremely reduced levels of circulating IGF-1 (Bartke 2005). Conversely, IGF-1 levels are elevated in IGF-1R[+/−] mice, probably because of lack of negative feedback on GH release, but intracellular signaling through AKT and p66[Shc] is dampened in these animals (Holzenberger et al. 2003). IIS mutant mice display higher levels and activity of antioxidant enzymes (Brown-Borg et al. 1999; Brown-Borg and Rakoczy 2000; Hauck and Bartke 2000) and their cells show increased resistance to oxidative stress in culture. In fact, reducing IIS in mice inhibits p66[Shc], activates the phase 2 antioxidant response via NFE2L2 (Nrf2) (Holzenberger et al. 2003; Murakami et al. 2003; Salmon et al. 2005; Leiser and Miller 2010; Sun et al. 2011), and possibly activates the FoxO transcription factor (Nemoto and Finkel 2002). Furthermore, reduced IIS protects mice against age-related pathologies, such as Alzheimer's disease (Cohen et al. 2009).

In addition to increased stress resistance, animals with defective GH signaling have altered metabolism, as evidenced by increased insulin sensitivity and glucose tolerance, reduced fat accumulation, resistance to high-fat diets, and increased oxidative metabolism and fatty acid oxidation, especially at older ages (Bartke and Westbrook 2012). Some of these phenotypes are recapitulated in IGF-1R$^{+/-}$ and IGF-1-deficient mice, such as resistance to high-fat diets and reduced fat accumulation, even though both of these models present slight impairments in glucose homeostasis (Holzenberger et al. 2003; Salmon et al. 2015). Because metabolism is regulated by a limited number of endocrine organs in metazoans, these observations suggest that IIS influences aging both via cell-autonomous mechanisms, such as stress resistance, and nonautonomous ones. This dual role for IIS is supported by several bodies of evidence. First, life-span extension and delayed aging are more prominent in pituitary-deficient and GHRKO mice, which have specific defects in the somatotropic axis, than in the general, broad-targeting, IGF-1R$^{+/-}$ and IGF-1-deficient mice, in which the beneficial effects of the mutations are reduced in magnitude and mostly restricted to female animals (Holzenberger et al. 2003; Lorenzini et al. 2013). Second, mutations in the IIS pathway can extend life span even when they occur only in select endocrine organs, such as adipose tissue in flies and mice, or intestine and neurons in *C. elegans*, thus suggesting that reducing specific endocrine signaling is as effective as a general suppression of growth-factor signaling (Kenyon 2005). This may explain the apparent dichotomy between the specific effects of GH and IGF-1 supplementation on one side, and the long-lived phenotype of reduced GH and IGF-1-signaling mutants on the other side. Although reducing systemic GH and IGF-1 signaling promotes longevity by improving metabolism and raising cellular defenses to stress, high local levels of IGF-1 could protect specific tissues against damage and apoptosis, which is especially detrimental in postmitotic tissues, such as muscle, heart, and brain (Bartke 2008). Interestingly, Ames mice show increased levels of IGF-1

in the brain (Sun et al. 2005a,b), and IGF-1 protects against age-related decline in heart function (Moellendorf et al. 2012), further supporting this hypothesis. This evidence notwithstanding, elevated local IGF-1 may still have detrimental effects for longevity, as indicated by the long-lived phenotype of mice lacking the pregnancy-associated plasma protein A (PAPP-A), a serum metalloprotease that increases peripheral activity of IGF-1 by cleaving IGF-1-binding proteins (IGF-BPs) (Conover and Bale 2007). The impact of GH and IGF-1 on longevity may thus be exquisitely tissue specific. GHR tissue-specific knockout mice have been developed, including fat, macrophage, liver, muscle, and β-cell-specific knockouts, and it will be of particular interest to see which of these animals have altered longevity and health span (Sun and Bartke 2013).

Mitochondrial Stress and Antioxidants

Mitochondrial function has long been thought to be important for healthy aging. Indeed, the first molecular theory of aging proposed that free radicals, produced in part through mitochondrial metabolic processes, were the primary cause of age-related damage and declines in cellular function (Harman 1956). In recent years, the idea that mitochondrial oxidative stress has a purely negative effect on age-related parameters has been revised, based on a growing body of evidence supporting the idea that mitochondrial stress and reactive oxygen species (ROS) can also induce protective mechanisms in addition to damage (Bennett and Kaeberlein 2014), an idea referred to as "mitohormesis" (Ristow and Zarse 2010). In addition, several life-span-extending mutations have been identified that impair mitochondrial function, suggesting a complex relationship between longevity and mitochondria.

Loss-of-function alleles of the genes encoding the Rieske iron sulfur protein *isp-1* and the coenzyme Q biosynthetic gene *clk-1* increase life span in worms (Felkai et al. 1999; Feng et al. 2001). Furthermore, genome-wide RNAi studies in *C. elegans* provide direct evidence that decreasing mitochondrial respiration enhances longevity (Hwang et al. 2012; Yanos et al. 2012). RNAi clones corresponding to numerous components of the mitochondrial electron transport chain were found to extend life span when knockdown was initiated during development (Dillin et al. 2002; Lee et al. 2003). Limited examples of life-span extension from inhibition of mitochondrial function have also been described in fruit flies (Copeland et al. 2009; Owusu-Ansah et al. 2013) and mice (Hughes and Hekimi 2011; Wang et al. 2012), although these appear to be far less common than in *C. elegans*. This may be partially explained by the lack of apoptosis in adult *C. elegans*, which allows worms to maintain somatic tissue in the presence of increased oxidative damage. In all cases, severe mitochondrial dysfunction leads to reduced survival, consistent with the central idea of mitohormesis that a little mitochondrial stress is protective, although a large degree of mitochondrial stress is detrimental. This concept has been further tested in studies showing that treatment with low doses of paraquat, which produce mitochondrial ROS in the form of superoxide, is sufficient to enhance longevity in worms (Yang and Hekimi 2010; Hwang et al. 2014).

Although inhibition of mitochondrial respiration appears to be a conserved prolongevity strategy, the mechanistic basis for this life-span extension remains obscure, and the degree to which similar mechanisms are engaged in different organisms is unclear (Bennett et al. 2014a). It was initially suggested that mitochondrial inhibition extends life span in *C. elegans* through generation of a neuronal signal that is propagated to the intestine and other tissues to induce the mitochondrial unfolded protein response (UPR^{mt}), which up-regulates mitochondrial-specific chaperones and proteases (Durieux et al. 2011). The neuronal mitochondrial signal has not yet been identified; however, one study in flies suggests that mitochondrial dysfunction in muscle can also induce a prolongevity signal in that organism (Owusu-Ansah et al. 2013). The model that the UPR^{mt} promotes longevity has since been weakened, however, with the finding that induction of the UPR^{mt} is neither necessary nor sufficient for

life-span extension in worms (Bennett et al. 2014b). Despite the lack of causal evidence that the UPR^{mt} mediates enhanced longevity, mitochondrial stress in long-lived flies and mice is capable of activating the UPR^{mt} in certain tissues, akin to *C. elegans* (Owusu-Ansah et al. 2013; Pulliam et al. 2014). Thus, the UPR^{mt} may be one response that is induced to regulate mitochondrial proteostasis in conjunction with others that directly promote longevity.

Several groups have attempted to define the mechanistic basis for life-span extension from electron transport chain (ETC) inhibition in *C. elegans*, resulting in the identification of multiple transcription factors that are required for enhanced longevity in a subset of cases. These include AHA-1, CEH-18, CEH-23, CEP-1, GCN-2, HIF-1, JUN-1, NHR-27, NHR-49, and TAF-4 (Ventura et al. 2009; Lee et al. 2010a; Walter et al. 2011; Baker et al. 2012; Khan et al. 2013). Of these, hypoxia-inducible factor 1 (HIF-1) is particularly interesting, as HIF-1 serves as a major transcriptional regulator of the hypoxic response, which is highly conserved in metazoans (Leiser and Kaeberlein 2010). HIF-1 is activated in response to low levels of ROS produced by some forms of mitochondrial inhibition (Lee et al. 2010a), and activation of HIF-1, either through genetic manipulation or hypoxic conditioning, is sufficient to extend the life span in *C. elegans* (Mehta et al. 2009; Muller et al. 2009; Zhang et al. 2009a; Leiser et al. 2011, 2013). In addition, HIF-1 can be activated in mammalian cells by metabolites that are enriched in *C. elegans* mitochondrial mutants, suggesting multiple mechanisms that may impinge on HIF-1 during mitochondrial dysfunction (Lu et al. 2005; Butler et al. 2013). Chronic activation of HIF-1 in humans causes Von Hippel–Lindau syndrome and is, therefore, unlikely to improve longevity or health span (Ivan and Kaelin 2001). However, it remains possible that tissue-specific activation of HIF-1 or activation of a subset of HIF-1 target genes could be beneficial for longevity in mammals. Recently, another mechanism for life-span extension from inhibition of respiration in *C. elegans* has been proposed involving the activation of the intrinsic apoptotic pathway

(Yee et al. 2014), but whether this mechanism plays any role in aging in other organisms is yet to be determined.

Another aspect of mitohormesis that should be considered is the role of antioxidant systems in coping with oxidative stress. Studies implicating mitohormesis with increased antioxidant capacity were first described in *C. elegans* using both inhibitors of glycolysis and knockdown of *daf-2* (Schulz et al. 2007; Zarse et al. 2012). These interventions were found to increase mitochondrial respiration, cause a burst of ROS, and increase cellular antioxidant activity. This response and the associated life-span extensions were proposed to be mediated by SKN-1 and PMK-1. SKN-1 is the worm homolog of the NFE2L2 (Nrf2) transcription factor of the phase 2 detoxification response, which controls expression of antioxidant and glutathione biosynthesis genes, whereas PMK-1 is a p38 MAPK homolog that regulates SKN-1 nuclear localization and activity under oxidative stress conditions (Inoue et al. 2005). Overexpression of SKN-1 is sufficient to extend life span in *C. elegans* (Tullet et al. 2008). Additionally, flies with reduced expression of the Nrf2 negative regulator Keap1 show enhanced paraquat resistance and life-span extension (Sykiotis and Bohmann 2008).

The role of Nrf2/SKN-1 in aging of mammals is correlative but compelling. In rodents, Nrf2 activity decreases with age (Suh et al. 2004; Shih and Yen 2007), conceivably leading to increased oxidative damage. In addition, DR and many DR mimetics activate Nrf2 and increase antioxidant capacity of mice (Martin-Montalvo et al. 2011). However, Nrf2 KO mice still receive benefits from DR, such as increased insulin sensitivity and similar increased life span compared with wild-type controls (Pearson et al. 2008). Whether Nrf2 mediates any of the health benefits of the long-lived mitochondrial mutant mice models is still an open question. For example, mice lacking the ETC chaperone Surf1 show Nrf2 activation in cardiac tissue (Pulliam et al. 2014), but whether this is beneficial to cardiac aging or is just an indicator of mitochondrial dysfunction in this background is unknown. Nevertheless, activation of Nrf2 and

other antioxidant pathways is a common feature of several long-lived mouse models (Leiser and Miller 2010; Sun et al. 2011; Michael et al. 2012), and Nrf2 mediates at least some of the prolongevity effects of rapamycin in vitro (Lerner et al. 2013), suggesting that this transcription factor may actually play an active role in enhancing health span.

A similar case could be drawn for p66[shc], a mediator of GH/IGF-1 signaling (see above) that regulates oxidative stress responses. Although its ablation increases resistance to oxidative and genotoxic stress, and was shown to extend life span in laboratory mice (Migliaccio et al. 1999), more recent studies have brought into question the long-lived phenotypes of p66[shc] knockout mice (Giorgio et al. 2012; Ramsey et al. 2014), suggesting a complex relationship between oxidative stress, stress resistance, and longevity.

For many years, it was conventional wisdom that antioxidants should have a beneficial effect on longevity and health. The mitohormesis model and supporting data would suggest this idea is overly simplistic, and the actual data showing a direct relationship between exogenous treatment with antioxidants and longevity or health span in any species is quite limited. In mammals, the strongest evidence in support of the idea that boosting antioxidant capacity can improve healthy aging comes from studies of mice transgenically expressing mitochondrial targeted human catalase (MCAT). Catalase catalyzes the breakdown of H_2O_2 and normally functions within the peroxisome. MCAT mice show increases in life span and decreases in cardiac pathology, inflammation, and tumor burden (Schriner et al. 2005; Dai et al. 2009; Lee et al. 2010b). Furthermore, MCAT mice maintain better muscle function with age owing to reduced oxidation of ryanodine receptor 1, the Ca^{2+} release channel of the sarcoplasmic reticulum involved in muscle contraction (Umanskaya et al. 2014). Ectopic expression of other antioxidant enzymes, such as thioredoxin-1, thioredoxin reductase, Mn superoxide dismutase (MnSOD), copper zinc superoxide dismutase (CuZnSOD), or nonmitochondrial catalase, does not cause robust life-span extension in fruit flies or mice, even in cases of combinatorial expression (Mitsui et al. 2002; Orr et al. 2003; Jang et al. 2009; Perez et al. 2009, 2011). In both mice and worms, deletion of antioxidant enzymes in many cases does not shorten life span, despite causing substantial increases in oxidative damage to cells and tissues and striking sensitivity to oxidative stress (Van Raamsdonk and Hekimi 2009; Zhang et al. 2009b). Taken together, the bulk of data would seem to suggest that antioxidant capacity does not seem to be generally limiting for life span of wild-type organisms. However, increased antioxidant capacity or improvements in redox homeostasis, perhaps in specific cellular compartments, such as mitochondria, can improve longevity and health during aging at least in some situations.

CONNECTIONS BETWEEN CONSERVED ANTIAGING PATHWAYS: TOWARD A NETWORK VIEW OF AGING

As mentioned in the introductory section, one of the general features of known conserved longevity pathways is that they tend to regulate the relationship between environmental cues and growth and reproduction. Therefore, it is not surprising that all of the major conserved longevity pathways have been proposed to play key roles in mediating the beneficial effects of DR. In nonmammalian models, the case can best be made for inhibition of mTOR acting downstream from DR, as mTOR is a direct sensor of amino acids (Bar-Peled and Sabatini 2014), and studies in yeast, worms, and flies have indicated that inhibition of mTOR, or downstream processes regulated by mTOR, such as autophagy, are both necessary and sufficient for lifespan extension from DR (Kapahi et al. 2004; Kaeberlein et al. 2005a; Jia and Levine 2007; Morck and Pilon 2007; Zid et al. 2009). In mammals, the situation is probably more complex, and initial studies suggest that DR and rapamycin induce overlapping but distinct changes, suggesting that rapamycin treatment does not mimic every aspect of DR or vice versa (Fok et al. 2014).

IIS and DR increase life span through overlapping but also distinct mechanisms. DR and reduced IIS affect growth, metabolism, endocrine signaling, and stress resistance in a similar way (Bartke et al. 2001; Yamaza et al. 2010), yet DR can further increase life span in Ames mice (Bartke et al. 2001), GHRHKO (Liou et al. 2013), and, to a lesser extent, in GHRKO mice (Bonkowski et al. 2006), suggesting that DR engages other pathways in addition to IIS. Similarly, DR increases life span in dFoxO$^{-/-}$ flies, although not to the extent of wild-type animals. In worms, several studies have reported that different methods of DR are able to extend life span in a *daf-16* null background, although removal of food is sufficient to activate DAF-16 (Henderson and Johnson 2001), suggesting that the IIS pathway is engaged by DR in worms.

Whether rapamycin or other DR mimetics further extend the life span of IIS-mutant or dietary-restricted mammals is still unknown. Preliminary evidence suggests that DR inhibits mTOR to promote the function of at least one stem-cell niche, the Paneth cells of the intestine (Yilmaz et al. 2012). In addition, S6K1$^{-/-}$ mice show several features typical of the IIS mutants or of mice under DR, such as reduced size, better insulin sensitivity, and improved glucose tolerance at old ages, and have a gene expression profile that overlap with IRS1$^{-/-}$ and dietary-restricted mice (Selman et al. 2009). Together with the observation that Ames dwarf mice have reduced mTOR signaling (Sharp and Bartke 2005), this evidence further suggests that DR, IIS, and TOR extend longevity through overlapping pathways in mice too.

A role for sirtuins in DR was first suggested based on studies in yeast showing that deletion of Sir2 prevents replicative life-span extension from DR (Lin et al. 2000). Based on this observation, Guarente and colleagues proposed that DR might activate sirtuins by increasing the availability of NAD$^+$, a substrate of sirtuin-mediated deacetylation (Guarente and Picard 2005). The generality of this model has been weakened by subsequent reports that neither Sir2 nor the other yeast sirtuins are absolutely required for replicative life-span extension from DR in yeast, and that overexpression of Sir2 further increases life span in combination with DR (Kaeberlein et al. 2004, 2006; Tsuchiya et al.

2006). Likewise, Sir2 is not required for chronological life-span extension from DR in yeast, and *sir-2.1* is not required for life-span extension from several different methods of DR in *C. elegans* (Kaeberlein and Powers 2007). In contrast, studies in mice have generally supported the idea that sirtuins, particularly SIRT1 and SIRT3 are activated in some tissues by DR and may be required for some of the health benefits associated with this dietary regimen (Giblin et al. 2014).

There is abundant evidence that mitochondrial metabolism is altered in response to DR, although whether this is mechanistically related to the life-span extension seen in respiratory-deficient animals is unclear. In *C. elegans*, inhibition of electron transport chain components only extends life span when knockdown is accomplished during development (Rea et al. 2007), whereas DR extends life span in *C. elegans* when initiated during adulthood (Smith et al. 2008b), suggesting that these are distinct longevity pathways. In yeast, mitochondrial biogenesis and mitochondrial respiration are increased in response to DR (Lin et al. 2002), and similar effects have been reported in some tissues of mice and in cultured mammalian cells (Nisoli et al. 2005; Lopez-Lluch et al. 2006), further supporting the idea that mitochondrial inhibition extends life span by a mechanism distinct from that of DR.

AMPK has also been implicated in mediating mitochondrial and metabolic adaptation to DR. In worms, most forms of DR are dependent on *aak-2* (Greer and Brunet 2009) and AMPK is necessary for the beneficial effects of DR on cardiac function in mouse (Chen et al. 2012). Furthermore, metformin, an AMPK activator, induces metabolic changes in mouse liver and skeletal muscle consistent with those seen under caloric restriction, although not completely overlapping (Martin-Montalvo et al. 2013). These observations indicate that DR may exert some of its beneficial effects through AMPK, although compelling evidence is still lacking, especially in mammals.

Importantly, all of the pathways described above interact with each other, making it difficult and perhaps unrealistic to pinpoint any individual factor as the sole or even the major mediator of DR. For example, both IGF-1 and AMPK can modulate the TOR pathway, and all three pathways can influence mitochondrial function and metabolism independently and in concert with one another (see above). Similarly, cross talk between the TOR pathway and several sirtuins has been found in vitro (Csibi et al. 2013; Hong et al. 2014), several proteins of the IGF-1R-signaling pathway are subject to deacetylation by sirtuins in vitro (Zhang 2007; Sundaresan et al. 2011), and AMPK can promote sirtuin activity by increasing the levels of NAD^+ in mouse skeletal muscle (Canto et al. 2009). In addition, mitochondrial dysfunction and ROS can activate TOR independent of growth factors and nutrients (Nacarelli et al. 2014). Together, these observations suggest that age-delaying interventions are likely to engage multiple pathways to some extent and that, although single gene perturbations may have sizable effects in term of health and longevity, no single factor is solely responsible for long-lived and healthy aging phenotypes. Importantly, all of the above age-related pathways increase the activity of homeostatic mechanisms, such as authophagy and stress-resistance and detoxification enzymes. The importance of these mechanisms for aging phenotypes, as well as the cells and tissue most affected by their activity, are currently object of intense research and will likely help fill in the blanks on the network of interaction that affect aging and age-related diseases.

CONCLUSIONS AND FUTURE DIRECTIONS

Over the past few decades, there have been numerous successes at identifying genetic factors that appear to modulate the rate of aging in evolutionarily divergent organisms. We now know of several conserved pathways and gene families that are key modulators of aging, and a picture of the network in which they interact is beginning to emerge. In the case of rapamycin, these advances have progressed to the point where we have a pharmacological intervention with the ability to extend life span and delay age-related declines in function in organisms from yeast to worms. Going forward, we anticipate these trends to continue, with the identification of additional longevity-promoting interventions and further advances in understanding the molecular mechanisms of aging.

A key challenge, however, will be developing strategies to successfully translate these discoveries to improve human health span. One approach is to attempt to validate the efficacy of interventions, such as rapamycin, to treat age-related diseases in patients. Although this type of approach may be successful, it is unclear whether any benefits will be obtained from delaying aging in individuals whose health has already deteriorated to the point where clinical diagnosis has occurred. A more promising approach is to test these interventions as preventative, or even rejuvenating, therapies in people. One good example of this is a recent placebo controlled study in which short-term treatment with low-dose rapamycin was found to improve the immune response to an influenza vaccine of otherwise healthy elderly people (Mannick et al. 2014). Although it remains unclear whether these effects are mechanistically related to the life-span extension seen in mice following rapamycin treatment, this parallels prior work showing a similar rejuvenation of immune function in aged mice (Chen et al. 2009). Although this study does not show that the other positive effects of rapamycin on longevity and health span in mice will also be seen in humans, it is highly suggestive.

In addition to examining the potential benefits of rapamycin as a preventative measure for select age-related phenotypes in humans, we have recently proposed that a more comprehensive evaluation of rapamycin in a large mammal can be accomplished using the domestic dog *Canis lupus familiaris* (Check Hayden 2014). Companion dogs, in particular, have several advantages as a model for human aging over laboratory organisms. These include a breadth of phenotypic and genetic diversity across breeds, a high quality of veterinary care and expertise, exposure to most of the same environmental conditions as people, susceptibility to many of the same diseases of aging as people, and life spans that are amenable to experimental time frames. For example, if rapamycin treatment initiated in middle age extends life span and health span in dogs similarly to the effects seen in mice, we might expect the average life span of a cohort of large-size dogs to increase from about 9 years to about 11 years if treatment were initiated at 6–7 years of age. Thus, a 5-year study would be more than sufficient to document these effects if sufficiently powered.

ACKNOWLEDGMENTS

Studies related to this topic in the Kaeberlein laboratory are supported by National Institutes of Health (NIH) Grants AG038518, AG039390, and AG033598 to M.K. A.B. and A.M.W. are supported by NIH Grant T32AG000057. C.F.B. is supported by NIH Training Grant T32ES007032.

REFERENCES

Aguilaniu H, Gustafsson L, Rigoulet M, Nystrom T. 2003. Asymmetric inheritance of oxidatively damaged proteins during cytokinesis. *Science* **299:** 1751–1753.

Alcendor RR, Gao S, Zhai P, Zablocki D, Holle E, Yu X, Tian B, Wagner T, Vatner SF, Sadoshima J. 2007. Sirt1 regulates aging and resistance to oxidative stress in the heart. *Circ Res* **100:** 1512–1521.

Anisimov VN, Semenchenko AV, Yashin AI. 2003. Insulin and longevity: Antidiabetic biguanides as geroprotectors. *Biogerontology* **4:** 297–307.

Anisimov VN, Zabezhinski MA, Popovich IG, Piskunova TS, Semenchenko AV, Tyndyk ML, Yurova MN, Rosenfeld SV, Blagosklonny MV. 2011. Rapamycin increases life-span and inhibits spontaneous tumorigenesis in inbred female mice. *Cell Cycle* **10:** 4230–4236.

Aparicio OM, Billington BL, Gottschling DE. 1991. Modifiers of position effect are shared between telomeric and silent mating-type loci in *S. cerevisiae*. *Cell* **66:** 1279–1287.

Apfeld J, O'Connor G, McDonagh T, DiStefano PS, Curtis R. 2004. The AMP-activated protein kinase AAK-2 links energy levels and insulin-like signals to lifespan in *C. elegans*. *Genes Dev* **18:** 3004–3009.

Ashrafi K, Sinclair D, Gordon JI, Guarente L. 1999. Passage through stationary phase advances replicative aging in *Saccharomyces cerevisiae*. *Proc Natl Acad Sci* **96:** 9100–9105.

Austad SN. 2009. Comparative biology of aging. *J Gerontol A Biol Sci Med Sci* **64:** 199–201.

Bai H, Kang P, Hernandez AM, Tatar M. 2013. Activin signaling targeted by insulin/dFOXO regulates aging and muscle proteostasis in *Drosophila*. *PLoS Genet* **9:** e1003941.

Baker BM, Nargund AM, Sun T, Haynes CM. 2012. Protective coupling of mitochondrial function and protein synthesis via the eIF2α kinase GCN-2. *PLoS Genet* **8**: e1002760.

Bannister CA, Holden SE, Jenkins-Jones S, Morgan CL, Halcox JP, Schernthaner G, Mukherjee J, Currie CJ. 2014. Can people with type 2 diabetes live longer than those without? A comparison of mortality in people initiated with metformin or sulphonylurea monotherapy and matched, non-diabetic controls. *Diabetes Obes Metab* **16**: 1165–1173.

Bar-Peled L, Sabatini DM. 2014. Regulation of mTORC1 by amino acids. *Trends Cell Biol* **24**: 400–406.

Bartke A. 2005. Minireview: Role of the growth hormone/insulin-like growth factor system in mammalian aging. *Endocrinology* **146**: 3718–3723.

Bartke A. 2008. Growth hormone and aging: A challenging controversy. *Clin Interv Aging* **3**: 659–665.

Bartke A, Westbrook R. 2012. Metabolic characteristics of long-lived mice. *Front Genet* **3**: 288.

Bartke A, Wright J, Mattison J, Ingram D, Miller R, Roth G. 2001. Extending the lifespan of long-lived mice. *Nature* **414**: 412.

Bass TM, Weinkove D, Houthoofd K, Gems D, Partridge L. 2007. Effects of resveratrol on lifespan in *Drosophila melanogaster* and *Caenorhabditis elegans*. *Mech Ageing Dev* **128**: 546–552.

Baur JA, Pearson KJ, Price NL, Jamieson HA, Lerin C, Kalra A, Prabhu VV, Allard JS, Lopez-Lluch G, Lewis K, et al. 2006. Resveratrol improves health and survival of mice on a high-calorie diet. *Nature* **444**: 337–342.

Bennett CF, Kaeberlein M. 2014. The mitochondrial unfolded protein response and increased longevity: Cause, consequence, or correlation? *Exp Gerontol* **56**: 142–146.

Bennett CF, Choi H, Kaeberlein M. 2014a. Searching for the elusive mitochondrial longevity signal in *C. elegans*. *Worm* **3**: e959404.

Bennett CF, Vander Wende H, Simko M, Klum S, Barfield S, Choi H, Pineda VV, Kaeberlein M. 2014b. Activation of the mitochondrial unfolded protein response does not predict longevity in *Caenorhabditis elegans*. *Nat Commun* **5**: 3483.

Bjedov I, Toivonen JM, Kerr F, Slack C, Jacobson J, Foley A, Partridge L. 2010. Mechanisms of life span extension by rapamycin in the fruit fly *Drosophila melanogaster*. *Cell Metab* **11**: 35–46.

Bonkowski M, Rocha J, Masternak M, Al Regaiey K, Bartke A. 2006. Targeted disruption of growth hormone receptor interferes with the beneficial actions of calorie restriction. *Proc Natl Acad Sci* **103**: 7901–7905.

Bordone L, Cohen D, Robinson A, Motta MC, van Veen E, Czopik A, Steele AD, Crowe H, Marmor S, Luo J, et al. 2007. SIRT1 transgenic mice show phenotypes resembling calorie restriction. *Aging Cell* **6**: 759–767.

Borra MT, Smith BC, Denu JM. 2005. Mechanism of human SIRT1 activation by resveratrol. *J Biol Chem* **280**: 17187–17195.

Brown-Borg H, Rakoczy S. 2000. Catalase expression in delayed and premature aging mouse models. *Exp Gerontol* **35**: 199–212.

Brown-Borg H, Borg K, Meliska C, Bartke A. 1996. Dwarf mice and the ageing process. *Nature* **384**: 33.

Brown-Borg H, Bode A, Bartke A. 1999. Antioxidative mechanisms and plasma growth hormone levels: Potential relationship in the aging process. *Endocrine* **11**: 41–48.

Budanov AV, Karin M. 2008. p53 target genes sestrin1 and sestrin2 connect genotoxic stress and mTOR signaling. *Cell* **134**: 451–460.

Burnett C, Valentini S, Cabreiro F, Goss M, Somogyvari M, Piper MD, Hoddinott M, Sutphin GL, Leko V, McElwee JJ, et al. 2011. Absence of effects of Sir2 overexpression on lifespan in *C. elegans* and *Drosophila*. *Nature* **477**: 482–485.

Burtner CR, Murakami CJ, Kennedy BK, Kaeberlein M. 2009. A molecular mechanism of chronological aging in yeast. *Cell Cycle* **8**: 1256–1270.

Burtner CR, Murakami CJ, Olsen B, Kennedy BK, Kaeberlein M. 2011. A genomic analysis of chronological longevity factors in budding yeast. *Cell Cycle* **10**: 1385–1396.

Butler JA, Mishur RJ, Bhaskaran S, Rea SL. 2013. A metabolic signature for long life in the *Caenorhabditis elegans* Mit mutants. *Aging Cell* **12**: 130–138.

Cabreiro F, Au C, Leung KY, Vergara-Irigaray N, Cocheme HM, Noori T, Weinkove D, Schuster E, Greene ND, Gems D. 2013. Metformin retards aging in *C. elegans* by altering microbial folate and methionine metabolism. *Cell* **153**: 228–239.

Canto C, Gerhart-Hines Z, Feige JN, Lagouge M, Noriega L, Milne JC, Elliott PJ, Puigserver P, Auwerx J. 2009. AMPK regulates energy expenditure by modulating NAD$^+$ metabolism and SIRT1 activity. *Nature* **458**: 1056–1060.

Carson CT, Lynne C, Heather RK, Tim W, Nathanael SG, David MS. 2012. A unifying model for mTORC1-mediated regulation of mRNA translation. *Nature* **485**: 109–113.

Chang HC, Guarente L. 2014. SIRT1 and other sirtuins in metabolism. *Trends Endocrinol Metab* **25**: 138–145.

Charest P, Shen Z, Lakoduk A, Sasaki A, Briggs S, Firtel R. 2010. A Ras signaling complex controls the RasC-TORC2 pathway and directed cell migration. *Dev Cell* **18**: 737–749.

Check Hayden E. 2014. Pet dogs set to test anti-ageing drug. *Nature* **514**: 546.

Chen C, Liu Y, Zheng P. 2009. mTOR regulation and therapeutic rejuvenation of aging hematopoietic stem cells. *Sci Signal* **2**: ra75.

Chen K, Kobayashi S, Xu X, Viollet B, Liang Q. 2012. AMP activated protein kinase is indispensable for myocardial adaptation to caloric restriction in mice. *PLoS ONE* **8**: e59682.

Cho J, Hur JH, Walker DW. 2011. The role of mitochondria in *Drosophila* aging. *Exp Gerontol* **46**: 331–334.

Clancy DJ, Gems D, Harshman LG, Oldham S, Stocker H, Hafen E, Leevers SJ, Partridge L. 2001. Extension of lifespan by loss of CHICO, a *Drosophila* insulin receptor substrate protein. *Science* **292**: 104–106.

Cohen E, Paulsson J, Blinder P, Burstyn-Cohen T, Du D, Estepa G, Adame A, Pham H, Holzenberger M, Kelly J, et al. 2009. Reduced IGF-1 signaling delays age-associated proteotoxicity in mice. *Cell* **139**: 1157–1169.

Cite this article as *Cold Spring Harb Perspect Med* doi: 10.1101/cshperspect.a025114

Conover CA, Bale LK. 2007. Loss of pregnancy-associated plasma protein A extends lifespan in mice. *Aging Cell* **6:** 727–729.

Copeland JM, Cho J, Lo T Jr, Hur JH, Bahadorani S, Arabyan T, Rabie J, Soh J, Walker DW. 2009. Extension of *Drosophila* life span by RNAi of the mitochondrial respiratory chain. *Curr Biol* **19:** 1591–1598.

Coschigano K, Clemmons D, Bellush L, Kopchick J. 2000. Assessment of growth parameters and life span of GHR/BP gene-disrupted mice. *Endocrinology* **141:** 2608–2613.

Coschigano KT, Holland AN, Riders ME, List EO, Flyvbjerg A, Kopchick JJ. 2003. Deletion, but not antagonism, of the mouse growth hormone receptor results in severely decreased body weights, insulin, and insulin-like growth factor I levels and increased life span. *Endocrinology* **144:** 3799–3810.

Csibi A, Fendt SMM, Li C, Poulogiannis G, Choo AY, Chapski DJ, Jeong SM, Dempsey JM, Parkhitko A, Morrison T, et al. 2013. The mTORC1 pathway stimulates glutamine metabolism and cell proliferation by repressing SIRT4. *Cell* **153:** 840–854.

Dai DF, Santana LF, Vermulst M, Tomazela DM, Emond MJ, MacCoss MJ, Gollahon K, Martin GM, Loeb LA, Ladiges WC, et al. 2009. Overexpression of catalase targeted to mitochondria attenuates murine cardiac aging. *Circulation* **119:** 2789–2797.

Dai DF, Karunadharma PP, Chiao YA, Basisty N, Crispin D, Hsieh EJ, Chen T, Gu H, Djukovic D, Raftery D, et al. 2014. Altered proteome turnover and remodeling by short-term caloric restriction or rapamycin rejuvenate the aging heart. *Aging Cell* **13:** 529–539.

Defossez PA, Park PU, Guarente L. 1998. Vicious circles: A mechanism for yeast aging. *Curr Opin Microbiol* **1:** 707–711.

Defossez PA, Prusty R, Kaeberlein M, Lin SJ, Ferrigno P, Silver PA, Keil RL, Guarente L. 1999. Elimination of replication block protein Fob1 extends the life span of yeast mother cells. *Mol Cell* **3:** 447–455.

De Haes W, Frooninckx L, Van Assche R, Smolders A, Depuydt G, Billen J, Braeckman BP, Schoofs L, Temmerman L. 2014. Metformin promotes lifespan through mitohormesis via the peroxiredoxin PRDX-2. *Proc Natl Acad Sci* **111:** E2501–E2509.

Delaney JR, Murakami C, Chou A, Carr D, Schleit J, Sutphin GL, An EH, Castanza AS, Fletcher M, Goswami S, et al. 2013. Dietary restriction and mitochondrial function link replicative and chronological aging in *Saccharomyces cerevisiae*. *Exp Gerontol* **48:** 1006–1013.

Dillin A, Hsu AL, Arantes-Oliveira N, Lehrer-Graiwer J, Hsin H, Fraser AG, Kamath RS, Ahringer J, Kenyon C. 2002. Rates of behavior and aging specified by mitochondrial function during development. *Science* **298:** 2398–2401.

Dilman VM, Anisimov VN. 1980. Effect of treatment with phenformin, diphenylhydantoin or L-dopa on life span and tumour incidence in C3H/Sn mice. *Gerontology* **26:** 241–246.

Dorman JB, Albinder B, Shroyer T, Kenyon C. 1995. The *age-1* and *daf-2* genes function in a common pathway to control the lifespan of *Caenorhabditis elegans*. *Genetics* **141:** 1399–1406.

Du J, Zhou Y, Su X, Yu JJ, Khan S, Jiang H, Kim J, Woo J, Kim JH, Choi BH, et al. 2011. Sirt5 is a NAD-dependent protein lysine demalonylase and desuccinylase. *Science* **334:** 806–809.

Durieux J, Wolff S, Dillin A. 2011. The cell-non-autonomous nature of electron transport chain-mediated longevity. *Cell* **144:** 79–91.

Edgar B. 2006. How flies get their size: Genetics meets physiology. *Nat Rev Genet* **7:** 907–916.

Egilmez NK, Jazwinski SM. 1989. Evidence for the involvement of a cytoplasmic factor in the aging of the yeast *Saccharomyces cerevisiae*. *J Bacteriol* **171:** 37–42.

El-Mir MY, Nogueira V, Fontaine E, Averet N, Rigoulet M, Leverve X. 2000. Dimethylbiguanide inhibits cell respiration via an indirect effect targeted on the respiratory chain complex I. *J Biol Chem* **275:** 223–228.

Erjavec N, Nystrom T. 2007. Sir2p-dependent protein segregation gives rise to a superior reactive oxygen species management in the progeny of *Saccharomyces cerevisiae*. *Proc Natl Acad Sci* **104:** 10877–10881.

Erjavec N, Larsson L, Grantham J, Nystrom T. 2007. Accelerated aging and failure to segregate damaged proteins in Sir2 mutants can be suppressed by overproducing the protein aggregation-remodeling factor Hsp104p. *Genes Dev* **21:** 2410–2421.

Fabrizio P, Longo VD. 2007. The chronological life span of *Saccharomyces cerevisiae*. *Methods Mol Biol* **371:** 89–95.

Fabrizio P, Pozza F, Pletcher SD, Gendron CM, Longo VD. 2001. Regulation of longevity and stress resistance by Sch9 in yeast. *Science* **292:** 288–290.

Felkai S, Ewbank JJ, Lemieux J, Labbe JC, Brown GG, Hekimi S. 1999. CLK-1 controls respiration, behavior and aging in the nematode *Caenorhabditis elegans*. *EMBO J* **18:** 1783–1792.

Feng J, Bussiere F, Hekimi S. 2001. Mitochondrial electron transport is a key determinant of life span in *Caenorhabditis elegans*. *Dev Cell* **1:** 633–644.

Fernandez R, Tabarini D, Azpiazu N, Frasch M, Schlessinger J. 1995. The *Drosophila* insulin receptor homolog: A gene essential for embryonic development encodes two receptor isoforms with different signaling potential. *EMBO J* **14:** 3373–3384.

Finkel T, Deng CX, Mostoslavsky R. 2009. Recent progress in the biology and physiology of sirtuins. *Nature* **460:** 587–591.

Firestein R, Blander G, Michan S, Oberdoerffer P, Ogino S, Campbell J, Bhimavarapu A, Luikenhuis S, de Cabo R, Fuchs C, et al. 2008. The SIRT1 deacetylase suppresses intestinal tumorigenesis and colon cancer growth. *PLoS ONE* **3:** e2020.

Flurkey K, Papaconstantinou J, Miller R, Harrison D. 2001. Lifespan extension and delayed immune and collagen aging in mutant mice with defects in growth hormone production. *Proc Natl Acad Sci* **98:** 6736–6741.

Flynn JM, O'Leary MN, Zambataro CA, Academia EC, Presley MP, Garrett BJ, Zykovich A, Mooney SD, Strong R, Rosen CJ, et al. 2013. Late-life rapamycin treatment reverses age-related heart dysfunction. *Aging Cell* **12:** 851–862.

Fok WC, Bokov A, Gelfond J, Yu Z, Zhang Y, Doderer M, Chen Y, Javors M, Wood WH III, Zhang Y, et al. 2014.

Combined treatment of rapamycin and dietary restriction has a larger effect on the transcriptome and metabolome of liver. *Aging Cell* **13:** 311–319.

Fontana L, Partridge L, Longo VD. 2010. Extending healthy life span—From yeast to humans. *Science* **328:** 321–326.

Forsyth I, Wallis M. 2002. Growth hormone and prolactin—Molecular and functional evolution. *J Mamm Gland Biol Neoplasia* **7:** 291–312.

Friedman DB, Johnson TE. 1988. A mutation in the *age-1* gene in *Caenorhabditis elegans* lengthens life and reduces hermaphrodite fertility. *Genetics* **118:** 75–86.

Fritze CE, Verschueren K, Strich R, Easton Esposito R. 1997. Direct evidence for SIR2 modulation of chromatin structure in yeast rDNA. *EMBO J* **16:** 6495–6509.

Frye RA. 2000. Phylogenetic classification of prokaryotic and eukaryotic Sir2-like proteins. *Biochem Biophys Res Commun* **273:** 793–798.

García-Martínez J, Alessi D. 2008. mTOR complex 2 (mTORC2) controls hydrophobic motif phosphorylation and activation of serum- and glucocorticoid-induced protein kinase 1 (SGK1). *Biochem J* **416:** 375–385.

Gelato M, Merriam G. 1986. Growth hormone releasing hormone. *Annu Rev Physiol* **48:** 569–591.

Gershon H, Gershon D. 2000. Paradigms in aging research: A critical review and assessment. *Mech Ageing Dev* **117:** 21–28.

Giblin W, Skinner ME, Lombard DB. 2014. Sirtuins: Guardians of mammalian healthspan. *Trends Genet* **30:** 271–286.

Giorgio M, Berry A, Berniakovich I, Poletaeva I, Trinei M, Stendardo M, Hagopian K, Ramsey J, Cortopassi G, Migliaccio E, et al. 2012. The p66[Shc] knocked out mice are short lived under natural condition. *Aging Cell* **11:** 162–168.

Gottlieb S, Esposito RE. 1989. A new role for a yeast transcriptional silencer gene, *SIR2*, in regulation of recombination in ribosomal DNA. *Cell* **56:** 771–776.

Greer EL, Brunet A. 2009. Different dietary restriction regimens extend lifespan by both independent and overlapping genetic pathways in *C. elegans*. *Aging Cell* **8:** 113–127.

Guarente L, Picard F. 2005. Calorie restriction—The *SIR2* connection. *Cell* **120:** 473–482.

Haigis MC, Mostoslavsky R, Haigis KM, Fahie K, Christodoulou DC, Murphy AJ, Valenzuela DM, Yancopoulos GD, Karow M, Blander G, et al. 2006. SIRT4 inhibits glutamate dehydrogenase and opposes the effects of calorie restriction in pancreatic β cells. *Cell* **126:** 941–954.

Halloran J, Hussong S, Burbank R, Podlutskaya N, Fischer K, Sloane L, Austad SN, Strong R, Richardson A, Hart M, et al. 2012. Chronic inhibition of mammalian target of rapamycin by rapamycin modulates cognitive and noncognitive components of behavior throughout lifespan in mice. *Neuroscience* **223:** 102–113.

Hallows WC, Yu W, Smith BC, Devries MK, Ellinger JJ, Someya S, Shortreed MR, Prolla T, Markley JL, Smith LM, et al. 2011. Sirt3 promotes the urea cycle and fatty acid oxidation during dietary restriction. *Mol Cell* **41:** 139–149.

Hardie D. 2007. AMP-activated/SNF1 protein kinases: Conserved guardians of cellular energy. *Nat Rev Mol Cell Biol* **8:** 774–785.

Harman D. 1956. Aging: A theory based on free radical and radiation chemistry. *J Gerontol* **11:** 298–300.

Harrison DE, Strong R, Sharp ZD, Nelson JF, Astle CM, Flurkey K, Nadon NL, Wilkinson JE, Frenkel K, Carter CS, et al. 2009. Rapamycin fed late in life extends lifespan in genetically heterogeneous mice. *Nature* **460:** 392–395.

Hauck S, Bartke A. 2000. Effects of growth hormone on hypothalamic catalase and Cu/Zn superoxide dismutase. *Free Radic Biol Med* **28:** 970–978.

Henderson ST, Johnson TE. 2001. *daf-16* integrates developmental and environmental inputs to mediate aging in the nematode *Caenorhabditis elegans*. *Curr Biol* **11:** 1975–1980.

Herndon LA, Schmeissner PJ, Dudaronek JM, Brown PA, Listner KM, Sakano Y, Paupard MC, Hall DH, Driscoll M. 2002. Stochastic and genetic factors influence tissue-specific decline in ageing *C. elegans*. *Nature* **419:** 808–814.

Herranz D, Munoz-Martin M, Canamero M, Mulero F, Martinez-Pastor B, Fernandez-Capetillo O, Serrano M. 2010. Sirt1 improves healthy ageing and protects from metabolic syndrome-associated cancer. *Nat Commun* **1:** 3.

Herskovits AZ, Guarente L. 2014. SIRT1 in neurodevelopment and brain senescence. *Neuron* **81:** 471–483.

Hirschey MD, Shimazu T, Goetzman E, Jing E, Schwer B, Lombard DB, Grueter CA, Harris C, Biddinger S, Ilkayeva OR, et al. 2010. SIRT3 regulates mitochondrial fatty-acid oxidation by reversible enzyme deacetylation. *Nature* **464:** 121–125.

Holzenberger M, Dupont J, Ducos B, Leneuve P, Géloën A, Even P, Cervera P, Le Bouc Y. 2003. IGF-1 receptor regulates lifespan and resistance to oxidative stress in mice. *Nature* **421:** 182–189.

Hong S, Zhao B, Lombard DB, Fingar DC, Inoki K. 2014. Cross-talk between sirtuin and mammalian target of rapamycin complex 1 (mTORC1) signaling in the regulation of S6 kinase 1 (S6K1) phosphorylation. *J Biol Chem* **289:** 13132–13141.

Howitz KT, Bitterman KJ, Cohen HY, Lamming DW, Lavu S, Wood JG, Zipkin RE, Chung P, Kisielewski A, Zhang LL, et al. 2003. Small molecule activators of sirtuins extend *Saccharomyces cerevisiae* lifespan. *Nature* **425:** 191–196.

Hsu CP, Odewale I, Alcendor RR, Sadoshima J. 2008. Sirt1 protects the heart from aging and stress. *Biol Chem* **389:** 221–231.

Huang K, Fingar DC. 2014. Growing knowledge of the mTOR signaling network. *Semin Cell Dev Biol* **36:** 79–90.

Huang C, Xiong C, Kornfeld K. 2004. Measurements of age-related changes of physiological processes that predict lifespan of *Caenorhabditis elegans*. *Proc Natl Acad Sci* **101:** 8084–8089.

Hughes BG, Hekimi S. 2011. A mild impairment of mitochondrial electron transport has sex-specific effects on lifespan and aging in mice. *PLoS ONE* **6:** e26116.

Hwang AB, Jeong DE, Lee SJ. 2012. Mitochondria and organismal longevity. *Curr Genomics* **13:** 519–532.

Hwang AB, Ryu EA, Artan M, Chang HW, Kabi MH, Nam HJ, Lee D, Yang JS, Kim S, Mair WB, et al. 2014. Feedback regulation via AMPK and HIF-1 mediates ROS-dependent longevity in *Caenorhabditis elegans*. *Proc Natl Acad Sci* **111**: E4458–E4467.

Ikenoue T, Inoki K, Yang Q, Zhou X, Guan KL. 2008. Essential function of TORC2 in PKC and Akt turn motif phosphorylation, maturation and signalling. *EMBO J* **27**: 1919–1931.

Imai S, Armstrong CM, Kaeberlein M, Guarente L. 2000. Transcriptional silencing and longevity protein Sir2 is an NAD-dependent histone deacetylase. *Nature* **403**: 795–800.

Inoue H, Hisamoto N, An JH, Oliveira RP, Nishida E, Blackwell TK, Matsumoto K. 2005. The *C. elegans* p38 MAPK pathway regulates nuclear localization of the transcription factor SKN-1 in oxidative stress response. *Genes Dev* **19**: 2278–2283.

Ivan M, Kaelin WG. 2001. The von Hippel–Lindau tumor suppressor protein. *Curr Opin Genet Dev* **11**: 27–34.

Ivy JM, Klar AJ, Hicks JB. 1986. Cloning and characterization of four *SIR* genes of Saccharomyces cerevisiae. *Mol Cell Biol* **6**: 688–702.

Jang YC, Perez VI, Song W, Lustgarten MS, Salmon AB, Mele J, Qi W, Liu Y, Liang H, Chaudhuri A, et al. 2009. Overexpression of Mn superoxide dismutase does not increase life span in mice. *J Gerontol A Biol Sci Med Sci* **64**: 1114–1125.

Jeong H, Cohen DE, Cui L, Supinski A, Savas JN, Mazzulli JR, Yates JR III, Bordone L, Guarente L, Krainc D. 2012. Sirt1 mediates neuroprotection from mutant huntingtin by activation of the TORC1 and CREB transcriptional pathway. *Nat Med* **18**: 159–165.

Jia K, Levine B. 2007. Autophagy is required for dietary restriction-mediated life span extension in *C. elegans*. *Autophagy* **3**: 597–599.

Jia K, Chen D, Riddle DL. 2004. The TOR pathway interacts with the insulin signaling pathway to regulate *C. elegans* larval development, metabolism and life span. *Development* **131**: 3897–3906.

Johnson SC, Sangesland M, Kaeberlein M, Rabinovitch PS. 2015. Modulating mTOR in aging and health. *Interdiscip Top Gerontol* **40**: 107–127.

Kaeberlein M. 2007. Longevity genomics across species. *Curr Genomics* **8**: 73–78.

Kaeberlein M. 2010. Lessons on longevity from budding yeast. *Nature* **464**: 513–519.

Kaeberlein M. 2013. Longevity and aging. *F1000Prime Rep* **5**: 5.

Kaeberlein M, Powers RW III. 2007. Sir2 and calorie restriction in yeast: A skeptical perspective. *Ageing Res Rev* **6**: 128–140.

Kaeberlein M, McVey M, Guarente L. 1999. The *SIR2/3/4* complex and *SIR2* alone promote longevity in *Saccharomyces cerevisiae* by two different mechanisms. *Genes Dev* **13**: 2570–2580.

Kaeberlein M, McVey M, Guarente L. 2001. Using yeast to discover the fountain of youth. *Sci Aging Knowledge Environ* **2001**: pe1.

Kaeberlein M, Kirkland KT, Fields S, Kennedy BK. 2004. Sir2-independent life span extension by calorie restriction in yeast. *PLoS Biol* **2**: E296.

Kaeberlein M, Powers RW III, Steffen KK, Westman EA, Hu D, Dang N, Kerr EO, Kirkland KT, Fields S, Kennedy BK.

2005a. Regulation of yeast replicative life span by TOR and Sch9 in response to nutrients. *Science* **310**: 1193–1196.

Kaeberlein M, McDonagh T, Heltweg B, Hixon J, Westman EA, Caldwell SD, Napper A, Curtis R, DiStefano PS, Fields S, et al. 2005b. Substrate-specific activation of sirtuins by resveratrol. *J Biol Chem* **280**: 17038–17045.

Kaeberlein M, Steffen KK, Hu D, Dang N, Kerr EO, Tsuchiya M, Fields S, Kennedy BK. 2006. Comment on "*HST2* mediates *SIR2*-independent life-span extension by calorie restriction." *Science* **312**: 1312; author reply 1312.

Kanfi Y, Naiman S, Amir G, Peshti V, Zinman G, Nahum L, Bar-Joseph Z, Cohen HY. 2012. The sirtuin SIRT6 regulates lifespan in male mice. *Nature* **483**: 218–221.

Kapahi P, Zid BM, Harper T, Koslover D, Sapin V, Benzer S. 2004. Regulation of lifespan in *Drosophila* by modulation of genes in the TOR signaling pathway. *Curr Biol* **14**: 885–890.

Keith CT, Schreiber SL. 1995. PIK-related kinases: DNA repair, recombination, and cell cycle checkpoints. *Science* **270**: 50–51.

Kennedy BK, Austriaco NR Jr, Guarente L. 1994. Daughter cells of *Saccharomyces cerevisiae* from old mothers display a reduced life span. *J Cell Biol* **127**: 1985–1993.

Kenyon C. 2005. The plasticity of aging: Insights from long-lived mutants. *Cell* **120**: 449–460.

Kenyon C. 2011. The first long-lived mutants: Discovery of the insulin/IGF-1 pathway for ageing. *Philos Trans R Soc Lond B Biol Sci* **366**: 9–16.

Kenyon C, Chang J, Gensch E, Rudner A, Tabtiang R. 1993. A *C. elegans* mutant that lives twice as long as wild type. *Nature* **366**: 461–464.

Khan MH, Ligon M, Hussey LR, Hufnal B, Farber R II, Munkacsy E, Rodriguez A, Dillow A, Kahlig E, Rea SL. 2013. TAF-4 is required for the life extension of *isp-1, clk-1* and *tpk-1* Mit mutants. *Aging (Albany NY)* **5**: 741–758.

Kimura K, Tissenbaum H, Liu Y, Ruvkun G. 1997. *daf-2*, an insulin receptor-like gene that regulates longevity and diapause in *Caenorhabditis elegans*. *Science* **277**: 942–946.

Klar AJ, Fogel S, Macleod K. 1979. *MAR1*—A regulator of the *HM*a and *HM*α loci in *Saccharomyces cerevisiae*. *Genetics* **93**: 37–50.

Kojima M, Hosoda H, Date Y, Nakazato M, Matsuo H, Kangawa K. 1999. Ghrelin is a growth-hormone-releasing acylated peptide from stomach. *Nature* **402**: 656–660.

Lai CY, Jaruga E, Borghouts C, Jazwinski SM. 2002. A mutation in the *ATP2* gene abrogates the age asymmetry between mother and daughter cells of the yeast *Saccharomyces cerevisiae*. *Genetics* **162**: 73–87.

Lamming DW, Ye L, Katajisto P, Goncalves MD, Saitoh M, Stevens DM, Davis JG, Salmon AB, Richardson A, Ahima RS, et al. 2012. Rapamycin-induced insulin resistance is mediated by mTORC2 loss and uncoupled from longevity. *Science* **335**: 1638–1643.

Landry J, Sutton A, Tafrov ST, Heller RC, Stebbins J, Pillus L, Sternglanz R. 2000. The silencing protein SIR2 and its homologs are NAD-dependent protein deacetylases. *Proc Natl Acad Sci* **97**: 5807–5811.

Laplante M, Sabatini DM. 2012. mTOR signaling in growth control and disease. *Cell* **149**: 274–293.

Lee SS, Lee RY, Fraser AG, Kamath RS, Ahringer J, Ruvkun G. 2003. A systematic RNAi screen identifies a critical role for mitochondria in *C. elegans* longevity. *Nat Genet* **33:** 40–48.

Lee S, Comer F, Sasaki A, McLeod I, Duong Y, Okumura K, Yates J, Parent C, Firtel R. 2005. TOR complex 2 integrates cell movement during chemotaxis and signal relay in *Dictyostelium*. *Mol Biol Cell* **16:** 4572–4583.

Lee SJ, Hwang AB, Kenyon C. 2010a. Inhibition of respiration extends *C. elegans* life span via reactive oxygen species that increase HIF-1 activity. *Curr Biol* **20:** 2131–2136.

Lee HY, Choi CS, Birkenfeld AL, Alves TC, Jornayvaz FR, Jurczak MJ, Zhang D, Woo DK, Shadel GS, Ladiges W, et al. 2010b. Targeted expression of catalase to mitochondria prevents age-associated reductions in mitochondrial function and insulin resistance. *Cell Metab* **12:** 668–674.

Leiser SF, Kaeberlein M. 2010. The hypoxia-inducible factor HIF-1 functions as both a positive and negative modulator of aging. *Biol Chem* **391:** 1131–1137.

Leiser S, Miller R. 2010. Nrf2 signaling, a mechanism for cellular stress resistance in long-lived mice. *Mol Cell Biol* **30:** 871–955.

Leiser SF, Begun A, Kaeberlein M. 2011. HIF-1 modulates longevity and healthspan in a temperature-dependent manner. *Aging Cell* **10:** 318–326.

Leiser SF, Fletcher M, Begun A, Kaeberlein M. 2013. Lifespan extension from hypoxia in *Caenorhabditis elegans* requires both HIF-1 and DAF-16 and is antagonized by SKN-1. *J Gerontol A Biol Sci Med Sci* **68:** 1135–1144.

Lerner C, Bitto A, Pulliam D, Nacarelli T, Konigsberg M, Van Remmen H, Torres C, Sell C. 2013. Reduced mammalian target of rapamycin activity facilitates mitochondrial retrograde signaling and increases life span in normal human fibroblasts. *Aging Cell* **12:** 966–977.

Li X. 2013. SIRT1 and energy metabolism. *Acta Biochim Biophys Sin (Shanghai)* **45:** 51–60.

Lim JH, Lee YM, Chun YS, Chen J, Kim JE, Park JW. 2010. Sirtuin 1 modulates cellular responses to hypoxia by deacetylating hypoxia-inducible factor 1α. *Mol Cell* **38:** 864–878.

Lin SJ, Defossez PA, Guarente L. 2000. Requirement of NAD and *SIR2* for life-span extension by calorie restriction in *Saccharomyces cerevisiae*. *Science* **289:** 2126–2128.

Lin SJ, Kaeberlein M, Andalis AA, Sturtz LA, Defossez PA, Culotta VC, Fink GR, Guarente L. 2002. Calorie restriction extends *Saccharomyces cerevisiae* lifespan by increasing respiration. *Nature* **418:** 344–348.

Liou YS, Adam S, William RS, Yimin F, Cristal H, Joshua AH, Jacob DB, Reyhan W, Roberto S, Andrzej B. 2013. Growth hormone-releasing hormone disruption extends lifespan and regulates response to caloric restriction in mice. *eLife* **2:** e01098.

Loeb J, Northrop JH. 1916. Is there a temperature coefficient for the duration of life? *Proc Natl Acad Sci* **2:** 456–457.

Loewith R, Jacinto E, Wullschleger S, Lorberg A, Crespo J, Bonenfant D, Oppliger W, Jenoe P, Hall M. 2002. Two TOR complexes, only one of which is rapamycin sensitive, have distinct roles in cell growth control. *Mol Cell* **10:** 457–468.

Lombard DB, Pletcher SD, Canto C, Auwerx J. 2011. Ageing: Longevity hits a roadblock. *Nature* **477:** 410–411.

Longo VD, Fabrizio P. 2012. Chronological aging in *Saccharomyces cerevisiae*. *Subcell Biochem* **57:** 101–121.

Longo VD, Shadel GS, Kaeberlein M, Kennedy B. 2012. Replicative and chronological aging in *Saccharomyces cerevisiae*. *Cell Metab* **16:** 18–31.

Lopez-Lluch G, Hunt N, Jones B, Zhu M, Jamieson H, Hilmer S, Cascajo MV, Allard J, Ingram DK, Navas P, et al. 2006. Calorie restriction induces mitochondrial biogenesis and bioenergetic efficiency. *Proc Natl Acad Sci* **103:** 1768–1773.

Lorenzini A, Salmon A, Lerner C, Torres C, Ikeno Y, Motch S, McCarter R, Sell C. 2013. Mice producing reduced levels of insulin-like growth factor type 1 display an increase in maximum, but not mean, lifespan. *J Gerontol A Biol Sci Med Sci* **69:** 410–419.

Lu H, Dalgard CL, Mohyeldin A, McFate T, Tait AS, Verma A. 2005. Reversible inactivation of HIF-1 prolyl hydroxylases allows cell metabolism to control basal HIF-1. *J Biol Chem* **280:** 41928–41939.

Luo XY, Qu SL, Tang ZH, Zhang Y, Liu MH, Peng J, Tang H, Yu KL, Zhang C, Ren Z, et al. 2014. SIRT1 in cardiovascular aging. *Clin Chim Acta* **437:** 106–114.

Majumder S, Caccamo A, Medina DX, Benavides AD, Javors MA, Kraig E, Strong R, Richardson A, Oddo S. 2012. Lifelong rapamycin administration ameliorates age-dependent cognitive deficits by reducing IL-1β and enhancing NMDA signaling. *Aging Cell* **11:** 326–335.

Mannick JB, Giudice GD, Lattanzi M, Valiante NM, Praestgaard J, Huang B, Lonetto MA, Maecker HT, Kovarik J, Carson S, et al. 2014. mTOR inhibition improves immune function in the elderly. *Sci Transl Med* **6:** 268ra179.

Martin-Montalvo A, Villalba JM, Navas P, de Cabo R. 2011. NRF2, cancer and calorie restriction. *Oncogene* **30:** 505–520.

Martin-Montalvo A, Mercken EM, Mitchell SJ, Palacios HH, Mote PL, Scheibye-Knudsen M, Gomes AP, Ward TM, Minor RK, Blouin MJ, et al. 2013. Metformin improves healthspan and lifespan in mice. *Nat Commun* **4:** 2192.

Medvedik O, Lamming DW, Kim KD, Sinclair DA. 2007. *MSN2* and *MSN4* link calorie restriction and TOR to sirtuin-mediated lifespan extension in *Saccharomyces cerevisiae*. *PLoS Biol* **5:** e261.

Mehta R, Steinkraus KA, Sutphin GL, Ramos FJ, Shamieh LS, Huh A, Davis C, Chandler-Brown D, Kaeberlein M. 2009. Proteasomal regulation of the hypoxic response modulates aging in *C. elegans*. *Science* **324:** 1196–1198.

Mercken EM, Mitchell SJ, Martin-Montalvo A, Minor RK, Almeida M, Gomes AP, Scheibye-Knudsen M, Palacios HH, Licata JJ, Zhang Y, et al. 2014. SRT2104 extends survival of male mice on a standard diet and preserves bone and muscle mass. *Aging Cell* **13:** 787–796.

Michael JS, Liou YS, Andrzej B, Richard AM. 2012. Activation of genes involved in xenobiotic metabolism is a shared signature of mouse models with extended lifespan. *Am J Physiol Endocrinol Metab* **303:** 95.

Migliaccio E, Giorgio M, Mele S, Pelicci G, Reboldi P, Pandolfi P, Lanfrancone L, Pelicci P. 1999. The p66shc adaptor protein controls oxidative stress response and life span in mammals. *Nature* **402:** 309–313.

Miller RA, Harrison DE, Astle CM, Floyd RA, Flurkey K, Hensley KL, Javors MA, Leeuwenburgh C, Nelson JF,

Ongini E, et al. 2007. An aging interventions testing program: Study design and interim report. *Aging Cell* **6:** 565–575.

Miller RA, Harrison DE, Astle CM, Baur JA, Boyd AR, de Cabo R, Fernandez E, Flurkey K, Javors MA, Nelson JF, et al. 2011. Rapamycin, but not resveratrol or simvastatin, extends life span of genetically heterogeneous mice. *J Gerontol A Biol Sci Med Sci* **66:** 191–201.

Miller RA, Harrison DE, Astle CM, Fernandez E, Flurkey K, Han M, Javors MA, Li X, Nadon NL, Nelson JF, et al. 2014. Rapamycin-mediated lifespan increase in mice is dose and sex dependent and metabolically distinct from dietary restriction. *Aging Cell* **13:** 468–477.

Minor RK, Baur JA, Gomes AP, Ward TM, Csiszar A, Mercken EM, Abdelmohsen K, Shin YK, Canto C, Scheibye-Knudsen M, et al. 2011. SRT1720 improves survival and healthspan of obese mice. *Sci Rep* **1:** 70.

Mitchell SJ, Martin-Montalvo A, Mercken EM, Palacios HH, Ward TM, Abulwerdi G, Minor RK, Vlasuk GP, Ellis JL, Sinclair DA, et al. 2014. The SIRT1 activator SRT1720 extends lifespan and improves health of mice fed a standard diet. *Cell Rep* **6:** 836–843.

Mitsui A, Hamuro J, Nakamura H, Kondo N, Hirabayashi Y, Ishizaki-Koizumi S, Hirakawa T, Inoue T, Yodoi J. 2002. Overexpression of human thioredoxin in transgenic mice controls oxidative stress and life span. *Antioxid Redox Signal* **4:** 693–696.

Moellendorf S, Kessels C, Peiseler L, Raupach A, Jacob C, Vogt N, Lindecke A, Koch L, Brüning J, Heger J, et al. 2012. IGF-IR signaling attenuates the age-related decline of diastolic cardiac function. *Am J Physiol Endocrinol Metab* **303:** 22.

Morck C, Pilon M. 2007. Caloric restriction and autophagy in *Caenorhabditis elegans*. *Autophagy* **3:** 51–53.

Morris JZ, Tissenbaum HA, Ruvkun G. 1996. A phosphatidylinositol-3-OH kinase family member regulating longevity and diapause in *Caenorhabditis elegans*. *Nature* **382:** 536–539.

Mortimer RK, Johnston JR. 1959. Life span of individual yeast cells. *Nature* **183:** 1751–1752.

Mostoslavsky R, Chua KF, Lombard DB, Pang WW, Fischer MR, Gellon L, Liu P, Mostoslavsky G, Franco S, Murphy MM, et al. 2006. Genomic instability and aging-like phenotype in the absence of mammalian SIRT6. *Cell* **124:** 315–329.

Muller RU, Fabretti F, Zank S, Burst V, Benzing T, Schermer B. 2009. The von Hippel Lindau tumor suppressor limits longevity. *J Am Soc Nephrol* **20:** 2513–2517.

Murakami C, Kaeberlein M. 2009. Quantifying yeast chronological life span by outgrowth of aged cells. *J Vis Exp* doi: 10.3791/1156.

Murakami S, Salmon A, Miller R. 2003. Multiplex stress resistance in cells from long-lived dwarf mice. *FASEB J* **17:** 1565–1566.

Murakami CJ, Burtner CR, Kennedy BK, Kaeberlein M. 2008. A method for high-throughput quantitative analysis of yeast chronological life span. *J Gerontol A Biol Sci Med Sci* **63:** 113–121.

Murakami CJ, Wall V, Basisty N, Kaeberlein M. 2011. Composition and acidification of the culture medium influ-

ences chronological aging similarly in vineyard and laboratory yeast. *PLoS ONE* **6:** e24530.

Murakami C, Delaney JR, Chou A, Carr D, Schleit J, Sutphin GL, An EH, Castanza AS, Fletcher M, Goswami S, et al. 2012. pH neutralization protects against reduction in replicative lifespan following chronological aging in yeast. *Cell Cycle* **11:** 3087–3096.

Nacarelli T, Azar A, Sell C. 2014. Aberrant mTOR activation in senescence and aging: A mitochondrial stress response? *Exp Gerontol* doi: 10.1016/j.exger.2014.11.004.

Nadon NL, Strong R, Miller RA, Nelson J, Javors M, Sharp ZD, Peralba JM, Harrison DE. 2008. Design of aging intervention studies: The NIA interventions testing program. *Age (Dordr)* **30:** 187–199.

Nakagawa T, Lomb DJ, Haigis MC, Guarente L. 2009. SIRT5 deacetylates carbamoyl phosphate synthetase 1 and regulates the urea cycle. *Cell* **137:** 560–570.

Neff F, Flores-Dominguez D, Ryan DP, Horsch M, Schroder S, Adler T, Afonso LC, Aguilar-Pimentel JA, Becker L, Garrett L, et al. 2013. Rapamycin extends murine lifespan but has limited effects on aging. *J Clin Invest* **123:** 3272–3291.

Nemoto S, Finkel T. 2002. Redox regulation of forkhead proteins through a *p66shc*–dependent signaling pathway. *Science* **295:** 2450–2452.

Nisoli E, Tonello C, Cardile A, Cozzi V, Bracale R, Tedesco L, Falcone S, Valerio A, Cantoni O, Clementi E, et al. 2005. Calorie restriction promotes mitochondrial biogenesis by inducing the expression of eNOS. *Science* **310:** 314–317.

North BJ, Rosenberg MA, Jeganathan KB, Hafner AV, Michan S, Dai J, Baker DJ, Cen Y, Wu LE, Sauve AA. et al. 2014. SIRT2 induces the checkpoint kinase BubR1 to increase lifespan. *EMBO J* **33:** 1438–1453.

Ogg S, Paradis S, Gottlieb S, Patterson GI, Lee L, Tissenbaum HA, Ruvkun G. 1997. The Fork head transcription factor DAF-16 transduces insulin-like metabolic and longevity signals in *C. elegans*. *Nature* **389:** 994–999.

Oldham S, Hafen E. 2003. Insulin/IGF and target of rapamycin signaling: ATOR de force in growth control. *Trends Cell Biol* **13:** 79–85.

Onken B, Driscoll M. 2010. Metformin induces a dietary restriction-like state and the oxidative stress response to extend *C. elegans* healthspan via AMPK, LKB1, and SKN-1. *PLoS ONE* **5:** e8758.

Orr WC, Mockett RJ, Benes JJ, Sohal RS. 2003. Effects of overexpression of copper-zinc and manganese superoxide dismutases, catalase, and thioredoxin reductase genes on longevity in *Drosophila melanogaster*. *J Biol Chem* **278:** 26418–26422.

Owen MR, Doran E, Halestrap AP. 2000. Evidence that metformin exerts its anti-diabetic effects through inhibition of complex 1 of the mitochondrial respiratory chain. *Biochem J* **348:** 607–614.

Owusu-Ansah E, Song W, Perrimon N. 2013. Muscle mitohormesis promotes longevity via systemic repression of insulin signaling. *Cell* **155:** 699–712.

Palacios OM, Carmona JJ, Michan S, Chen KY, Manabe Y, Ward JL III, Goodyear LJ, Tong Q. 2009. Diet and exercise signals regulate SIRT3 and activate AMPK and PGC-1α in skeletal muscle. *Aging (Albany NY)* **1:** 771–783.

Park J, Chen Y, Tishkoff DX, Peng C, Tan M, Dai L, Xie Z, Zhang Y, Zwaans BM, Skinner ME, et al. 2013. SIRT5-mediated lysine desuccinylation impacts diverse metabolic pathways. *Mol Cell* **50:** 919–930.

Pearson KJ, Lewis KN, Price NL, Chang JW, Perez E, Cascajo MV, Tamashiro KL, Poosala S, Csiszar A, Ungvari Z, et al. 2008. Nrf2 mediates cancer protection but not prolongevity induced by caloric restriction. *Proc Natl Acad Sci* **105:** 2325–2330.

Peng C, Lu Z, Xie Z, Cheng Z, Chen Y, Tan M, Luo H, Zhang Y, He W, Yang K, et al. 2011. The first identification of lysine malonylation substrates and its regulatory enzyme. *Mol Cell Proteomics* **10:** M111.012658.

Perez VI, Van Remmen H, Bokov A, Epstein CJ, Vijg J, Richardson A. 2009. The overexpression of major antioxidant enzymes does not extend the lifespan of mice. *Aging Cell* **8:** 73–75.

Perez VI, Cortez LA, Lew CM, Rodriguez M, Webb CR, Van Remmen H, Chaudhuri A, Qi W, Lee S, Bokov A, et al. 2011. Thioredoxin 1 overexpression extends mainly the earlier part of life span in mice. *J Gerontol A Biol Sci Med Sci* **66:** 1286–1299.

Petes TD, Botstein D. 1977. Simple Mendelian inheritance of the reiterated ribosomal DNA of yeast. *Proc Natl Acad Sci* **74:** 5091–5095.

Petralia RS, Mattson MP, Yao PJ. 2014. Aging and longevity in the simplest animals and the quest for immortality. *Ageing Res Rev* **16:** 66–82.

Philippsen P, Thomas M, Kramer RA, Davis RW. 1978. Unique arrangement of coding sequences for 5 S, 5.8 S, 18 S and 25 S ribosomal RNA in *Saccharomyces cerevisiae* as determined by R-loop and hybridization analysis. *J Mol Biol* **123:** 387–404.

Piper PW, Harris NL, MacLean M. 2006. Preadaptation to efficient respiratory maintenance is essential both for maximal longevity and the retention of replicative potential in chronologically ageing yeast. *Mech Ageing Dev* **127:** 733–740.

Powers RW III, Kaeberlein M, Caldwell SD, Kennedy BK, Fields S. 2006. Extension of chronological life span in yeast by decreased TOR pathway signaling. *Genes Dev* **20:** 174–184.

Pulliam DA, Deepa SS, Liu Y, Hill S, Lin AL, Bhattacharya A, Shi Y, Sloane L, Viscomi C, Zeviani M, et al. 2014. Complex IV-deficient *Surf1⁻/⁻* mice initiate mitochondrial stress responses. *Biochem J* **462:** 359–371.

Ramsey JJ, Tran D, Giorgio M, Griffey SM, Koehne A, Laing ST, Taylor SL, Kim K, Cortopassi GA, Lloyd KC, et al. 2014. The influence of Shc proteins on life span in mice. *J Gerontol A Biol Sci Med Sci* **69:** 1177–1185.

Rardin MJ, He W, Nishida Y, Newman JC, Carrico C, Danielson SR, Guo A, Gut P, Sahu AK, Li B, et al. 2013. SIRT5 regulates the mitochondrial lysine succinylome and metabolic networks. *Cell Metab* **18:** 920–933.

Rea SL, Ventura N, Johnson TE. 2007. Relationship between mitochondrial electron transport chain dysfunction, development, and life extension in *Caenorhabditis elegans*. *PLoS Biol* **5:** e259.

Reinecke M, Collet C. 1998. The phylogeny of the insulin-like growth factors. *Int Rev Cytol* **183:** 1–94.

Rera M, Bahadorani S, Cho J, Koehler CL, Ulgherait M, Hur JH, Ansari WS, Lo T Jr, Jones DL, Walker DW. 2011. Modulation of longevity and tissue homeostasis by the *Drosophila* PGC-1 homolog. *Cell Metab* **14:** 623–634.

Ridgway ID, Richardson CA, Austad SN. 2011. Maximum shell size, growth rate, and maturation age correlate with longevity in bivalve molluscs. *J Gerontol A Biol Sci Med Sci* **66:** 183–190.

Rine J, Herskowitz I. 1987. Four genes responsible for a position effect on expression from *HML* and *HMR* in *Saccharomyces cerevisiae*. *Genetics* **116:** 9–22.

Ristow M, Zarse K. 2010. How increased oxidative stress promotes longevity and metabolic health: The concept of mitochondrial hormesis (mitohormesis). *Exp Gerontol* **45:** 410–418.

Robida-Stubbs S, Glover-Cutter K, Lamming DW, Mizunuma M, Narasimhan SD, Neumann-Haefelin E, Sabatini DM, Blackwell TK. 2012. TOR signaling and rapamycin influence longevity by regulating SKN-1/Nrf and DAF-16/FoxO. *Cell Metab* **15:** 713–724.

Rogina B, Helfand SL. 2004. Sir2 mediates longevity in the fly through a pathway related to calorie restriction. *Proc Natl Acad Sci* **101:** 15998–16003.

Salmon A, Murakami S, Bartke A, Kopchick J, Yasumura K, Miller R. 2005. Fibroblast cell lines from young adult mice of long-lived mutant strains are resistant to multiple forms of stress. *Am J Physiol Endocrinol Metab* **289:** E23–E29.

Salmon AB, Lerner C, Ikeno Y, Motch Perrine SM, McCarter R, Sell C. 2015. Altered metabolism and resistance to obesity in long-lived mice producing reduced levels of IGF-1. *Am J Physiol Endocrinol Metab* doi: 10.1152/ajpendo.00558.2014.

Sarbassov D, Ali S, Kim DH, Guertin D, Latek R, Erdjument-Bromage H, Tempst P, Sabatini D. 2004. Rictor, a novel binding partner of mTOR, defines a rapamycin-insensitive and raptor-independent pathway that regulates the cytoskeleton. *Curr Biol* **14:** 1296–1302.

Sarbassov D, Guertin D, Ali S, Sabatini D. 2005. Phosphorylation and regulation of Akt/PKB by the rictor-mTOR complex. *Science* **307:** 1098–1101.

Satoh A, Stein L, Imai S. 2011. The role of mammalian sirtuins in the regulation of metabolism, aging, and longevity. *Handb Exp Pharmacol* **206:** 125–162.

Satoh A, Brace CS, Rensing N, Cliften P, Wozniak DF, Herzog ED, Yamada KA, Imai S. 2013. Sirt1 extends life span and delays aging in mice through the regulation of Nk2 homeobox 1 in the DMH and LH. *Cell Metab* **18:** 416–430.

Schriner SE, Linford NJ, Martin GM, Treuting P, Ogburn CE, Emond M, Coskun PE, Ladiges W, Wolf N, Van Remmen H, et al. 2005. Extension of murine life span by overexpression of catalase targeted to mitochondria. *Science* **308:** 1909–1911.

Schulz TJ, Zarse K, Voigt A, Urban N, Birringer M, Ristow M. 2007. Glucose restriction extends *Caenorhabditis elegans* life span by inducing mitochondrial respiration and increasing oxidative stress. *Cell Metab* **6:** 280–293.

Schwer B, Eckersdorff M, Li Y, Silva JC, Fermin D, Kurtev MV, Giallourakis C, Comb MJ, Alt FW, Lombard DB.

2009. Calorie restriction alters mitochondrial protein acetylation. *Aging Cell* **8:** 604–606.

Selman C, Tullet JM, Wieser D, Irvine E, Lingard SJ, Choudhury AI, Claret M, Al-Qassab H, Carmignac D, Ramadani F, et al. 2009. Ribosomal protein S6 kinase 1 signaling regulates mammalian life span. *Science* **326:** 140–144.

Sharp Z, Bartke A. 2005. Evidence for down-regulation of phosphoinositide 3-kinase/Akt/mammalian target of rapamycin (PI3K/Akt/mTOR)-dependent translation regulatory signaling pathways in Ames dwarf mice. *J Gerontol A Biol Sci Med Sci* **60:** 293–300.

Shi T, Wang F, Stieren E, Tong Q. 2005. SIRT3, a mitochondrial sirtuin deacetylase, regulates mitochondrial function and thermogenesis in brown adipocytes. *J Biol Chem* **280:** 13560–13567.

Shih PH, Yen GC. 2007. Differential expressions of antioxidant status in aging rats: The role of transcriptional factor Nrf2 and MAPK signaling pathway. *Biogerontology* **8:** 71–80.

Shimobayashi M, Hall MN. 2014. Making new contacts: The mTOR network in metabolism and signalling crosstalk. *Nat Rev Mol Cell Biol* **15:** 155–162.

Sinclair DA, Guarente L. 1997. Extrachromosomal rDNA circles—A cause of aging in yeast. *Cell* **91:** 1033–1042.

Smith JS, Boeke JD. 1997. An unusual form of transcriptional silencing in yeast ribosomal DNA. *Genes Dev* **11:** 241–254.

Smith JS, Brachmann CB, Celic I, Kenna MA, Muhammad S, Starai VJ, Avalos JL, Escalante-Semerena JC, Grubmeyer C, Wolberger C, et al. 2000. A phylogenetically conserved NAD$^+$-dependent protein deacetylase activity in the Sir2 protein family. *Proc Natl Acad Sci* **97:** 6658–6663.

Smith ED, Tsuchiya M, Fox LA, Dang N, Hu D, Kerr EO, Johnston ED, Tchao BN, Pak DN, Welton KL, et al. 2008a. Quantitative evidence for conserved longevity pathways between divergent eukaryotic species. *Genome Res* **18:** 564–570.

Smith ED, Kaeberlein TL, Lydum BT, Sager J, Welton KL, Kennedy BK, Kaeberlein M. 2008b. Age- and calorie-independent life span extension from dietary restriction by bacterial deprivation in *Caenorhabditis elegans*. *BMC Dev Biol* **8:** 49.

Someya S, Yu W, Hallows WC, Xu J, Vann JM, Leeuwenburgh C, Tanokura M, Denu JM, Prolla TA. 2010. Sirt3 mediates reduction of oxidative damage and prevention of age-related hearing loss under caloric restriction. *Cell* **143:** 802–812.

Song G, Ouyang G, Bao S. 2005. The activation of Akt/PKB signaling pathway and cell survival. *J Cell Mol Med* **9:** 59–71.

Srisuttee R, Koh SS, Kim SJ, Malilas W, Boonying W, Cho IR, Jhun BH, Ito M, Horio Y, Seto E, et al. 2012. Hepatitis B virus X (HBX) protein upregulates β-catenin in a human hepatic cell line by sequestering SIRT1 deacetylase. *Oncol Rep* **28:** 276–282.

Stanfel MN, Shamieh LS, Kaeberlein M, Kennedy BK. 2009. The TOR pathway comes of age. *Biochim Biophys Acta* **1790:** 1067–1074.

Steffen KK, Kennedy BK, Kaeberlein M. 2009. Measuring replicative life span in the budding yeast. *J Vis Exp* doi: 10.3791/1209.

Steinkraus KA, Kaeberlein M, Kennedy BK. 2008. Replicative aging in yeast: The means to the end. *Annu Rev Cell Dev Biol* **24:** 29–54.

Stitt T, Drujan D, Clarke B, Panaro F, Timofeyva Y, Kline W, Gonzalez M, Yancopoulos G, Glass D. 2004. The IGF-1/PI3K/Akt pathway prevents expression of muscle atrophy-induced ubiquitin ligases by inhibiting FOXO transcription factors. *Mol Cell* **14:** 395–403.

Suh JH, Shenvi SV, Dixon BM, Liu H, Jaiswal AK, Liu RM, Hagen TM. 2004. Decline in transcriptional activity of Nrf2 causes age-related loss of glutathione synthesis, which is reversible with lipoic acid. *Proc Natl Acad Sci* **101:** 3381–3386.

Sun LY, Bartke A. 2013. Tissue-specific GHR knockout mice: Metabolic phenotypes. *Front Endocrinol (Lausanne)* **5:** 243.

Sun L, Evans M, Hsieh J, Panici J, Bartke A. 2005a. Increased neurogenesis in dentate gyrus of long-lived Ames dwarf mice. *Endocrinology* **146:** 1138–1144.

Sun L, Al-Regaiey K, Masternak M, Wang J, Bartke A. 2005b. Local expression of GH and IGF-1 in the hippocampus of GH-deficient long-lived mice. *Neurobiol Aging* **26:** 929–937.

Sun L, Bokov A, Richardson A, Miller R. 2011. Hepatic response to oxidative injury in long-lived Ames dwarf mice. *FASEB J* **25:** 398–806.

Sundaresan NR, Pillai VB, Wolfgeher D, Samant S, Vasudevan P, Parekh V, Raghuraman H, Cunningham JM, Gupta M, Gupta MP. 2011. The deacetylase SIRT1 promotes membrane localization and activation of Akt and PDK1 during tumorigenesis and cardiac hypertrophy. *Sci Signal* **4:** ra46.

Sutphin GL, Kaeberlein M. 2009. Measuring *Caenorhabditis elegans* life span on solid media. *J Vis Exp* doi: 10.3791/1152.

Sutphin G, Kaeberlein M. 2011. Comparative genetics of aging. In *Handbook of the biology of aging* (ed. Masoro E, Austad S), pp. 215–241. Elsevier, Amsterdam.

Sykiotis GP, Bohmann D. 2008. Keap1/Nrf2 signaling regulates oxidative stress tolerance and lifespan in *Drosophila*. *Dev Cell* **14:** 76–85.

Takahara T, Maeda T. 2013. Evolutionarily conserved regulation of TOR signalling. *J Biochem* **154:** 1–10.

Tan M, Peng C, Anderson Kristin A, Chhoy P, Xie Z, Dai L, Park J, Chen Y, Huang H, et al. 2014. Lysine glutarylation is a protein posttranslational modification regulated by SIRT5. *Cell Metab* **19:** 605–617.

Tatar M, Kopelman A, Epstein D, Tu MP, Yin CM, Garofalo RS. 2001. A mutant *Drosophila* insulin receptor homolog that extends life-span and impairs neuroendocrine function. *Science* **292:** 107–110.

Tatar M, Post S, Yu K. 2014. Nutrient control of *Drosophila* longevity. *Trends Endocrinol Metab* **25:** 509–517.

Tissenbaum HA, Guarente L. 2001. Increased dosage of a *sir-2* gene extends lifespan in *Caenorhabditis elegans*. *Nature* **410:** 227–230.

Tissenbaum HA, Ruvkun G. 1998. An insulin-like signaling pathway affects both longevity and reproduction in *Caenorhabditis elegans*. *Genetics* **148:** 703–717.

Treuting PM, Linford NJ, Knoblaugh SE, Emond MJ, Morton JF, Martin GM, Rabinovitch PS, Ladiges WC. 2008. Reduction of age-associated pathology in old mice by overexpression of catalase in mitochondria. *J Gerontol A Biol Sci Med Sci* **63:** 813–822.

Tsuchiya M, Dang N, Kerr EO, Hu D, Steffen KK, Oakes JA, Kennedy BK, Kaeberlein M. 2006. Sirtuin-independent effects of nicotinamide on lifespan extension from calorie restriction in yeast. *Aging Cell* **5:** 505–514.

Tu MP, Epstein D, Tatar M. 2002. The demography of slow aging in male and female *Drosophila* mutant for the insulin-receptor substrate homologue chico. *Aging Cell* **1:** 75–80.

Tullet JM, Hertweck M, An JH, Baker J, Hwang JY, Liu S, Oliveira RP, Baumeister R, Blackwell TK. 2008. Direct inhibition of the longevity-promoting factor SKN-1 by insulin-like signaling in *C. elegans*. *Cell* **132:** 1025–1038.

Ulgherait M, Rana A, Rera M, Graniel J, Walker DW. 2014. AMPK modulates tissue and organismal aging in a non-cell-autonomous manner. *Cell Rep* **8:** 1767–1780.

Um SH, Frigerio F, Watanabe M, Picard F, Joaquin M, Sticker M, Fumagalli S, Allegrini PR, Kozma SC, Auwerx J, et al. 2004. Absence of S6K1 protects against age- and diet-induced obesity while enhancing insulin sensitivity. *Nature* **431:** 200–205.

Umanskaya A, Santulli G, Xie W, Andersson DC, Reiken SR, Marks AR. 2014. Genetically enhancing mitochondrial antioxidant activity improves muscle function in aging. *Proc Natl Acad Sci* **111:** 15250–15255.

Upton Z, Francis G, Chan S, Steiner D, Wallace J, Ballard F. 1997. Evolution of insulin-like growth factor (IGF) function: Production and characterization of recombinant hagfish IGF. *Gen Comp Endocrinol* **105:** 79–90.

Valenzano DR, Kirschner J, Kamber RA, Zhang E, Weber D, Cellerino A, Englert C, Platzer M, Reichwald K, Brunet A. 2009. Mapping loci associated with tail color and sex determination in the short-lived fish *Nothobranchius furzeri*. *Genetics* **183:** 1385–1395.

Van Raamsdonk JM, Hekimi S. 2009. Deletion of the mitochondrial superoxide dismutase *sod-2* extends lifespan in *Caenorhabditis elegans*. *PLoS Genet* **5:** e1000361.

Vellai T, Takacs-Vellai K, Zhang Y, Kovacs AL, Orosz L, Muller F. 2003. Genetics: Influence of TOR kinase on lifespan in *C. elegans*. *Nature* **426:** 620.

Ventura N, Rea SL, Schiavi A, Torgovnick A, Testi R, Johnson TE. 2009. p53/CEP-1 increases or decreases lifespan, depending on level of mitochondrial bioenergetic stress. *Aging Cell* **8:** 380–393.

Viswanathan M, Guarente L. 2011. Regulation of *Caenorhabditis elegans* lifespan by *sir-2.1* transgenes. *Nature* **477:** E1–E2.

Vowels JJ, Thomas JH. 1992. Genetic analysis of chemosensory control of dauer formation in *Caenorhabditis elegans*. *Genetics* **130:** 105–123.

Walter L, Baruah A, Chang HW, Pace HM, Lee SS. 2011. The homeobox protein CEH-23 mediates prolonged longevity in response to impaired mitochondrial

electron transport chain in *C. elegans*. *PLoS Biol* **9:** e1001084.

Wang D, Wang Y, Argyriou C, Carriere A, Malo D, Hekimi S. 2012. An enhanced immune response of *Mclk1*$^{+/-}$ mutant mice is associated with partial protection from fibrosis, cancer and the development of biomarkers of aging. *PLoS ONE* **7:** e49606.

Warner HR. 2003. Subfield history: Use of model organisms in the search for human aging genes. *Sci Aging Knowledge Environ* **2003:** RE1.

Wasko BM, Carr DT, Tung H, Doan H, Schurman N, Neault JR, Feng J, Lee J, Zipkin B, Mouser J, et al. 2013. Buffering the pH of the culture medium does not extend yeast replicative lifespan. *F1000Res* **2:** 216.

Weber JD, Gutmann DH. 2012. Deconvoluting mTOR biology. *Cell Cycle* **11:** 236–248.

Weindruch R, Walford RL. 1988. *The retardation of aging and disease by dietary restriction*. C.C. Thomas, Springfield, IL.

Wilkinson JE, Burmeister L, Brooks SV, Chan CC, Friedline S, Harrison DE, Hejtmancik JF, Nadon N, Strong R, Wood LK, et al. 2012. Rapamycin slows aging in mice. *Aging Cell* **11:** 675–682.

Wood JG, Rogina B, Lavu S, Howitz K, Helfand SL, Tatar M, Sinclair D. 2004. Sirtuin activators mimic caloric restriction and delay ageing in metazoans. *Nature* **430:** 686–689.

Wu JJ, Liu J, Chen EB, Wang JJ, Cao L, Narayan N, Fergusson MM, Rovira II, Allen M, Springer DA, et al. 2013. Increased mammalian lifespan and a segmental and tissue-specific slowing of aging after genetic reduction of mTOR expression. *Cell Rep* **4:** 913–920.

Yamaza H, Komatsu T, Wakita S, Kijogi C, Park S, Hayashi H, Chiba T, Mori R, Furuyama T, Mori N, et al. 2010. FoxO1 is involved in the antineoplastic effect of calorie restriction. *Aging Cell* **9:** 372–382.

Yang W, Hekimi S. 2010. A mitochondrial superoxide signal triggers increased longevity in *Caenorhabditis elegans*. *PLoS Biol* **8:** e1000556.

Yang J, Maika S, Craddock L, King JA, Liu ZM. 2008. Chronic activation of AMP-activated protein kinase-α1 in liver leads to decreased adiposity in mice. *Biochem Biophys Res Commun* **370:** 248–253.

Yanos ME, Bennett CF, Kaeberlein M. 2012. Genome-wide RNAi longevity screens in *Caenorhabditis elegans*. *Curr Genomics* **13:** 508–518.

Yee C, Yang W, Hekimi S. 2014. The intrinsic apoptosis pathway mediates the pro-longevity response to mitochondrial ROS in *C. elegans*. *Cell* **157:** 897–909.

Yilmaz ÖH, Katajisto P, Lamming D, Gültekin Y, Bauer-Rowe K, Sengupta S, Birsoy K, Dursun A, Yilmaz V, Selig M, et al. 2012. mTORC1 in the Paneth cell niche couples intestinal stem-cell function to calorie intake. *Nature* **486:** 490–495.

Yuan H, Su L, Chen WY. 2013. The emerging and diverse roles of sirtuins in cancer: A clinical perspective. *Onco Targets Ther* **6:** 1399–1416.

Zarse K, Schmeisser S, Groth M, Priebe S, Beuster G, Kuhlow D, Guthke R, Platzer M, Kahn CR, Ristow M. 2012. Impaired insulin/IGF1 signaling extends life span by promoting mitochondrial L-proline catabolism to induce a transient ROS signal. *Cell Metab* **15**: 451–465.

Zhang J. 2007. The direct involvement of SirT1 in insulin-induced insulin receptor substrate-2 tyrosine phosphorylation. *J Biol Chem* **282**: 34356–34364.

Zhang Y, Shao Z, Zhai Z, Shen C, Powell-Coffman JA. 2009a. The HIF-1 hypoxia-inducible factor modulates lifespan in C. elegans. *PLoS ONE* **4**: e6348.

Zhang Y, Ikeno Y, Qi W, Chaudhuri A, Li Y, Bokov A, Thorpe SR, Baynes JW, Epstein C, Richardson A, et al. 2009b. Mice deficient in both Mn superoxide dismutase and glutathione peroxidase-1 have increased oxidative dam-age and a greater incidence of pathology but no reduc-tion in longevity. *J Gerontol A Biol Sci Med Sci* **64**: 1212–1220.

Zhang Y, Bokov A, Gelfond J, Soto V, Ikeno Y, Hubbard G, Diaz V, Sloane L, Maslin K, Treaster S, et al. 2014. Rapa-mycin extends life and health in C57BL/6 mice. *J Gerontol A Biol Sci Med Sci* **69**: 119–130.

Zid BM, Rogers AN, Katewa SD, Vargas MA, Kolipinski MC, Lu TA, Benzer S, Kapahi P. 2009. 4E-BP ex-tends lifespan upon dietary restriction by enhancing mitochondrial activity in *Drosophila*. *Cell* **139**: 149–160.

DNA Damage, DNA Repair, Aging, and Neurodegeneration

Scott Maynard[1], Evandro Fei Fang[2], Morten Scheibye-Knudsen[2], Deborah L. Croteau[2], and Vilhelm A. Bohr[1,2]

[1]Department of Cellular and Molecular Medicine, Center for Healthy Aging, University of Copenhagen, DK-2200 Copenhagen, Denmark

[2]Laboratory of Molecular Gerontology, National Institute on Aging, National Institutes of Health, Baltimore, Maryland 21224

Correspondence: vbohr@nih.gov

Aging in mammals is accompanied by a progressive atrophy of tissues and organs, and stochastic damage accumulation to the macromolecules DNA, RNA, proteins, and lipids. The sequence of the human genome represents our genetic blueprint, and accumulating evidence suggests that loss of genomic maintenance may causally contribute to aging. Distinct evidence for a role of imperfect DNA repair in aging is that several premature aging syndromes have underlying genetic DNA repair defects. Accumulation of DNA damage may be particularly prevalent in the central nervous system owing to the low DNA repair capacity in postmitotic brain tissue. It is generally believed that the cumulative effects of the deleterious changes that occur in aging, mostly after the reproductive phase, contribute to species-specific rates of aging. In addition to nuclear DNA damage contributions to aging, there is also abundant evidence for a causative link between mitochondrial DNA damage and the major phenotypes associated with aging. Understanding the mechanistic basis for the association of DNA damage and DNA repair with aging and age-related diseases, such as neurodegeneration, would give insight into contravening age-related diseases and promoting a healthy life span.

Aging is a major risk factor for neurodegeneration, cancer, and other chronic diseases (Hoeijmakers 2009). No single molecular mechanism appears to account for the functional decline in different organ systems in older humans; however, one dominant theory is that molecular damage, including DNA damage and mutations, accumulate over time, and that this damage has phenotypic consequences in adult organisms (Kirkwood 2005). This article discusses our current understanding of the role of DNA repair in counteracting aging-associated disease, the mechanisms by which DNA damage leads to aging and disease, and recent efforts to use this knowledge as a basis for therapeutic approaches to prevent cancer and neurodegenerative disease. Although the focus of aging research is on understanding human aging, many pieces of the puzzle have been revealed through use of animal model systems, which are also discussed here.

DNA DAMAGE AND DNA REPAIR PATHWAYS

Nucleic acids, proteins, and lipids are continually damaged by physical and chemical agents. Exogenous sources of DNA damage include radiation, diet, and environmental chemicals. Endogenous sources of DNA damage include chemical instability, such as depurination, spontaneous errors during DNA replication and repair, and reactive oxygen species (ROS), as by-products of normal metabolism. ROS are thought to generate as many as 50,000 DNA lesions per human cell per day (Lindahl 1993),

including base modifications, single-strand breaks (SSBs), double strand breaks (DSBs), and interstrand cross-links (ICLs). Prominent DNA repair pathways in mammalian cells are base excision repair (BER), nucleotide excision repair (NER), mismatch repair (MMR), and double-strand break repair (DSBR) (Fig. 1). BER excises mostly oxidative and alkylation DNA damage, NER removes bulky, helix-distorting lesions from DNA (e.g., ultraviolet [UV] photodimers), MMR reverses replication errors, and DSBR is specific for repairing DSBs, mainly by either error-prone rejoining of the broken DNA ends (nonhomologous end joining

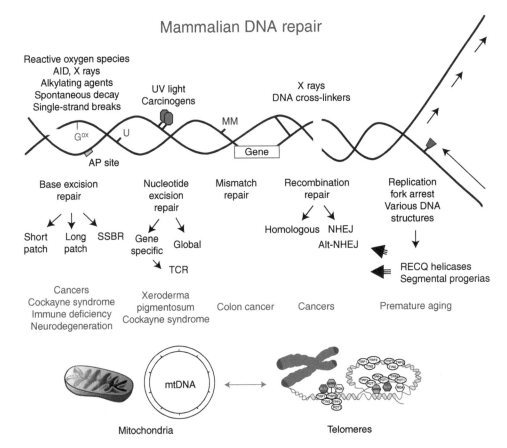

Figure 1. Mammalian DNA repair pathways. Various types of genotoxic agents result in specific DNA damage lesions repaired by specific DNA repair pathways. Defects in each repair pathway are associated with various diseases (shown in red). Subpathways are also referred to, which are described elsewhere (Khakhar et al. 2003; Iyama and Wilson 2013). The black arrows extending from "RecQ helicases" signify that these enzymes are involved in several DNA repair pathways (Croteau et al. 2014). SSBR, single-stand break repair; TCR, transcription coupled repair; Alt-NHEJ, alternative nonhomologous end-joining pathway; AID, activation-induced cytidine deaminase; MM, mismatch of DNA bases.

[NHEJ]) or accurately repairing the DSB using information on the undamaged sister chromatid (homologous recombination [HR]) (Wyman and Kanaar 2006; Iyama and Wilson 2013). Telomere maintenance is a form of targeted DNA repair that is critical for genome stability; defects in telomere maintenance are associated with cellular senescence and aging (von Zglinicki 2000; Rodier et al. 2005; Singh et al. 2011; Calado and Dumitriu 2013).

Unrepaired DNA damage can give rise to genomic instability and induce signaling cascades leading to cell senescence or cell death, which are cellular phenotypes associated with aging (Rodier et al. 2009). Indeed, the capacity to repair DNA damage is thought to decline as cells age (Moriwaki et al. 1996; Muiras et al. 1998; Li and Vijg 2012). DNA lesions can lead to mutations, some of which can be oncogenic (Bohr 2002; Maynard et al. 2009; Cha and Yim 2013). Apoptosis, senescence, and DNA repair are mechanisms that counteract oncogenesis. In addition, unrepaired DNA damage may reduce the capacity for tissue self-renewal, thus inhibiting recovery from acute stress or injury (Rossi et al. 2007). The genetic integrity of stem cells is especially important. Embryonic stems cells (ESCs) have the capacity to differentiate into all cell types, including germ cells; thus, any genetic alterations that are not corrected can compromise the genome stability and functionality of entire cell lineages. Adult stem cells are important for the long-term maintenance of tissues throughout life (Kenyon and Gerson 2007). Indeed, stem cells appear to be very proficient at DNA repair (Maynard et al. 2008; Rocha et al. 2013).

NEURODEGENERATIVE DISEASE AND PREMATURE AGING

Mitochondrial DNA (mtDNA) damage, mitochondrial dysfunction, and defects in BER can adversely affect neuronal functions, thus increasing the risk of neurodegenerative disease (de Souza-Pinto et al. 2008; Fernandez-Checa et al. 2010; Wang and Michaelis 2010). In fact, neurological dysfunction is found in individuals and mouse models with genetic errors in

DNA repair genes (McKinnon 2009; Jeppesen et al. 2011). This is consistent with accumulating evidence that DNA repair pathways and other components of the DNA damage response play a role in preventing neuropathology (Fishel et al. 2007; Rulten and Caldecott 2013; Madabhushi et al. 2014). Oxidative damage to neuronal cells may be an important component of neurodegeneration, as suggested by reports that BER is important in preventing neurodegeneration (Liu et al. 2011; Bosshard et al. 2012; Sheng et al. 2012; Canugovi et al. 2013; Lillenes et al. 2013). The mammalian brain consumes oxygen at a relatively high rate, leading to high exposure of neurons to the associated ROS byproducts; if antioxidants are depleted in the brain, neurons become susceptible to ROS-induced DNA damage (Sai et al. 1992; Hirano et al. 1996; Kaneko et al. 1996; Nakae et al. 2000; Barja 2004).

Defects in DNA repair contribute to genomic instability, which increases with age. Interestingly, *age1* and other long-lived mutants of the nematode *Caenorhabditis elegans* show increased DNA repair capacity, whereas DNA repair–deficient nematodes have a significantly shorter life span, supporting the hypothesis that DNA repair capacity influences longevity (Hyun et al. 2008). In humans, premature aging and early death are characteristics of several rare heritable diseases linked to defects in DNA repair or the processing of DNA damage (Brosh and Bohr 2007; Vijg 2008; Campisi and Vijg 2009; Martin 2011). In many cases, genetically modified mice with comparable defects in DNA repair show similar disease phenotypes that resemble normal aging, suggesting a causal relationship between DNA repair defects and premature aging (de Boer et al. 2002; Andressoo et al. 2006; Garinis et al. 2008; Gredilla et al. 2012). Human diseases of premature or accelerated aging include Werner syndrome (WS), Cockayne syndrome (CS), and Hutchinson–Gilford progeria syndrome (HGPS). All three are termed "segmental progerias" because patients prematurely display some but not all features of normal aging. WS is the most well-characterized premature aging disorder in humans and appears to most closely resemble an acceleration of normal aging (Goto 1997; Kyng et al. 2009). The Werner protein (WRN), mutated in WS patients, is a member of the highly conserved RecQ helicase family, which consists of enzymes that unwind double-stranded DNA and play important roles in DNA replication, recombination, and repair (Rossi et al. 2010). Several studies implicate RecQ helicases in critical processes of DNA replication and repair that influence genomic stability (Sun et al. 1998; Fry and Loeb 1999; Machwe et al. 2006; Bachrati and Hickson 2008; Croteau et al. 2014). CS is a premature aging disorder associated with specific defects in DNA repair and transcription. CS patients show severe developmental and neurological abnormalities (Jeppesen et al. 2011). Some clinicians view CS as a distinct early aging phenotype (Weidenheim et al. 2009; Natale 2011). Mutations in *CSB* (Cockayne syndrome complementation group B) account for ~80% of CS cases, with the majority of remaining CS patients carrying mutations in the *CSA* gene. In addition to its well-established role in transcription-coupled NER, *CSB* also plays a role in BER and maintenance of mitochondrial function (Stevnsner et al. 2008; Osenbroch et al. 2009; Aamann et al. 2010; Scheibye-Knudsen et al. 2012). HGPS is commonly caused by a point mutation in the lamin A gene, leading to a dysfunctional truncated version of the nuclear scaffolding protein lamin A (this aberrant protein is termed progerin). Unlike WS and CS, the defect is not presumed to be directly associated with a defect in DNA repair or processing; however, recent studies suggest that lamin A promotes DNA repair, especially DSBR (Redwood et al. 2011; Gonzalo 2014).

THEORIES OF AGING

Aging is often defined as the accumulation of deleterious biological changes over time, which increases an organism's vulnerability to disease and renders it more likely to die. However, the causal relationship between the biological changes that occur with time and aging is not fully understood. Numerous theories of aging have been suggested, but none of these fully explain all aspects of aging. In this section, we summarize common theories of aging (Fig. 2).

These processes are not mutually exclusive, they may interact in complex ways, and they lead to DNA damage accumulation.

Evolutionary Theories of Aging

The programmed (adaptive) theory of aging states that a genetic program drives the aging process, and that organisms have evolved mechanisms to limit the organism's life span beyond a specific age to benefit subsequent generations. In contrast, many other theories state that aging is a stochastic or random process, with no specific evolutionary value or force. As to why we age, the evolutionary theory of aging states that aging is the result of a decline in the force of natural selection (Tosato et al. 2007; Robert et al. 2010; Goldsmith 2012). One proposed mechanism (or subtheory) of the evolutionary theory is referred to as mutational accumulation, which suggests that the evolutionary effects of adverse events decline following the peak of reproduction (Charlesworth 2001; Martin 2011). Another proposed mechanism is referred to as antagonistic pleiotropy (or trade-off). It suggests that gene variants that enhance reproductive fitness early in life show deleterious effects later in life, after the peak of reproduction (Ljubuncic and Reznick 2009; Martin 2011). Yet another evolutionary theory is the so-called disposable soma theory, which proposes that the failure to repair accumulated stochastic damage is a consequence of evolved limitations in somatic maintenance and repair functions. It predicts that organisms with expected high survival and low reproductive rates should use more metabolic resources in protecting their soma than organisms that expect a shorter life span and to reproduce rapidly (Kirkwood 2005). It has been suggested that the life-span extension induced by caloric restriction may represent the adaptive readjustment of the organism's metabolic resources away from growth and reproduction and toward somatic maintenance (Shanley and Kirkwood 2000). These theories suggest that the cumulative effects of late-acting deleterious changes contribute to species-specific rates of aging (Brunet-Rossinni and Austad 2004).

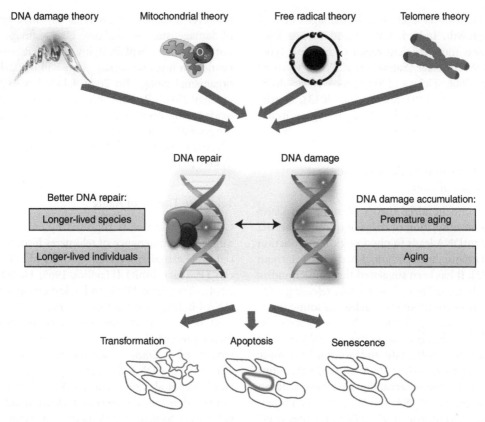

Figure 2. Four major theories of aging. Each theory has DNA damage accumulation and DNA repair as a major component. A variety of evidence, not always consistent (Moskalev et al. 2013), indicates that DNA damage accumulation is associated with aging.

DNA Damage/Repair Theory of Aging

This theory states that unrepaired DNA damage contributes to genomic instability and the aging process (Fig. 2). Although which specific DNA lesions contribute to aging is still debated, age-associated DNA damage could include DNA breaks, cross-links, and modified bases (e.g., oxidative lesions). This theory is part of a broader concept that aging results from a general loss of molecular fidelity (Hayflick 2007). In this context, it has been postulated that natural selection has allowed us to maintain optimal biomolecular fidelity through the peak period of reproductive potential. After this period, survival of the individual is superfluous to survival of the species, and molecular fidelity declines. Genomic maintenance has been described as a double-edged sword (Vijg 2014)—DNA dam-

age by exogenous and endogenous sources is often not perfectly repaired, thus leading to mutations. In germline cells, these mutations drive evolutionary change through natural selection. In somatic cells of multicellular organisms, these mutations could contribute to aging.

Some DNA repair pathways, such as MMR, HR, and NHEJ, are associated with replication, and thus are attenuated in nondividing cells, such as neuronal cells; BER, NER, and transcription-coupled repair appear to play important roles in neurons (Fishel et al. 2007; Nouspikel 2007; Iyama and Wilson 2013). Accumulation of DNA mutations with age is accompanied by an increase in probability of tumor formation. Germline cells, and the ESCs that they originate from, may avoid the buildup of stochastic DNA damage by either more efficient DNA repair systems or replacement of cells that have lower levels of DNA damage by clonal outgrowth. Indeed, human and mouse ESCs possess more efficient repair of multiple types of DNA damage (Saretzki et al. 2004; Maynard et al. 2008; Tichy and Stambrook 2008; Momcilovic et al. 2010; Rocha et al. 2013), thereby decreasing the likelihood of passing on mutations to daughter cells.

The Mitochondrial and Free Radical Theories of Aging

The mitochondrial theory of aging postulates that accumulation of damage to mitochondria and mtDNA leads to physiological dysfunction and eventually to pathological disease (Harman 1972). It has been suggested that mitochondria also become "leaky" over time, releasing ROS that may contribute to nuclear genomic instability (Samper et al. 2003). This theory is consistent with the observation that mtDNA mutates at a much faster rate and accumulates more damage than nuclear DNA, and is further supported by the observations that mice with an error-prone mtDNA polymerase γ age prematurely (Trifunovic et al. 2004) and that overexpression of mitochondrial-targeted catalase (an antioxidant) in mice leads to increased life span (Schriner et al. 2005).

The free radical theory of aging proposes that aging is caused by the accumulation of damage inflicted by free radicals (Park and Yeo 2013; Vina et al. 2013; Gladyshev 2014). This has obvious strong overlap with the mitochondrial theory (Fig. 2), because much of the endogenous ROS comes from imperfect ("leaky") mitochondrial respiration. There is abundant evidence for a role of mitochondrial dysfunction and ROS in age-associated diseases (Lee et al. 2010; Victor et al. 2011; Ramamoorthy et al. 2012; Lagouge and Larsson 2013; Marzetti et al. 2013; Maynard et al. 2013). Indeed, our group has recently reported associations of ROS and DNA damage with self-reported fatigue/vitality in peripheral blood mononuclear cells of human participants (Maynard et al. 2013, 2014). When aging in the various organ systems is considered, the theory that often appears is the oxidative stress/free radical theory (Cefalu 2011).

However, it has been argued that any single type of damage, such as oxidative DNA damage, is not enough to explain aging and that intervention in just one damage type will not delay organismal aging (Jin 2010; Gladyshev 2014; Liochev 2014). Perhaps this explains why antioxidant supplementation strategies, with the purpose of promoting longevity, have not been effective (Fusco et al. 2007; Vina et al. 2013; Gladyshev 2014).

Telomere Theory of Aging

Cellular senescence is triggered by erosion or improper maintenance of telomeres leading to cell-cycle exit after a certain number of cell cycles (Hayflick limit) (Hayflick 1965; Holliday 2014). Telomeric DNA and telomere-specific DNA-binding proteins form a structurally distinct domain at chromosome termini. Telomeres prevent chromosome ends from being recognized as DSBs. When telomeres shorten, cells induce a DNA damage response (Fig. 2) (Karlseder et al. 1999). Furthermore, depletion of DNA damage–response factors can result in defective telomere maintenance; for example, studies have shown that cells lacking or deficient in certain helicases display telomere attrition and/or replication defects (Crabbe et al. 2004; Sfeir et al. 2009; Ghosh et al. 2011). Human telomeric DNA includes 2–15 kb of tandem repeats of TTAGGG, plus a terminal $3'$-protruding G-rich single-stranded DNA (ssDNA) tail greater than 100 nucleotides long (Lin et al. 2014). The ssDNA tail folds back and invades the telomeric double-stranded DNA (dsDNA) forming a telomeric T-loop that is critical for telomere capping (Hanish et al. 1994). Telomerase is a specialized DNA polymerase that is responsible for telomere replication. The somatic expression of telomerase is insufficient to compensate for telomere loss. Even though mice have longer telomeres, telomere dysfunction is thought to contribute significantly to aging in mice. Telomerase reactivation (and telomere elongation) reverses age-related pathology in mice (Jaskelioff et al. 2011); however, early death often occurs because of tumor formation (Gonzalez-Suarez et al. 2002). DNA glycosylases are critical in the removal of oxidative base damage at telomeres, but they do so imperfectly, and, thus, oxidative DNA damage still accumulates in telomeres of older mice and disrupts telomere length homoeostasis (Lu and Liu 2010; Wang et al. 2010; Rhee et al. 2011). The loss of telomeres is a classic example of the imperfect homeostasis that may contribute to aging (Teplyuk 2012).

A new research area of great interest is the association between telomere shortening and mitochondrial dysfunction (Sahin et al. 2011). Although the mechanism of such "cross talk" is not yet well understood, there are examples of proteins that participate in both compartments, such as RECQL4, which interacts with human telomeric DNA (Ghosh et al. 2011) and is present in mitochondria (Croteau et al. 2012). A number of recent observations (von Zglinicki et al. 2001; Passos et al. 2007; Sahin et al. 2011) have indicated that mitochondrial dysfunction can lead to telomere attrition and vice versa.

Cellular Senescence during Aging

Stochastic damage events can lead to cellular senescence. The senescence cellular stress response is now considered one of the major drivers of aging. Consistent with this concept, in most cases, fibroblasts from patients with progeriod syndromes have accelerated senescence (van de Ven et al. 2006). Conversely, fibroblasts from long-lived Snell dwarf mice are resistant to the oxidative damage that contributes to growth arrest in vitro (Maynard and Miller 2006). Senescent cells now appear to be a major player in the aging process by the acquisition of the senescence-associated secretory phenotype, which impacts the cellular milieu; this property of senescent cells is apparently a response to genotoxic stress (Campisi and Robert 2014). Senescent cells accumulate in many tissues over time (Fig. 2) (Dimri et al. 1995; Ressler et al. 2006), indicating that their formation occurs at a faster rate than their death or removal. The senescence cellular stress response is also an important anticancer mechanism (Campisi 2013). Indeed, the tumor suppressor p53 plays an important role in the regulatory mechanisms between DNA repair, apoptosis, and senescence (Erol

2011). However, evidence indicates that senescence drives both degenerative and hyperplasic pathologies; thus, it has been proposed that the senescence response may have features that are antagonistically pleotropic (Campisi 2013). Thus, senescence appears to be another demonstration of imperfect homeostasis as a basis of aging (Teplyuk 2012). Notably, germ cells/ESCs do not undergo senescence; this is a manifestation of the germline evasion of imperfect homeostasis (Evans and Kaufman 1981; Teplyuk 2012).

Stem-Cell Depletion during Aging

Most tissues of multicellular organisms have the ability for regeneration and self-renewal, by repopulating adult stem cells; this ability declines with age. The exhaustion of stem-cell pools with age is likely a result of several of the processes discussed above, such as accumulation of DNA mutations, telomere attrition, apoptosis, and senescence (Rossi et al. 2007; Beltrami et al. 2011). Bone marrow, intestines, and other highly proliferative tissues may be particularly sensitive to loss of functional stem cells. Consequently, not all tissues in an organism age at the same rate. It has been suggested that the general reason for the decline in stem-cell pools is the tissue-dependent imperfect balance between stem-cell self-renewal and differentiation (Teplyuk 2012).

MITOCHONDRIAL DYSFUNCTION AND ENERGY HOMEOSTASIS

Mitochondria are emerging as central players in aging, neurodegeneration, and metabolic diseases (Wallace 1999; Wallace et al. 2010). This dynamic organelle is central in ATP generation through oxidative phosphorylation (OXPHOS); however, mitochondria are also involved in other processes, such as biomolecule synthesis, apoptosis, and calcium regulation. The importance of mitochondria is highlighted by the elaborate and conserved maintenance pathways that ensure proper function of this organelle (Fig. 3). These include redox regulation, mtDNA repair, and autophagy. In addition,

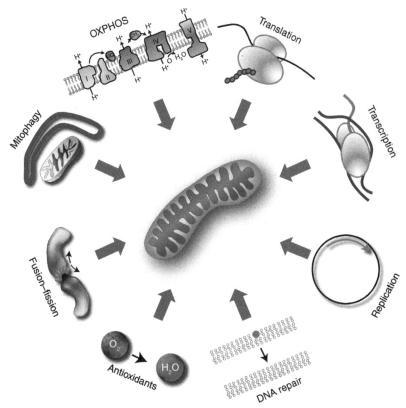

Figure 3. Pathways involved in mitochondrial maintenance.

mtDNA transcription, translation, and replication as well as proper function of OXPHOS are essential for organismal survival. Mitochondria are the primary source of superoxide, a ROS that acts both as a signaling molecule and as a source of damage. To remove ROS, aerobic organisms have an antioxidant defense system, which includes enzymes, such as superoxide dismutase, catalase, and peroxiredoxins. In addition, there are many dietary antioxidants, including glutathione, vitamin E, and vitamin C. As a feedback loop, ROS can induce the expression of antioxidant genes through transcription factors, such as PGC-1α and NRF2 (Ventura-Clapier et al. 2008). If ROS is not scavenged, it can oxidize molecules in the cells, such as lipids, proteins, and DNA.

mtDNA may be particularly prone to damage in part because of its vicinity to the source of ROS, the OXPHOS machinery. Oxidatively damaged mtDNA is repaired primarily through BER, a repair pathway that deals with single-base damage (Gredilla et al. 2010). This repair pathway entails several enzymatic steps. First, damage is recognized by a glycosylase, such as 8-oxoguanine glycosylase 1 (OGG1), that removes the damaged base. The resultant abasic site is then recognized by the apurinic/apyrimidinic endonuclease 1 (APE1) that removes the ribose leaving a gap in the DNA strand. The gap is then subsequently filled by the mtDNA polymerase γ (POL-γ) and the DNA is sealed by ligase III (Gredilla et al. 2010).

If mitochondria become damaged beyond repair, the whole organelle can be degraded through a subpathway of autophagy termed mitophagy (Campello et al. 2014). Mitophagy is the process by which a double lipid bilayer is formed around the damaged mitochondria, such that it is engulfed in a vesicle, the autophagosome. The autophagosome fuses with a lysosome facilitating the degradation of its content.

At least two mitophagy pathways can lead to mitochondrial degradation: (1) programmed mitophagy through up-regulation of the mitochondrial receptor NIX, leading to removal of all mitochondria, a pathway necessary for erythropoiesis; and (2) selective mitophagy whereby single mitochondria are degraded. The second pathway entails the initiation of mitophagy at damaged mitochondria through loss of inner mitochondrial membrane potential. Inner membrane depolarization leads to the accumulation of the kinase PINK1 on the outer membrane. PINK1 phosphorylates a number of proteins in the outer mitochondrial membrane, including MFN2 and ubiquitin, which leads to the recruitment, phosphorylation, and activation of parkin, an E3-ubiquitin ligase. Parkin ubiquitinates outer membrane proteins and facilitates the association of the mitochondria with a growing autophagosome membrane. After mitochondria become engulfed, fusion with a lysosome ensures degradation of the mitochondria (Youle and Narendra 2011; Youle and van der Bliek 2012).

It is becoming apparent that these mitochondrial-associated pathways are important for maintenance of organismal health. For example, loss of antioxidative capacity, such as vitamin E or glutathione synthase deficiency, can lead to neurodegeneration. In addition, defects in mtDNA repair can lead to neurodegeneration, as is the case for ataxia with oculomotor apraxia 1 (Sykora et al. 2011). Defects in mitophagy are associated with Parkinson's disease through mutations in PINK1 and parkin (Narendra et al. 2008). It is therefore clear that preserving mitochondrial health is important for maintaining organismal health.

Approximately one in 5000 individuals suffer from a mitochondrial disorder (Haas et al. 2007). Furthermore, several relatively common aging-related diseases appear to have a mitochondrial component. For example, mitochondrial dysfunction is a hallmark of β-amyloid-induced neural toxicity in Alzheimer's disease (Lustbader et al. 2004; Tillement et al. 2011); Parkinson's disease patients, as well as elderly individuals, have the burden of mtDNA deletions within substantia nigra neurons (Bender et al. 2006; Kraytsberg et al. 2006); cardiovascular disease is associated with increased production of ROS in mitochondria, accumulation of mtDNA damage, and progressive respiratory chain dysfunction (Madamanchi and Runge 2007; Mercer et al. 2010); and mitochondrial dysfunction characterized by reduced ATP generation appears to have a causal role in features of type 2 diabetes (insulin resistance and hyperglycemia) (Lowell and Shulman 2005). Declining mitochondrial function over the lifetime of an organism has been shown by several lines of evidence (Corral-Debrinski et al. 1992; Vendelbo and Nair 2011; Chistiakov et al. 2014). For example, it has been reported that mtDNA deletions and oxidative damage accumulate with aging (Corral-Debrinski et al. 1992; Hudson et al. 1998).

If mitochondrial dysfunction is pivotal in aging, diseases displaying accelerated aging should phenocopy the signs and symptoms seen in mitochondrial diseases. To test this hypothesis, we recently generated an online database of signs and symptoms seen in human mitochondrial disorders, www.mitodb.com (Scheibye-Knudsen et al. 2013). We then created a number of online advanced bioinformatics tools to test whether a disorder can be characterized as mitochondrial, based on its clinical signs and symptoms. We believe that this database will be useful to physicians and researchers who are studying diseases of unknown etiology. As a proof of principle, the database segregated CS and xeroderma pigmentosum group A ([XPA], a disorder with deficient nucleotide excision repair) with mitochondrial diseases. We recently reported mitochondrial and bioenergetic changes in CS cell lines and mouse tissues (Scheibye-Knudsen et al. 2012), and, thus, we had expected that this condition would cluster with mitochondrial diseases. The XPA segregation with mitochondrial diseases, however, was unexpected, and we then investigated whether XPA cells displayed altered mitochondrial properties. Both XPA-knockdown and XPA-deficient patient cells did indeed show distinct mitochondrial changes, including higher membrane potential, altered mitophagy, and higher basal oxygen and ATP consumption rates (Fang et al. 2014). Interestingly, the ataxia telangiectasia mutated (ATM) disorder also segregated with mitochondrial diseases. Recently, ATM cells that were deficient in a kinase important in the DNA damage–response cascade were shown to have higher mitochondrial membrane potential, decreased mitophagy, and higher respiration (Valentin-Vega et al. 2012). These observations suggest that www.mitodb.com is a useful tool for studying diseases linked to DNA repair defects and premature aging. Interestingly, normal aging shares many features of mitochondrial dysfunction, corroborating the mitochondrial theory of aging.

Cellular DNA damage triggers activation of the DNA damage sensor poly(ADP-ribose) polymerase 1 (PARP1) to recruit DNA repair proteins and fix the damaged DNA. PARP1 is an important protein that is involved in a variety of intracellular processes. It is one of 18 members in the PARP family and plays a major role in PARylation, the process by which PARP is covalently linked to and regulate cellular proteins. PARylation regulates a variety of intracellular processes, including DNA repair, transcription, replication, chromatin modification and cell death (Rouleau et al. 2004; Erdelyi et al. 2005). PARP1 is involved in both BER and DSBR. In addition, PARP1 appears to be involved in NER, as evidenced by the activation of this protein after UV damage, a classic NER substrate (Robu et al. 2013). In addition, PARP1 interacts with core NER proteins, such as CSB (Thorslund et al. 2005), XPA (Fan and Luo 2010; Fischer et al. 2014), and XPC (Robu et al. 2013). Although a predominantly nuclear enzyme, a portion of PARP1 proteins localizes to mitochondria and interacts with the mitochondrial protein mitofilin. The mitochondrial PARP1 may play a role in maintaining mtDNA integrity (Rossi et al. 2009).

Even though PARP1 plays a role in genome maintenance, hyperactivation of PARP1 may be detrimental to the cell. Indeed, increased activation of PARP1 has been associated with aging, abnormal metabolism, neurodegeneration, and a specific form of cell death called parthanatos. Parthanatos is a caspase-independent pathway of programmed cell death dependent on the nuclear translocation of the mitochondrial-associated apoptosis-inducing factor (AIF) (Fatokun et al. 2014). Hyperactivation of PARP1 is associated with stroke (Andrabi et al. 2011) and neurodegeneration in some premature aging disorders, such as XPA, CSB, and ATM (Fang et al. 2014). Increased PARP1 activation occurs with age in wild-type *C. elegans*, and DNA damage expedites this process; this is evidenced by the findings that supplementation with PARP inhibitors extends life span in wild-type and DNA repair–deficient (*xpa-1, csb-1*) *C. elegans* (Mouchiroud et al. 2013; Fang et al. 2014; Scheibye-Knudsen et al. 2014b). Additionally, in a PARP1 knockout mouse model, there is an increased NAD^+ content that leads to increased SIRT1 activity and cellular metabolism in brown adipose tissue and muscle (Bai et al. 2011).

The side effects of PARP1 hyperactivation may be partially attributed to a reduction of the NAD^+-SIRT1 pathway because both PARP1 and SIRT1 compete for NAD^+. The consequences of lower NAD^+ levels because of PARP1 hyperactivation has been shown in the DNA repair defect disease model XPA (Fang et al. 2014). XPA is a 40-kDa nuclear protein, essential for NER. XPA physically interacts with both PARP1 and PAR, and this interaction may be of importance for the repair of UV-induced damage (Fan and Luo 2010; Fischer et al. 2014). Furthermore, SIRT1 physically interacts with, and deacetylates, XPA to promote NER (Fan and Luo 2010). Because both PARP1 and SIRT1 consume NAD^+ on activation, PARP1 hyperactivation in XPA leads to reduced SIRT1 deacetylation because of loss of NAD^+ (Fang et al. 2014; Scheibye-Knudsen et al. 2014a). PGC-1α is a master regulator of mitochondrial function, and SIRT1 positively regulates the activity of PGC-1α through deacetylation of this transcriptional coactivator. Loss of SIRT1 consequently leads to inactivation of PGC-1α and mitochondrial dysfunction. Thus, a hitherto unrecognized impairment of the nuclear-mitochondrial signaling may be involved in the pathogenesis of XPA and other neurodegenerative DNA repair–deficient disorders.

In an effort to synthesize results of our research on mitochondrial bioenergetics and aging, we have developed a working model. Persistent DNA damage activates the nuclear DNA-damage response, which includes kinases and PARP. These enzymes have the capacity to consume high levels of ATP and NAD$^+$. In an attempt to meet the ATP demands of the cell, the mitochondria become more coupled, which leads to a higher membrane potential and to lower mitophagy. However, a side effect of more coupled mitochondria is an increase in ROS production. Independently, NAD$^+$ depletion may arise from DNA damage–dependent PARP activation. Recycling NAD$^+$ from these polymers is energy demanding, but necessary for maintaining homeostatic NADH/NAD$^+$ levels and alternatively may account for the increased oxygen and ATP consumption observed in cells. It appears that deficiencies in CSB, XPA, or ATM can lead to the mitochondrial phenotypes described above; thus, uncovering common causes of mitochondrial dysfunction is an important strategy for investigating potential mechanisms of neurodegeneration in these disorders, as well as in normal brain aging (Weidenheim et al. 2009; Niedernhofer et al. 2011; Fang et al. 2014).

FUTURE DIRECTIONS: INTERVENTIONS

There are a number of proposed sites of intervention in DNA-damage accumulation, as highlighted in Figure 4. Agents that alter DNA repair may have therapeutic potential. For example, in the context of cancer chemotherapy with genotoxic drugs, inhibition of endogenous DNA repair is expected to act synergistically to increase the lethal impact on rapidly dividing cancer cells. However, genotoxic agents also damage healthy cells, and so mechanisms are needed to protect normal dividing cells from the toxic impact of the drugs. Therefore, it has been proposed that agents that stimulate DNA repair in normal cells would be useful as a component of cancer chemotherapy. Such agents might also correct deficient or defective DNA repair in other contexts, such as after acute ischemic events.

Historically, researchers have developed inhibitors of DNA repair enzymes rather than activators; however, these enzymatic pathways are highly complex, and it is not a trivial matter to enhance the overall rate of a multistep enzymatic process. One complication is that some reaction intermediates, such as certain BER intermediate products, can be toxic (Sobol et al. 2003; Rinne et al. 2004; Trivedi et al. 2005). One could envision targeting the rate-limiting step in these pathways to avoid buildup of toxic intermediates. However, although the rate-limiting step of a DNA repair pathway can be determined under defined conditions in the lab, translating this data to a living system may yield unpredictable results. Nevertheless, there is considerable interest in exploring the possibility that agents that stimulate or inhibit DNA repair can be developed as therapeutic options for cancer and other aging-associated diseases. The following paragraphs discuss some of the most promising approaches involving agents that stimulate DNA repair.

BER

Ligases (LIG) are proteins that perform the final step in the DNA repair process by sealing the ends of DNA. The prominent ligases are LIGI III and IV. LIGIII is the only ligase in mitochondria; it also operates in the nuclear compartment in complex with XRCC1 and is a component of the BER pathway. LIGIII is the only ligase essential for life and it is also a component of the mtDNA replication and DNA repair machinery (Tomkinson et al. 2013). Reduced levels and activity of LIGIII have been detected in major human neurodegenerative diseases including ataxia telangiectasia (in ATM patient cells and ATM-KO mice) (Sharma et al. 2014) and Alzheimer's disease (Canugovi et al. 2013, 2014). LIGIII activity has been reported to be the rate-limiting step of BER in mitochondria (Akbari et al. 2014). Thus, regulation of this step should be the best target for stimulation of mitochondrial BER. mtDNA repair is critical for cell survival, and stimulation of mitochondrial BER could have benefits for not only mitochondrial functions but also general cellular functions. Because LIGIII levels are decreased in ataxia telangiectasia and Alzheimer's disease, application of an LIGIII stim-

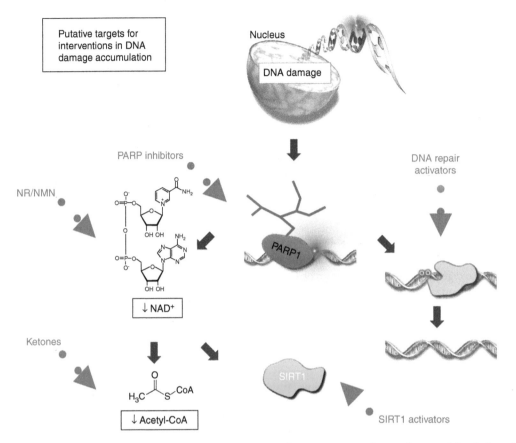

Figure 4. Possible sites of intervention that may ameliorate the consequence of DNA damage. Stimulation of DNA repair may be a feasible strategy for attenuating the effect of DNA damage. Recent research suggests that hyperactivation of the DNA damage–responsive enzyme poly(ADP-ribose) polymerase 1 (PARP1) may be involved in the aging process leading to downstream loss of SIRT1 and pleiotropic mitochondrial dysfunction. PARP1 hyperactivation leads to loss of NAD^+, and reconstitution of NAD^+ by inhibiting PARP1 or through supplementation with NAD^+ precursors nicotinamide riboside (NR) or nicotinamide mononucleotide (NMN) can rescue age-associated consequences of PARP1 activation. Loss of NAD^+ leads to downstream metabolic dysregulation and treatment with ketone bodies may ameliorate those changes.

ulator may be particularly beneficial in these disorders. Furthermore, other diseases might also benefit from enhanced mitochondrial BER activity, such as diseases with altered antioxidant defenses and increased oxidative stress.

Although it is pharmacologically easier to target a single enzymatic reaction, a strategy could be developed to enhance the whole BER process by enhancing all enzymatic steps at the same time. One future approach to this might be to alter posttranslational modifications in the process. For example, it was observed that one protein USP47, a de-ubiquitylating en-

zyme, can regulate the whole BER process, including the DNA polymerase β enzymatic step (Parsons et al. 2011). USP47 regulation could thus be an interesting druggable target and DNA repair pathways might then be regulated at this level.

PARP1

NAD^+ metabolism is of great importance in health and disease as it plays a key role in many molecular processes. Boosting NAD^+ levels could be efficacious in antiaging studies as well

Cite this article as *Cold Spring Harb Perspect Med* doi: 10.1101/cshperspect.a025130

as for the treatment of metabolic diseases and some neurodegenerative DNA repair–deficient disorders. Pharmacological approaches to increase intracellular NAD^+ pools include decreasing NAD^+ consumption by inhibition of PARPs and increasing synthesis of NAD^+ by supplementation with NAD^+ precursors, such as nicotinamide riboside (NR) and nicotinamide mononucleotide (NMN) (Fang et al. 2014; Imai and Guarente 2014). Indeed, both NR and NMN corrected mitochondrial dysfunction in XPA cells and in a $XPA^{-/-}/CSA^{-/-}$ double knockout mouse (Fang et al. 2014). Supplementation with NR may be a general strategy for improving overall mitochondrial fitness because this compound has been found to rescue respiratory chain defects and exercise intolerance in a mouse model of mitochondrial disease (Cerutti et al. 2014). Indeed, in mice, raising NAD^+ levels rescues age-related decline in mitochondrial-encoded OXPHOS subunits (Gomes et al. 2013).

Thus, NAD^+ supplementation appears to be a promising intervention. In addition, NR is stable at room temperature and is water soluble, making it a good candidate for further clinical trials in both healthy population and some specified disorders. Other current important topics/questions in this field include the possibility of synergistic effects of NAD^+ precursors with SIRT1 activators or NAD^+ precursors with PARP inhibitors, and efficacy of these compounds in both human diseases and even healthy individuals.

CONCLUDING REMARKS

Throughout this article, we have discussed the current understanding of the role of DNA repair in preventing aging-associated disease, the mechanisms by which DNA damage may lead to aging, and recent efforts to use this knowledge as a basis for therapeutic approaches to prevent aging, cancer, and neurodegenerative disease. Taken together, there is clear and compelling evidence to suggest that DNA repair mechanisms, both nuclear and mitochondrial, are essential for a long and healthy life. Regulation of DNA repair and of mitochondrial health as dis-

cussed above could be a promising strategic intervention in the future.

REFERENCES

Aamann MD, Sorensen MM, Hvitby C, Berquist BR, Muftuoglu M, Tian J, de Souza-Pinto NC, Scheibye-Knudsen M, Wilson DM III, Stevnsner T, et al. 2010. Cockayne syndrome group B protein promotes mitochondrial DNA stability by supporting the DNA repair association with the mitochondrial membrane. *FASEB J* **24:** 2334–2346.

Akbari M, Keijzers G, Maynard S, Scheibye-Knudsen M, Desler C, Hickson ID, Bohr VA. 2014. Overexpression of DNA ligase III in mitochondria protects cells against oxidative stress and improves mitochondrial DNA base excision repair. *DNA Repair (Amst)* **16:** 44–53.

Andrabi SA, Kang HC, Haince JF, Lee YI, Zhang J, Chi Z, West AB, Koehler RC, Poirier GG, Dawson TM, et al. 2011. Iduna protects the brain from glutamate excitotoxicity and stroke by interfering with poly(ADP-ribose) polymer-induced cell death. *Nat Med* **17:** 692–699.

Andressoo JO, Hoeijmakers JH, Mitchell JR. 2006. Nucleotide excision repair disorders and the balance between cancer and aging. *Cell Cycle* **5:** 2886–2888.

Bachrati CZ, Hickson ID. 2008. RecQ helicases: Guardian angels of the DNA replication fork. *Chromosoma* **117:** 219–233.

Bai P, Canto C, Oudart H, Brunyanszki A, Cen Y, Thomas C, Yamamoto H, Huber A, Kiss B, Houtkooper RH, et al. 2011. PARP-1 inhibition increases mitochondrial metabolism through SIRT1 activation. *Cell Metab* **13:** 461–468.

Barja G. 2004. Free radicals and aging. *Trends Neurosci* **27:** 595–600.

Beltrami AP, Cesselli D, Beltrami CA. 2011. At the stem of youth and health. *Pharmacol Ther* **129:** 3–20.

Bender A, Krishnan KJ, Morris CM, Taylor GA, Reeve AK, Perry RH, Jaros E, Hersheson JS, Betts J, Klopstock T, et al. 2006. High levels of mitochondrial DNA deletions in substantia nigra neurons in aging and Parkinson disease. *Nat Genet* **38:** 515–517.

Bohr VA. 2002. DNA damage and its processing. Relation to human disease. *J Inherit Metab Dis* **25:** 215–222.

Bosshard M, Markkanen E, van LB. 2012. Base excision repair in physiology and pathology of the central nervous system. *Int J Mol Sci* **13:** 16172–16222.

Brosh RM Jr, Bohr VA. 2007. Human premature aging, DNA repair and RecQ helicases. *Nucleic Acids Res* **35:** 7527–7544.

Brunet-Rossinni AK, Austad SN. 2004. Ageing studies on bats: A review. *Biogerontology* **5:** 211–222.

Calado RT, Dumitriu B. 2013. Telomere dynamics in mice and humans. *Semin Hematol* **50:** 165–174.

Campello S, Strappazzon F, Cecconi F. 2014. Mitochondrial dismissal in mammals, from protein degradation to mitophagy. *Biochim Biophys Acta* **1837:** 451–460.

Campisi J. 2013. Aging, cellular senescence, and cancer. *Annu Rev Physiol* **75:** 685–705.

Campisi J, Robert L. 2014. Cell senescence: Role in aging and age-related diseases. *Interdiscip Top Gerontol* **39:** 45–61.

Campisi J, Vijg J. 2009. Does damage to DNA and other macromolecules play a role in aging? If so, how? *J Gerontol A Biol Sci Med Sci* **64:** 175–178.

Canugovi C, Misiak M, Ferrarelli LK, Croteau DL, Bohr VA. 2013. The role of DNA repair in brain-related disease pathology. *DNA Repair (Amst)* **12:** 578–587.

Canugovi C, Shamanna RA, Croteau DL, Bohr VA. 2014. Base excision DNA repair levels in mitochondrial lysates of Alzheimer's disease. *Neurobiol Aging* **35:** 1293–1300.

Cefalu CA. 2011. Theories and mechanisms of aging. *Clin Geriatr Med* **27:** 491–506.

Cerutti R, Pirinen E, Lamperti C, Marchet S, Sauve AA, Li W, Leoni V, Schon EA, Dantzer F, Auwerx J, et al. 2014. NAD$^+$-dependent activation of Sirt1 corrects the phenotype in a mouse model of mitochondrial disease. *Cell Metab* **19:** 1042–1049.

Cha HJ, Yim H. 2013. The accumulation of DNA repair defects is the molecular origin of carcinogenesis. *Tumour Biol* **34:** 3293–3302.

Charlesworth B. 2001. Patterns of age-specific means and genetic variances of mortality rates predicted by the mutation-accumulation theory of ageing. *J Theor Biol* **210:** 47–65.

Chistiakov DA, Sobenin IA, Revin VV, Orekhov AN, Bobryshev YV. 2014. Mitochondrial aging and age-related dysfunction of mitochondria. *Biomed Res Int* **2014:** 238463.

Corral-Debrinski M, Horton T, Lott MT, Shoffner JM, Beal MF, Wallace DC. 1992. Mitochondrial DNA deletions in human brain: Regional variability and increase with advanced age. *Nat Genet* **2:** 324–329.

Crabbe L, Verdun RE, Haggblom CI, Karlseder J. 2004. Defective telomere lagging strand synthesis in cells lacking WRN helicase activity. *Science* **306:** 1951–1953.

Croteau DL, Rossi ML, Canugovi C, Tian J, Sykora P, Ramamoorthy M, Wang Z, Singh DK, Akbari M, Kasiviswanathan R, et al. 2012. RECQL4 localizes to mitochondria and preserves mitochondrial DNA integrity. *Aging Cell* **11:** 456–466.

Croteau DL, Popuri V, Opresko PL, Bohr VA. 2014. Human RecQ helicases in DNA repair, recombination, and replication. *Annu Rev Biochem* **83:** 519–552.

de Boer J, Andressoo JO, de WJ, Huijmans J, Beems RB, van SH, Weeda G, van der Horst GT, van LW, Themmen AP, et al. 2002. Premature aging in mice deficient in DNA repair and transcription. *Science* **296:** 1276–1279.

de Souza-Pinto NC, Wilson DM III, Stevnsner TV, Bohr VA. 2008. Mitochondrial DNA, base excision repair and neurodegeneration. *DNA Repair (Amst)* **7:** 1098–1109.

Dimri GP, Lee X, Basile G, Acosta M, Scott G, Roskelley C, Medrano EE, Linskens M, Rubelj I, Pereira-Smith O, et al. 1995. A biomarker that identifies senescent human cells in culture and in aging skin in vivo. *Proc Natl Acad Sci* **92:** 9363–9367.

Erdelyi K, Bakondi E, Gergely P, Szabo C, Virag L. 2005. Pathophysiologic role of oxidative stress-induced poly (ADP-ribose) polymerase 1 activation: Focus on cell death and transcriptional regulation. *Cell Mol Life Sci* **62:** 751–759.

Erol A. 2011. Deciphering the intricate regulatory mechanisms for the cellular choice between cell repair, apoptosis or senescence in response to damaging signals. *Cell Signal* **23:** 1076–1081.

Evans MJ, Kaufman MH. 1981. Establishment in culture of pluripotential cells from mouse embryos. *Nature* **292:** 154–156.

Fan W, Luo J. 2010. SIRT1 regulates UV-induced DNA repair through deacetylating XPA. *Mol Cell* **39:** 247–258.

Fang EF, Scheibye-Knudsen M, Brace LE, Kassahun H, SenGupta T, Nilsen H, Mitchell JR, Croteau DL, Bohr VA. 2014. Defective mitophagy in XPA via PARP-1 hyperactivation and NAD$^+$/SIRT1 reduction. *Cell* **157:** 882–896.

Fatokun AA, Dawson VL, Dawson TM. 2014. Parthanatos: Mitochondrial-linked mechanisms and therapeutic opportunities. *Br J Pharmacol* **171:** 2000–2016.

Fernandez-Checa JC, Fernandez A, Morales A, Mari M, Garcia-Ruiz C, Colell A. 2010. Oxidative stress and altered mitochondrial function in neurodegenerative diseases: Lessons from mouse models. *CNS Neurol Disord Drug Targets* **9:** 439–454.

Fischer JM, Popp O, Gebhard D, Veith S, Fischbach A, Beneke S, Leitenstorfer A, Bergemann J, Scheffner M, Ferrando-May E, et al. 2014. Poly(ADP-ribose)-mediated interplay of XPA and PARP1 leads to reciprocal regulation of protein function. *FEBS J* **281:** 3625–3641.

Fishel ML, Vasko MR, Kelley MR. 2007. DNA repair in neurons: So if they don't divide what's to repair? *Mutat Res* **614:** 24–36.

Fry M, Loeb LA. 1999. Human Werner syndrome DNA helicase unwinds tetrahelical structures of the fragile X syndrome repeat sequence d(CGG)$_n$. *J Biol Chem* **274:** 12797–12802.

Fusco D, Colloca G, Lo Monaco MR, Cesari M. 2007. Effects of antioxidant supplementation on the aging process. *Clin Interv Aging* **2:** 377–387.

Garinis GA, van der Horst GT, Vijg J, Hoeijmakers JH. 2008. DNA damage and ageing: New-age ideas for an age-old problem. *Nat Cell Biol* **10:** 1241–1247.

Ghosh AK, Rossi ML, Singh DK, Dunn C, Ramamoorthy M, Croteau DL, Liu Y, Bohr VA. 2011. RECQL4, the protein mutated in Rothmund–Thomson syndrome, functions in telomere maintenance. *J Biol Chem* **287:** 196–209.

Gladyshev VN. 2014. The free radical theory of aging is dead. Long live the damage theory! *Antioxid Redox Signal* **20:** 727–731.

Goldsmith TC. 2012. On the programmed/non-programmed aging controversy. *Biochemistry (Mosc)* **77:** 729–732.

Gomes AP, Price NL, Ling AJ, Moslehi JJ, Montgomery MK, Rajman L, White JP, Teodoro JS, Wrann CD, Hubbard BP, et al. 2013. Declining NAD$^+$ induces a pseudohypoxic state disrupting nuclear-mitochondrial communication during aging. *Cell* **155:** 1624–1638.

Gonzalez-Suarez E, Flores JM, Blasco MA. 2002. Cooperation between p53 mutation and high telomerase transgenic expression in spontaneous cancer development. *Mol Cell Biol* **22:** 7291–7301.

Gonzalo S. 2014. DNA damage and lamins. *Adv Exp Med Biol* **773:** 377–399.

Cite this article as *Cold Spring Harb Perspect Med* doi: 10.1101/cshperspect.a025130

Goto M. 1997. Hierarchical deterioration of body systems in Werner's syndrome: Implications for normal ageing. *Mech Ageing Dev* **98:** 239–254.

Gredilla R, Bohr VA, Stevnsner T. 2010. Mitochondrial DNA repair and association with aging—An update. *Exp Gerontol* **45:** 478–488.

Gredilla R, Garm C, Stevnsner T. 2012. Nuclear and mitochondrial DNA repair in selected eukaryotic aging model systems. *Oxid Med Cell Longev* **2012:** 282438.

Haas RH, Parikh S, Falk MJ, Saneto RP, Wolf NI, Darin N, Cohen BH. 2007. Mitochondrial disease: A practical approach for primary care physicians. *Pediatrics* **120:** 1326–1333.

Hanish JP, Yanowitz JL, de LT. 1994. Stringent sequence requirements for the formation of human telomeres. *Proc Natl Acad Sci* **91:** 8861–8865.

Harman D. 1972. The biologic clock: The mitochondria? *J Am Geriatr Soc* **20:** 145–147.

Hayflick L. 1965. The limited in vitro lifetime of human diploid cell strains. *Exp Cell Res* **37:** 614–636.

Hayflick L. 2007. Biological aging is no longer an unsolved problem. *Ann NY Acad Sci* **1100:** 1–13.

Hirano T, Yamaguchi R, Asami S, Iwamoto N, Kasai H. 1996. 8-Hydroxyguanine levels in nuclear DNA and its repair activity in rat organs associated with age. *J Gerontol A Biol Sci Med Sci* **51:** 303–307.

Hoeijmakers JH. 2009. DNA damage, aging, and cancer. *N Engl J Med* **361:** 1475–1485.

Holliday R. 2014. The commitment of human cells to senescence. *Interdiscip Top Gerontol* **39:** 1–7.

Hudson EK, Hogue B, Souza-Pinto N, Croteau DL, Anson RM, Bohr VA, Hansford RG. 1998. Age associated change in mitochondrial DNA. *Free Radic Res Commun* **29:** 573–579.

Hyun M, Lee J, Lee K, May A, Bohr VA, Ahn B. 2008. Longevity and resistance to stress correlate with DNA repair capacity in *Caenorhabditis elegans*. *Nucleic Acids Res* **36:** 1380–1389.

Imai S, Guarente L. 2014. NAD$^+$ and sirtuins in aging and disease. *Trends Cell Biol* **24:** 464–471.

Iyama T, Wilson DM III. 2013. DNA repair mechanisms in dividing and non-dividing cells. *DNA Repair (Amst)* **12:** 620–636.

Jaskelioff M, Muller FL, Paik JH, Jiang S, Adams AC, Sahin E, Kost-Alimova M, Protopopov A, Cadinanos J, Horner JW, et al. 2011. Telomerase reactivation reverses tissue degeneration in aged telomerase-deficient mice. *Nature* **469:** 102–106.

Jeppesen DK, Bohr VA, Stevnsner T. 2011. DNA repair deficiency in neurodegeneration. *Prog Neurobiol* **94:** 166–200.

Jin K. 2010. Modern biological theories of aging. *Aging Dis* **1:** 72–74.

Kaneko T, Tahara S, Matsuo M. 1996. Non-linear accumulation of 8-hydroxy-2′-deoxyguanosine, a marker of oxidized DNA damage, during aging. *Mutat Res* **316:** 277–285.

Karlseder J, Broccoli D, Dai Y, Hardy S, De Lange T. 1999. p53- and ATM-dependent apoptosis induced by telomeres lacking TRF2. *Science* **283:** 1321–1325.

Kenyon J, Gerson SL. 2007. The role of DNA damage repair in aging of adult stem cells. *Nucleic Acids Res* **35:** 7557–7565.

Khakhar RR, Cobb JA, Bjergbaek L, Hickson ID, Gasser SM. 2003. RecQ helicases: Multiple roles in genome maintenance. *Trends Cell Biol* **13:** 493–501.

Kirkwood TB. 2005. Understanding the odd science of aging. *Cell* **120:** 437–447.

Kraytsberg Y, Kudryavtseva E, McKee AC, Geula C, Kowall NW, Khrapko K. 2006. Mitochondrial DNA deletions are abundant and cause functional impairment in aged human substantia nigra neurons. *Nat Genet* **38:** 518–520.

Kyng K, Croteau DL, Bohr VA. 2009. Werner syndrome resembles normal aging. *Cell Cycle* **8:** 2323.

Lagouge M, Larsson NG. 2013. The role of mitochondrial DNA mutations and free radicals in disease and ageing. *J Intern Med* **273:** 529–543.

Lee HC, Chang CM, Chi CW. 2010. Somatic mutations of mitochondrial DNA in aging and cancer progression. *Ageing Res Rev* **9:** S47–S58.

Li W, Vijg J. 2012. Measuring genome instability in aging—A mini-review. *Gerontology* **58:** 129–138.

Lillenes MS, Stoen M, Gomez-Munoz M, Torp R, Gunther CC, Nilsson LN, Tonjum T. 2013. Transient OGG1, APE1, PARP1 and Polβ expression in an Alzheimer's disease mouse model. *Mech Ageing Dev* **134:** 467–477.

Lin J, Kaur P, Countryman P, Opresko PL, Wang H. 2014. Unraveling secrets of telomeres: One molecule at a time. *DNA Repair (Amst)* **20:** 142–153.

Lindahl T. 1993. Instability and decay of the primary structure of DNA. *Nature* **362:** 709–715.

Liochev SI. 2014. Reflections on the theories of aging, of oxidative stress, and of science in general. Is it time to abandon the free radical (oxidative stress) theory of aging? *Antioxid Redox Signal* doi: 10.1089/ars.2014.5928.

Liu D, Croteau DL, Souza-Pinto N, Pitta M, Tian J, Wu C, Jiang H, Mustafa K, Keijzers G, Bohr VA, et al. 2011. Evidence that OGG1 glycosylase protects neurons against oxidative DNA damage and cell death under ischemic conditions. *J Cereb Blood Flow Metab* **31:** 680–692.

Ljubuncic P, Reznick AZ. 2009. The evolutionary theories of aging revisited—A mini-review. *Gerontology* **55:** 205–216.

Lowell BB, Shulman GI. 2005. Mitochondrial dysfunction and type 2 diabetes. *Science* **307:** 384–387.

Lu J, Liu Y. 2010. Deletion of Ogg1 DNA glycosylase results in telomere base damage and length alteration in yeast. *EMBO J* **29:** 398–409.

Lustbader JW, Cirilli M, Lin C, Xu HW, Takuma K, Wang N, Caspersen C, Chen X, Pollak S, Chaney M, et al. 2004. ABAD directly links Aβ to mitochondrial toxicity in Alzheimer's disease. *Science* **304:** 448–452.

Machwe A, Xiao L, Groden J, Orren DK. 2006. The Werner and Bloom syndrome proteins catalyze regression of a model replication fork. *Biochemistry* **45:** 13939–13946.

Madamanchi NR, Runge MS. 2007. Mitochondrial dysfunction in atherosclerosis. *Circ Res* **100:** 460–473.

Madabhushi R, Pan L, Tsai LH. 2014. DNA damage and its links to neurodegeneration. *Neuron* **83:** 266–282.

Martin GM. 2011. The biology of aging: 1985-2010 and beyond. *FASEB J* **25**: 3756–3762.

Marzetti E, Csiszar A, Dutta D, Balagopal G, Calvani R, Leeuwenburgh C. 2013. Role of mitochondrial dysfunction and altered autophagy in cardiovascular aging and disease: From mechanisms to therapeutics. *Am J Physiol Heart Circ Physiol* **305**: H459–H476.

Maynard SP, Miller RA. 2006. Fibroblasts from long-lived Snell dwarf mice are resistant to oxygen-induced in vitro growth arrest. *Aging Cell* **5**: 89–96.

Maynard S, Swistowska AM, Lee JW, Liu Y, Liu ST, Da Cruz AB, Rao M, de Souza-Pinto NC, Zeng X, Bohr VA. 2008. Human embryonic stem cells have enhanced repair of multiple forms of DNA damage. *Stem Cells* **26**: 2266–2274.

Maynard S, Schurman SH, Harboe C, de Souza-Pinto NC, Bohr VA. 2009. Base excision repair of oxidative DNA damage and association with cancer and aging. *Carcinogenesis* **30**: 2–10.

Maynard S, Keijzers G, Gram M, Desler C, Bendix L, Budtz-Jorgensen E, Molbo D, Croteau DL, Osler M, Stevnsner T, et al. 2013. Relationships between human vitality and mitochondrial respiratory parameters, reactive oxygen species production and dNTP levels in peripheral blood mononuclear cells. *Aging (Albany NY)* **5**: 850–864.

Maynard S, Keijzers G, Hansen AM, Osler M, Molbo D, Bendix L, Moller P, Loft S, Moreno-Villanueva M, Burkle A, et al. 2014. Associations of subjective vitality with DNA damage, cardiovascular risk factors and physical performance. *Acta Physiol (Oxf)* **213**: 156–170.

McKinnon PJ. 2009. DNA repair deficiency and neurological disease. *Nat Rev Neurosci* **10**: 100–112.

Mercer JR, Cheng KK, Figg N, Gorenne I, Mahmoudi M, Griffin J, Vidal-Puig A, Logan A, Murphy MP, Bennett M. 2010. DNA damage links mitochondrial dysfunction to atherosclerosis and the metabolic syndrome. *Circ Res* **107**: 1021–1031.

Momcilovic O, Knobloch L, Fornsaglio J, Varum S, Easley C, Schatten G. 2010. DNA damage responses in human induced pluripotent stem cells and embryonic stem cells. *PLoS ONE* **5**: e13410.

Moriwaki S, Ray S, Tarone RE, Kraemer KH, Grossman L. 1996. The effect of donor age on the processing of UV-damaged DNA by cultured human cells: Reduced DNA repair capacity and increased DNA mutability. *Mutat Res* **364**: 117–123.

Moskalev AA, Shaposhnikov MV, Plyusnina EN, Zhavoronkov A, Budovsky A, Yanai H, Fraifeld VE. 2013. The role of DNA damage and repair in aging through the prism of Koch-like criteria. *Ageing Res Rev* **12**: 661–684.

Mouchiroud L, Houtkooper RH, Moullan N, Katsyuba E, Ryu D, Canto C, Mottis A, Jo YS, Viswanathan M, Schoonjans k, et al. 2013. The NAD$^+$/sirtuin pathway modulates longevity through activation of mitochondrial UPR and FOXO signaling. *Cell* **154**: 430–441.

Muiras ML, Muller M, Schachter F, Burkle A. 1998. Increased poly(ADP-ribose) polymerase activity in lymphoblastoid cell lines from centenarians. *J Mol Med* **76**: 346–354.

Nakae D, Akai H, Kishida H, Kusuoka O, Tsutsumi M, Konishi y. 2000. Age and organ dependent spontaneous generation of nuclear 8-hydroxydeoxyguanosine in male Fischer 344 rats. *Lab Invest* **80**: 249–261.

Narendra D, Tanaka A, Suen DF, Youle RJ. 2008. Parkin is recruited selectively to impaired mitochondria and promotes their autophagy. *J Cell Biol* **183**: 795–803.

Natale V. 2011. A comprehensive description of the severity groups in Cockayne syndrome. *Am J Med Genet A* **155A**: 1081–1095.

Niedernhofer LJ, Bohr VA, Sander M, Kraemer KH. 2011. Xeroderma pigmentosum and other diseases of human premature aging and DNA repair: Molecules to patients. *Mech Ageing Dev* **132**: 340–347.

Nouspikel T. 2007. DNA repair in differentiated cells: Some new answers to old questions. *Neuroscience* **145**: 1213–1221.

Osenbroch PØ, Auk-Emblem P, Halsne R, Strand J, Forstrøm RJ, van der Pluijm I, Eide L. 2009. Accumulation of mitochondrial DNA damage and bioenergetic dysfunction in CSB defective cells. *FEBS J* **276**: 2811–2821.

Park DC, Yeo SG. 2013. Aging. *Korean J Audiol* **17**: 39–44.

Parsons JL, Dianova II, Khoronenkova SV, Edelmann MJ, Kessler BM, Dianov GL. 2011. USP47 is a deubiquitylating enzyme that regulates base excision repair by controlling steady-state levels of DNA polymerase β. *Mol Cell* **41**: 609–615.

Passos JF, Saretzki G, von Zglinicki T. 2007. DNA damage in telomeres and mitochondria during cellular senescence: Is there a connection? *Nucleic Acids Res* **35**: 7505–7513.

Ramamoorthy M, Sykora P, Scheibye-Knudsen M, Dunn C, Kasmer C, Zhang Y, Becker KG, Croteau DL, Bohr VA. 2012. Sporadic Alzheimer disease fibroblasts display an oxidative stress phenotype. *Free Radic Biol Med* **53**: 1371–1380.

Redwood AB, Perkins SM, Vanderwaal RP, Feng Z, Biehl KJ, Gonzalez-Suarez I, Morgado-Palacin L, Shi W, Sage J, Roti-Roti JL, et al. 2011. A dual role for A-type lamins in DNA double-strand break repair. *Cell Cycle* **10**: 2549–2560.

Ressler S, Bartkova J, Niederegger H, Bartek J, Scharffetter-Kochanek K, Jansen-Durr P, Wlaschek M. 2006. p16INK4A is a robust in vivo biomarker of cellular aging in human skin. *Aging Cell* **5**: 379–389.

Rhee DB, Ghosh A, Lu J, Bohr VA, Liu Y. 2011. Factors that influence telomeric oxidative base damage and repair by DNA glycosylase OGG1. *DNA Repair (Amst)* **10**: 34–44.

Rinne M, Caldwell D, Kelley MR. 2004. Transient adenoviral N-methylpurine DNA glycosylase overexpression imparts chemotherapeutic sensitivity to human breast cancer cells. *Mol Cancer Ther* **3**: 955–967.

Robert L, Labat-Robert J, Robert AM. 2010. Genetic, epigenetic and posttranslational mechanisms of aging. *Biogerontology* **11**: 387–399.

Robu M, Shah RG, Petitclerc N, Brind'Amour J, Kandan-Kulangara F, Shah GM. 2013. Role of poly(ADP-ribose) polymerase-1 in the removal of UV-induced DNA lesions by nucleotide excision repair. *Proc Natl Acad Sci* **110**: 1658–1663.

Rocha CR, Lerner LK, Okamoto OK, Marchetto MC, Menck CF. 2013. The role of DNA repair in the pluripotency and differentiation of human stem cells. *Mutat Res* **752**: 25–35.

Rodier F, Kim SH, Nijjar T, Yaswen P, Campisi J. 2005. Cancer and aging: The importance of telomeres in genome maintenance. *Int J Biochem Cell Biol* **37:** 977–990.

Rodier F, Coppe JP, Patil CK, Hoeijmakers WA, Munoz DP, Raza SR, Freund A, Campeau E, Davalos AR, Campisi J. 2009. Persistent DNA damage signalling triggers senescence-associated inflammatory cytokine secretion. *Nat Cell Biol* **11:** 973–979.

Rossi DJ, Bryder D, Seita J, Nussenzweig A, Hoeijmakers J, Weissman IL. 2007. Deficiencies in DNA damage repair limit the function of haematopoietic stem cells with age. *Nature* **447:** 725–729.

Rossi MN, Carbone M, Mostocotto C, Mancone C, Tripodi M, Maione R, Amati P. 2009. Mitochondrial localization of PARP-1 requires interaction with mitofilin and is involved in the maintenance of mitochondrial DNA integrity. *J Biol Chem* **284:** 31616–31624.

Rossi ML, Ghosh AK, Bohr VA. 2010. Roles of Werner syndrome protein in protection of genome integrity. *DNA Repair (Amst)* **9:** 331–344.

Rouleau M, Aubin RA, Poirier GG. 2004. Poly(ADP-ribosyl)ated chromatin domains: Access granted. *J Cell Sci* **117:** 815–825.

Rulten SL, Caldecott KW. 2013. DNA strand break repair and neurodegeneration. *DNA Repair (Amst)* **12:** 558–567.

Sahin E, Colla S, Liesa M, Moslehi J, Muller FL, Guo M, Cooper M, Kotton D, Fabian AJ, Walkey C, et al. 2011. Telomere dysfunction induces metabolic and mitochondrial compromise. *Nature* **470:** 359–365.

Sai K, Takagi A, Umemura T, Hasegawa R, Kurokawa Y. 1992. Changes of 8-hydroxydeoxyguanosine levels in rat organ DNA during the aging process. *J Environ Pathol Toxicol Oncol* **11:** 139–143.

Samper E, Nicholls DG, Melov S. 2003. Mitochondrial oxidative stress causes chromosomal instability of mouse embryonic fibroblasts. *Aging Cell* **2:** 277–285.

Saretzki G, Armstrong L, Leake A, Lako M, von Zglinicki T. 2004. Stress defense in murine embryonic stem cells is superior to that of various differentiated murine cells. *Stem Cells* **22:** 962–971.

Scheibye-Knudsen M, Ramamoorthy M, Sykora P, Maynard S, Lin PC, Minor RK, Wilson DM III, Cooper M, Spencer R, de Cabo R, et al. 2012. Cockayne syndrome group B protein prevents the accumulation of damaged mitochondria by promoting mitochondrial autophagy. *J Exp Med* **209:** 855–869.

Scheibye-Knudsen M, Scheibye-Alsing K, Canugovi C, Croteau DL, Bohr VA. 2013. A novel diagnostic tool reveals mitochondrial pathology in human diseases and aging. *Aging (Albany NY)* **5:** 192–208.

Scheibye-Knudsen M, Fang EF, Croteau DL, Bohr VA. 2014a. Contribution of defective mitophagy to the neurodegeneration in DNA repair-deficient disorders. *Autophagy* **10:** 1468–1469.

Scheibye-Knudsen M, Mitchell SJ, Fang EF, Iyama T, Ward T, Wang J, Dunn CA, Singh N, Veith S, Hasan-Olive MM, et al. 2014b. A high-fat diet and NAD$^+$ activate Sirt1 to rescue premature aging in cockayne syndrome. *Cell Metab* **20:** 840–855.

Schriner SE, Linford NJ, Martin GM, Treuting P, Ogburn CE, Emond M, Coskun PE, Ladiges W, Wolf N, Van Remmen H, et al. 2005. Extension of murine life span by overexpression of catalase targeted to mitochondria. *Science* **308:** 1909–1911.

Sfeir A, Kosiyatrakul ST, Hockemeyer D, MacRae SL, Karlseder J, Schildkraut CL, de LT. 2009. Mammalian telomeres resemble fragile sites and require TRF1 for efficient replication. *Cell* **138:** 90–103.

Shanley DP, Kirkwood TB. 2000. Calorie restriction and aging: A life-history analysis. *Evolution* **54:** 740–750.

Sharma NK, Lebedeva M, Thomas T, Kovalenko OS, Stumpf JD, Shadel GS, Santos JH. 2014. Intrinsic mitochondrial DNA repair defects in Ataxia Telangiectasia. *DNA Repair (Amst)* **13:** 22–31.

Sheng Z, Oka S, Tsuchimoto D, Abolhassani N, Nomaru H, Sakumi K, Yamada H, Nakabeppu Y. 2012. 8-Oxoguanine causes neurodegeneration during MUTYH-mediated DNA base excision repair. *J Clin Invest* **122:** 4344–4361.

Singh DK, Ghosh AK, Croteau DL, Bohr VA. 2011. RecQ helicases in DNA double strand break repair and telomere maintenance. *Mutat Res* **736:** 15–24.

Sobol RW, Kartalou M, Almeida KH, Joyce DF, Engelward BP, Horton JK, Prasad R, Samson LD, Wilson SH. 2003. Base excision repair intermediates induce p53-independent cytotoxic and genotoxic responses. *J Biol Chem* **278:** 39951–39959.

Stevnsner T, Muftuoglu M, Aamann MD, Bohr VA. 2008. The role of Cockayne syndrome group B (CSB) protein in base excision repair and aging. *Mech Ageing Dev* **129:** 441–448.

Sun H, Karow JK, Hickson ID, Maizels N. 1998. The Bloom's syndrome helicase unwinds G_4 DNA. *J Biol Chem* **273:** 27587–27592.

Sykora P, Croteau DL, Bohr VA, Wilson DM III. 2011. Aprataxin localizes to mitochondria and preserves mitochondrial function. *Proc Natl Acad Sci* **108:** 7437–7442.

Teplyuk NM. 2012. Near-to-perfect homeostasis: Examples of universal aging rule which germline evades. *J Cell Biochem* **113:** 388–396.

Thorslund T, von KC, Harrigan JA, Indig FE, Christiansen M, Stevnsner T, Bohr VA. 2005. Cooperation of the Cockayne syndrome group B protein and poly(ADP-ribose) polymerase 1 in the response to oxidative stress. *Mol Cell Biol* **25:** 7625–7636.

Tichy ED, Stambrook PJ. 2008. DNA repair in murine embryonic stem cells and differentiated cells. *Exp Cell Res* **314:** 1929–1936.

Tillement L, Lecanu L, Papadopoulos V. 2011. Alzheimer's disease: Effects of β-amyloid on mitochondria. *Mitochondrion* **11:** 13–21.

Tomkinson AE, Howes TR, Wiest NE. 2013. DNA ligases as therapeutic targets. *Transl Cancer Res* **2:** 1219.

Tosato M, Zamboni V, Ferrini A, Cesari M. 2007. The aging process and potential interventions to extend life expectancy. *Clin Interv Aging* **2:** 401–412.

Trifunovic A, Wredenberg A, Falkenberg M, Spelbrink JN, Rovio AT, Bruder CE, Bohlooly Y, Gidlof S, Oldfors A, Wibom R, et al. 2004. Premature ageing in mice expressing defective mitochondrial DNA polymerase. *Nature* **429:** 417–423.

Trivedi RN, Almeida KH, Fornsaglio JL, Schamus S, Sobol RW. 2005. The role of base excision repair in the sensitivity and resistance to temozolomide-mediated cell death. *Cancer Res* **65:** 6394–6400.

Valentin-Vega YA, Maclean KH, Tait-Mulder J, Milasta S, Steeves M, Dorsey FC, Cleveland JL, Green DR, Kastan MB. 2012. Mitochondrial dysfunction in ataxia-telangiectasia. *Blood* **119:** 1490–1500.

van de Ven M, Andressoo JO, Holcomb VB, von LM, Jong WM, De Zeeuw CI, Suh Y, Hasty P, Hoeijmakers JH, van der Horst GT, et al. 2006. Adaptive stress response in segmental progeria resembles long-lived dwarfism and calorie restriction in mice. *PLoS Genet* **2:** 192.

Vendelbo MH, Nair KS. 2011. Mitochondrial longevity pathways. *Biochim Biophys Acta* **1813:** 634–644.

Ventura-Clapier R, Garnier A, Veksler V. 2008. Transcriptional control of mitochondrial biogenesis: The central role of PGC-1α. *Cardiovasc Res* **79:** 208–217.

Victor VM, Rocha M, Herance R, Hernandez-Mijares A. 2011. Oxidative stress and mitochondrial dysfunction in type 2 diabetes. *Curr Pharm Des* **17:** 3947–3958.

Vijg J. 2008. The role of DNA damage and repair in aging: New approaches to an old problem. *Mech Ageing Dev* **129:** 498–502.

Vijg J. 2014. Aging genomes: A necessary evil in the logic of life. *Bioessays* **36:** 282–292.

Vina J, Borras C, Abdelaziz KM, Garcia-Valles R, Gomez-Cabrera MC. 2013. The free radical theory of aging revisited: The cell signaling disruption theory of aging. *Antioxid Redox Signal* **19:** 779–787.

von Zglinicki T. 2000. Role of oxidative stress in telomere length regulation and replicative senescence. *Ann NY Acad Sci* **908:** 99–110.

von Zglinicki T, Burkle A, Kirkwood TB. 2001. Stress, DNA damage and ageing—An integrative approach. *Exp Gerontol* **36:** 1049–1062.

Wallace DC. 1999. Mitochondrial diseases in man and mouse. *Science* **283:** 1482–1488.

Wallace DC, Fan W, Procaccio V. 2010. Mitochondrial energetics and therapeutics. *Annu Rev Pathol* **5:** 297–348.

Wang X, Michaelis EK. 2010. Selective neuronal vulnerability to oxidative stress in the brain. *Front Aging Neurosci* **2:** 12.

Wang Z, Rhee DB, Lu J, Bohr CT, Zhou F, Vallabhaneni H, de Souza-Pinto NC, Liu Y. 2010. Characterization of oxidative guanine damage and repair in mammalian telomeres. *PLoS Genet* **6:** e1000951.

Weidenheim KM, Dickson DW, Rapin I. 2009. Neuropathology of Cockayne syndrome: Evidence for impaired development, premature aging, and neurodegeneration. *Mech Ageing Dev* **130:** 619–636.

Wyman C, Kanaar R. 2006. DNA double-strand break repair: All's well that ends well. *Annu Rev Genet* **40:** 363–383.

Youle RJ, Narendra DP. 2011. Mechanisms of mitophagy. *Nat Rev Mol Cell Biol* **12:** 9–14.

Youle RJ, van der Bliek AM. 2012. Mitochondrial fission, fusion, and stress. *Science* **337:** 1062–1065.

Cite this article as *Cold Spring Harb Perspect Med* doi: 10.1101/cshperspect.a025130

How Research on Human Progeroid and Antigeroid Syndromes Can Contribute to the Longevity Dividend Initiative

Fuki M. Hisama[1,3], Junko Oshima[2,3,4], and George M. Martin[2,3]

[1]Division of Medical Genetics, Department of Medicine, University of Washington School of Medicine, Seattle, Washington 98195

[2]Department of Pathology, University of Washington School of Medicine, Seattle, Washington 98195

[3]International Registry of Werner Syndrome, University of Washington School of Medicine, Seattle, Washington 98195

[4]Department of Medicine, Chiba University, Chiba 260-8670, Japan

Correspondence: gmmartin@uw.edu

Although translational applications derived from research on basic mechanisms of aging are likely to enhance health spans and life spans for most of us (the longevity dividend), there will remain subsets of individuals with special vulnerabilities. Medical genetics is a discipline that describes such "private" patterns of aging and can reveal underlying mechanisms, many of which support genomic instability as a major mechanism of aging. We review examples of three classes of informative disorders: "segmental progeroid syndromes" (those that appear to accelerate multiple features of aging), "unimodal progeroid syndromes" (those that impact on a single disorder of aging), and "unimodal antigeroid syndromes," variants that provide enhanced protection against specific disorders of aging; we urge our colleagues to expand our meager research efforts on the latter, including ancillary somatic cell genetic approaches.

A successful translation of the "longevity dividend" concept will require a fuller understanding of the fundamental mechanisms of intrinsic and extrinsic mechanisms of biological aging. In this review, we summarize what could be characterized as a medical genetics approach to the discovery of at least a subset of such mechanisms. That approach takes advantage of what might be called "experiments of nature"—mutations whose phenotypes appear to accelerate the ages of onset and the rates of progression of aging and diseases of aging (i.e., the "pathobiology of aging"). These can be clas-sified in two broad subtypes—those that mainly impact a single predominant phenotype associated with aging, "unimodal progeroid syndromes" (Martin 1982; Martin 2005) and those that impact multiple phenotypes, "segmental progeroid syndromes" (Table 1) (Martin 1978; Martin 2005). The latter class of syndromes has been a major focus of our research and has in fact revealed a fundamental underlying general mechanism of biological aging—genomic instability. We shall also briefly review the status of research on examples of "unimodal progeroid syndromes." Although these studies

Table 1. Summary of clinical features of segmental progeroid syndromes

Syndrome	Inheritance	Onset	Short stature	Skin/hair	Eyes	Vascular	Malignancy	Neurologic	Other
Werner	AR	Adult	+	Atrophy/graying	Cataract	+	+	–	DM osteoporosis
Bloom	AR	Child	++	Erythema, telangiectasia/graying	–	–	+	–	DM immunodeficiency
Rothmund–Thompson	AR	Child	+	Poikiloderma/graying	Cataract	–	+	–	Radial defect
Laminopathies	AD	Adult	+	Atrophy	–	++	–	–	
MDPL	AD	Child	+	Lipodystrophy	–	–	–	–	Deafness, mandible hypoplasia
Ruijs–Aalfs	AR	Child	+	–	Cataract	–	+	–	Kyphoscoliosis
Aicardi–Goutierres	AR AD	Child	+	Chillblains	–	–	–	ID, WM disease, BG calcification	Microcephaly, throbocytopenia
Mosaic trisomy 8	AD	Child	+	Palmar, plantar grooves	Corneal opacity	–	–	ID	Stiff joints
Hutchinson–Gilford	AD	Child	++	Alopecia	–	++	Rare	–	Loss of fat
Ataxia-telangiectasia	AR	Child	+	Telangiectasia	–	–	+	Ataxia	Immunodeficiency
Cockayne syndrome	AR	Child	++	Thin, dry skin and hair	Cataract, Pig. Ret.	–	No	Spasticity, ataxia, ID, CNS calcification	Microcephaly, SNHL, renal insufficiency
Fanconi anemia	AR XL	Child	+	Pigmentation abnormalities	–	–	+	ID	Radial defect, pancytopenia, deafness, heart or kidney birth defect
Myotonic dystrophy	AD	Child or adult	–	Frontal balding	Cataract	–	–	ID	Muscle atrophy and myotonia, diabetes, cardiac conduction abnormality
BSCL	AR	Child	–	Acanthosis nigricans	–	–	–	ID	Generalized lipodystrophy, DM, hypertrophic cardiomyopathy, skeletal muscle hypertrophy

MDPL, Mandibular hypoplasia, deafness, progeroid features, lipodystrophy; BSCL, Berardinelli–Seip congenital lipodystrophy; AR, autosomal recessive; AD, autosomal dominant; XL, X-linked; ID, intellectual disability; BG, basal ganglia; WM, white matter; DM, diabetes mellitus; Pig. Ret., pigmentary retinopathy; SNHL, sensorineural hearing loss; CNS, central nervous system.

are much less well developed, they are of potentially much greater relevance to the longevity dividend initiative. Given the biochemical genetic uniqueness of individual members of our species, one can anticipate that, despite major benefits from interventions that slow the intrinsic rates of aging, thus postponing multiple diseases of aging, many or perhaps most of us may still be handicapped by one or more "Achilles heels" of susceptibility to specific age-related disorders—susceptibilities that are functions of both constitutional genetic vulnerabilities, their interactions with environment, as well as a range of genetic and epigenetic stochastic events (Martin 2012). Although the disorders we are reviewing result from mutations with very large effects, we still have a great deal to learn about less penetrant mutations, polymorphic variants, and polygenic contributions to what may present as more subtle "Achilles heels," particularly given deleterious interactions with environmental agents. The same may be true for heterozygous carriers of autosomal recessive disorders. Our criteria for the definitions of segmental and unimodal progeroid syndromes have therefore been biased in favor of ascertainments of strongly expressed phenotypes consistent with accelerated rates of biological aging in one or more spheres of pathophysiology. There is much work to be performed by medical geneticists to enhance our understanding of more subtle phenotypic outliers.

This review also makes a plea for a new medical genetics research initiative. By definition, our clinics only enroll individuals and families in which the mutational experiments of nature lead to diminished structure and function. We will argue that it is time to also turn our attention to the discovery of alleles that provide remarkable degrees of enhanced structure and function for a range of relevant physiologies. Although there are colleagues engaged in the genetic analysis of human subjects with exceptionally long health spans and life spans, as reviewed by Milman and Barzilai (2015), the genetic analysis of individuals displaying exceptionally robust retention of function in specific physiological domains (cardiovascular, pulmonary, renal, skeletal muscular, hematopoietic, etc.) has

been comparatively neglected, perhaps with the exception of studies that have examined the genetic basis of resistance to dementias of the Alzheimer type (DATs). We shall refer to such yet-to-be-discovered medical genetic entities as "unimodal antigeroid syndromes," in keeping with the nomenclature of "antigeroid alleles" (Finch and Kirkwood 2000; Martin and Oshima 2000).

EXAMPLES OF SEGMENTAL PROGEROID SYNDROMES

Werner Syndrome—The Prototypic Segmental Progeroid Syndrome

Werner syndrome (WS) is perhaps the best example of segmental progeroid syndromes as the affected individuals undergo multiple features of accelerated aging (Fig. 1A) (Oshima et al. 2014). It is one of the few examples of adult-onset progeroid syndromes in which aged-appearances and common age-related disorders begin to manifest in the third decade of life. Short stature, which starts in the early teens, is often the first recognized sign. Typically, patients develop skin atrophy with loss of subcutaneous fat and graying and loss of hair in their 20s, bilateral cataracts in their late 20s or early 30s, hypogonadism, and, subsequently, a series of age-related diseases including type 2 diabetes mellitus, osteoporosis, atherosclerosis, and a variety of malignancies. WS patients have a disproportionally higher incidence of sarcomas, however (Goto et al. 1996), suggesting the importance of genomic instabilities of mesenchymal cells. Severe ulcerations occur in the lower extremities; these often lead to amputations. They are primarily related to a form of arteriosclerosis known as medial calcinosis. Interestingly, DATs are not well-recognized features of WS. The most recent data indicate a median age of death of 54 yr, usually caused by either cancer or myocardial infarction (Oshima et al. 2014).

WS is a rare autosomal recessive disorder caused by null mutations at the *WRN* locus, a member of the RecQ family of helicases. It encodes a nuclear protein with a $3' \rightarrow 5'$ exonuclease domain in its amino-terminal region, a

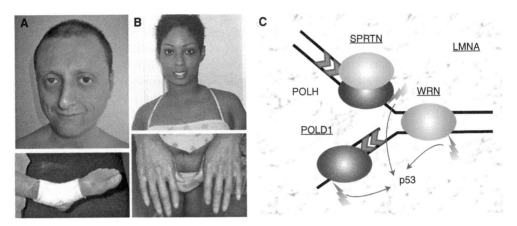

Figure 1. Examples of progeroid patients and selected causal genes for segmental progeroid syndromes. (*A*) A 35-yr-old Italian Caucasian male, registry #ROMA1010, with a homozygous WRN mutation, c.867_874del-AGAAAATC (p.Glu290fs) (Friedrich et al. 2010). He had a characteristic ankle ulcer (with bandage). (*B*) A 23-yr-old African–American Caucasian female, registry #ATLAN1010 with a heterozygous LMNA mutation, c.398G>T (p.Arg133Leu) (Chen et al. 2003). Atrophic skin of hands is shown. (*C*) Examples of major causal genes of segmental progeroid syndromes associated with DNA damage repair and response. Mutated loci so far identified among cases of atypical Werner syndrome (WS) within the International Registry of WS are underlined.

$3' \rightarrow 5'$ helicase in its central region, and a nuclear localization signal in its carboxy-terminal region (Croteau et al. 2014). The majority of disease mutations are nonsense mutations and small indels that truncate nuclear localization signals and/or cause nonsense-mediated RNA decay (Friedrich et al. 2010). Genomic rearrangements with or without involvement of neighboring loci have also been identified. With the advancement of next-generation DNA sequencing, *WRN* heterozygous disease mutations have been identified in association with at least one other known genetic disorder (see below, Aicardi–Goutieres syndrome) (Lessel et al. 2014a).

mTOR inhibitors have been shown to improve cellular growth and to reduce the accumulation of DNA damage (Talaei et al. 2013; Saha et al. 2014). Astaxanthin, an antioxidant, has shown improvement of fatty deposition in the liver of one case (Takemoto et al. 2015). WS-induced pluripotent stem cells have been generated in two independent laboratories and should be very useful for further research on the phenotypic alterations in a range of somatic cells (Cheung et al. 2014; Shimamoto et al. 2014).

More research is needed on the potential impacts of polymorphic variants and of potential deleterious interactions in heterozygotic carriers with genotoxic environmental agents (Ogburn et al. 1997; Blank et al. 2004). Several population studies have already linked the p.Leu1075Phe polymorphism with longevity, where a minor allele was associated with decreased longevity (Sebastiani et al. 2012).

Other RecQ Helicase Disorders

Bloom Syndrome

Bloom syndrome is a rare, autosomal recessive disorder recognized by a dermatologist, Dr. David Bloom who described three children with short stature and a sun-sensitive telangiectatic facial erythema (Bloom 1954). Additional clinical features include prenatal growth restriction, hypo- or hyperpigmentation elsewhere on the body, an immunodeficiency, and the following features that overlap with common features of aging: a paucity of subcutaneous fat, reduced male and female fertility, and increased risk of diabetes and cancers, particularly types of leukemias, non-Hodgkin's lymphomas, and carcinomas seen in the general population

(German et al. 1984; Kaneko and Kondo 2004). Bloom syndrome is more frequent in the Ashkenazi Jewish population owing to a founder effect.

Bloom syndrome is caused by mutations in one of the five human *RecQ* helicase genes—namely, *BLM*, also known as *RECQL3*. Biallelic mutations result in detectable chromosomal aberrations, including abnormal nuclei in interphase cells and excessive chromosome breaks, gaps, translocations, and quadriradials. In cells treated with bromodeoxyuridine (BrdU), the frequency of sister chromatid exchanges in cells from Bloom syndrome subjects is increased 10-fold compared with cells from controls, and is considered diagnostic (Chaganti et al. 1974). Thus, BLM functions to suppress hyperrecombination, and its absence results in genomic instability.

BLM is the only gene that, when mutated, causes Bloom syndrome. The majority of mutations result in loss-of-function caused by premature truncating, frameshift, splice-site or deletion mutations. The few causative missense substitutions so far reported are predicted to affect its helicase activity.

Expression of BLM increases in the S phase and G_2/M phase, and is localized to promyelocytic nuclear (PML) bodies and telomeres in the nucleus. BLM is rapidly recruited to sites of DNA damage, and can unwind forked double-stranded DNA (dsDNA) and anneal single-stranded DNA (ssDNA). In vitro, BLM stimulates DNA end resection, an exonucleiolytic process that generates ssDNA tails at a DNA double-strand break (DSB). BLM has been found to interact in vitro with replication protein A (RPA), which increases the efficiency of BLM for unwinding DNA substrates. Through direct and indirect protein interactions, BLM functions in DNA repair, recombination, replication, and thereby contributes to the maintenance of genomic stability (Wu et al. 2001; Machwe et al. 2011).

Rothmund–Thomson Syndrome

Rothmund–Thomson syndrome (RTS) was reported in 1887 by a German ophthalmologist,

August Rothmund, and in 1936 by a British dermatologist, Sidney Thomson. Most patients develop erythematous plaques on the cheeks, which spread to the forehead, ears, and neck. Telangiectasias similar to Bloom syndrome are common. Later, poikiloderma appears, resulting in dry, atrophic skin. The skin shows a marbleized appearance of hyperpigmentation and hypopigmentation in sun-exposed areas. Variable features include short stature, graying hair with sparse eyebrows/eyelashes, childhood cataracts, small saddle-shaped nose, and dystrophic teeth and nails (Vennos and James 1995). RTS patients are prone to develop osteosarcoma and squamous cell carcinoma (Drouin et al. 1993). Features that overlap with aging in the general population include graying hair, cataracts, and cancer risk. There is overlap among the three RecQ helicase-associated human diseases in terms of autosomal recessive inheritance, short stature, early onset of cataracts, skin changes, and increased risk of malignancy. In contrast to the other two diseases associated with RecQ helicase dysfunction (WS and Bloom syndrome), RTS patients do not have increased risk of type 2 diabetes mellitus. Hypoplastic thumbs are another occasional distinguishing feature.

In 1999, mutations were identified in *RECQL4* in a subset of patients with RTS (Kitao et al. 1999). In contrast to Bloom syndrome and WS, in which >85% of patients have mutations in the designated member of the RecQ helicase family, mutations in *RECQL4* can be identified in approximately two-thirds of RTS cases (Wang et al. 2001). No other germline mutations causing RTS have been identified to date. After the identification of RECQL4 mutations as the cause of RTS, two other disorders with radial defects and short stature were found to be caused in some cases by RECQL4 mutations: the Baller–Gerold syndrome and RAPADILINO (Siitonen et al. 2003; Van Maldergem et al. 2006).

Although the *RECQL4* helicase has the helicase domain common to all members of the RecQ family, it lacks the exonuclease domain of *WRN*, and lacks the clear nuclear localization signal of *WRN* and *BLM*. Nevertheless, by fluorescence immunostaining, RECQL4 is localized mainly in the nucleoplasm, but also exists in the nucleolus. *RECQL4* participates in nonhomologous end-joining. It is also essential for transport of p53 to mitochondria, where they potentiate the activity of polymerase γ (De et al. 2012; Gupta et al. 2014). In contrast to *WRN* knockout mice, which are born without an apparent phenotype, the *RECQL4* knockout results in early murine embryonic lethality; however, a mouse model with a deletion of exon 13 has features of RTS (Hoki et al. 2003; Lord et al. 2014). These results underscore differences in the biological functions of the RecQ helicases. The fact that these proteins are unable to compensate for one another's functions further underscores their special differentiations.

Atypical Werner Syndrome

Laminopathies

After the identification of mutations in *WRN* in subjects with WS, it became clear that, although most affected individuals carried biallelic mutations, often identical by descent, ∼15% of individuals thought to have WS on clinical grounds, had no detectible *WRN* mutations in coding domains and had normal levels and electrophoretic mobilities of WRN protein by Western blots.

Operationally, we classified these individuals as having "atypical WS." These subjects often did not meet clinical criteria for definite or even probable WS (www.wernersyndrome.org). For example, the patient might have prematurely gray hair and an aged appearance, but might lack bilateral cataracts, the most penetrant feature of classical WS. One of the first candidate genes examined in patients with atypical WS was the lamin A/C gene (*LMNA*), dominant mutations at which had recently been discovered to be the cause of childhood-onset progeria, also known as the Hutchinson–Gilford progeria syndrome (HGPS). In four out of 26 (15%) of atypical WS subjects within the International Registry of WS, a deleterious *LMNA* mutation was found, but the mutations differed from the classic mutation of HGPS (Fig 1B) (Chen et al. 2003). Two families studied in more detail were found to have short stature,

nonprogeroid appearance in childhood, but with the onset of progeroid features and the appearance of severe cardiovascular disease in adulthood, in sharp contrast to the much earlier onsets seen in HGPS. Each subject had an *LMNA* mutation that was associated with the production of the pathognomonic "progerin" protein of HGPS, a novel isoform, which, in HGPS, is the result of a 50-amino-acid deletion. The levels of progerin in these atypical cases, however, were substantially less than what is observed in HGPS (Hisama et al. 2011). The relevance of these observations for our understanding of the pathogenesis of atherosclerosis is highlighted by observations of increasing amounts of progerin in the media and adventitia of arterial walls and in atherosclerotic plaques as during human aging (Olive et al. 2010). The unusual degree of adventitial fibrosis in the atherosclerotic arteries of HGPS patients, however, is a caveat regarding the relevance of this disorder for the usual pathogenetic mechanisms of the disease (Olive et al. 2010).

LMNA encodes intermediate filament proteins of the nuclear lamina and nucleoplasmic structures, which are tightly associated with chromatin. Mature lamin A is produced by a series of posttranslational modification steps including farnesylation, methylation, and cleavage of 18 residues of the precursor protein, termed prelamin. Mutations in zmpste24, a metalloproteinase, which acts in the cleavage of prelamin, cause a neonatal progeroid restrictive dermopathy. *LMNA* is a "gene with many faces"; it causes more than ten distinct human diseases. These can be grouped into those with a neuromuscular phenotype (inherited recessive peripheral neuropathy, Emery–Dreifuss, and other muscular dystrophies), a phenotype of dilated cardiomyopathy with cardiac conduction disease, and progeroid or lipodystrophy disorders (atypical WS, mandibuloacral dysplasia, or familial partial lipodystrophy type 2) (reviewed in Capell and Collins 2006).

MDPL Syndrome

Germline mutations of the *POLD1* gene can cause an autosomal dominant multisystem

Cite this article as *Cold Spring Harb Perspect Med* doi: 10.1101/cshperspect.a025882

disorder known as the MDPL syndrome (mandibular hypoplasia, deafness, progeroid features, lipodystrophy) (Weedon et al. 2013). MDPL patients begin to develop prominent loss of subcutaneous fat, a characteristic facial appearance, metabolic abnormalities, including diabetes mellitus, and progeroid features usually during the first to second decades of life. Sensorineural deafness is seen in most, but not all cases. Undescended testes and hypogonadism have been reported in males but females may be fertile.

The *POLD1* gene encodes one of the main replicative polymerases, which contains an intrinsic exonuclease domain and interacts with WRN protein (Kamath-Loeb et al. 2012). Its primary role is in lagging strand synthesis and also functions as a translesion synthesis (TLS) polymerase. Unlike *POLH*, which catalyzes the TLS of the leading strand, *POLD1* functions in postreplication repair of the lagging strand during DNA replication. Most heterozygous *POLD1* mutations found in MDPL patients are a deletion (p.S605del) within the polymerase domain. A single missense mutation located in its exonuclease domain has also been identified (Weedon et al. 2013; Pelosini et al. 2014). Several germline mutations in POLD1 exonuclease domains are also known to predispose to cancers, particularly familial colorectal cancers (Palles et al. 2013).

Ruijs–Aalfs Syndrome

The Ruijs–Aalfs syndrome (RAS) is an early onset genomic instability syndrome characterized by developmental retardation, skeletal abnormalities, and a progeroid appearance (Ruijs et al. 2003). It is a rare autosomal recessive disorder caused by homozygous or compound heterozygous mutations in *SPRTN* (SprT-like amino-terminal domain). RAS patients are unusually sensitive to a specific type of malignancy, hepatocarcinoma, which can develop as early as the late teens (Lessel et al. 2014b).

SPRTN was originally identified as an interacting protein of ubiquitinated proliferating cell nuclear antigen (PCNA), a protein that is recruited to the sites of UV-induced DNA damage (Centore et al. 2012). The *SPRTN* associates with a TLS polymerase η (eta; encoded by *POLH*) on UV damage to stabilize the repair complex and enhance TLS. Cells carrying *SPRTN* mutations show hypersensitivities to certain genotoxic agents, consistent with the role of *SPRTN* in TLS and in G_2/M-checkpoint regulation (Lessel et al. 2014b).

Aicardi–Goutieres Syndrome

The Aicardi–Goutieres syndrome (AGS) is a progressive neurological disorder characterized by early onset encephalopathy, intellectual disability, and intracranial calcification involving the basal ganglia (Stephenson 2008). Hepatomegaly, thrombocytopenia, and chilblains are frequent. AGS is genetically heterogeneous. Causative genes include *TREX1*, which encodes a $3' \rightarrow 5'$ TREX1 exonuclease, components of the RNASEH2 endonuclease complex (*RNASEH2B, RNASEH2C, RNASEH2D*) and *SAMHD1*, which encodes a deoxynucleotide triphosphohydrolase (Stephenson 2008; Rice et al. 2009). The *SAMHD1* protein regulates intracellular pools of deoxynucleotide triphosphates (dNTPs) and also plays a role in regulation of cell proliferation and survival, and part of the response to DNA damage (Clifford et al. 2013).

One of the puzzling features of AGS is its pronounced phenotypic variability, which is quite unusual for autosomal recessive conditions. In particular, phenotypes associated with *SAMHD1* mutations range from "classical" presentations to those with mild intellectual disability or a nonspecific chronic inflammatory skin condition (Rice et al. 2009; Dale et al. 2010). A Turkish pedigree with a homozygous AGS mutation was identified among subjects enrolled in the International Registry of WS as having atypical WS (Lessel et al. 2014a). The patient also carried a heterozygous *WRN* mutation, raising the possibility that heterozygosity at the *WRN* locus may modify AGS phenotypes. Future research should explore the role of *WRN* heterozygosity in the modulation of other genomic instability syndromes.

Mosaic Trisomy 8

Trisomy 8 mosaicism is seen in individuals whose complement is mosaic for chromosome 8, meaning they have a chromosomally normal cell line, in addition to a cell line that is trisomic for chromosome 8. Full (nonmosaic) trisomy 8 is generally considered a lethal condition.

Common features in mosaic trisomy 8 patients include mild to moderate intellectual disability, agenesis of the corpus callosum, corneal opacities, strabismus, low set ears, a broad bulbous nose, congenital palate defect, congenital heart defect, hydronephrosis, cryptorchidism, and stiff joints. Deep vertical grooves of the palms and soles of the feet are also characteristic (Fineman et al. 1975). The features of mosaic trisomy 8, which overlap with aging in the normal population include cataracts and joint stiffness.

A remarkable feature of mosaic trisomy 8 is the variability in the clinical manifestations. The mechanisms that underlie this variability are not fully understood but must be determined in large degree by the extent and patterns of mosaicism, including tissue-specific effects. An online support group for rare chromosomal disorders (www.rarechromo.org) reports 23 adult members with trisomy 8 mosaicism. These subjects show a wide range of degrees of intellectual disability, from none (one is a college professor) to those requiring assistance for activities of daily living (Baidas et al. 2004).

One subject referred to the International Registry of WS was found to have trisomy 8 mosaicism in cultured fibroblast cells (Oshima and Hisama 2014). The *WRN* gene is located on chromosome 8p12, which means that individuals mosaic for trisomy 8 are predicted to have an additional functional copy of *WRN*. In contrast to the prototypic biallelic loss-of-function mutations that cause WS, whether and how trisomy for *WRN* contributes to the features of trisomy 8 is not yet known.

Hutchinson–Gilford Progeria Syndrome

The two segmental progeroid disorders with the most striking features of accelerated aging are WS (also termed adult-onset progeria) and the Hutchinson–Gilford progeria syndrome (HGPS, also termed childhood-onset progeria). Infants born with HGPS are normal appearing at birth but experience severe growth restriction in the first year of life. The affected children are always short and thin, and reach a final weight of 11–18 kg (25–40 lb) as teenagers (Gordon et al. 1993).

HGPS patients have a characteristic facial appearance, which is distinctive and pathognomonic to the informed clinician. This constellation of features include a bald head with loss of eyelashes and eyebrows, prominent eyes, beaked nose, and small jaw. Widespread loss of "baby fat" leads to prominent veins and an "aged" appearance to the skin. Additional features include acro-osteolysis, stiff finger joints, enlarged elbow and knee joints, coxa valga, aseptic necrosis of the head of the femur, as well as medical complications from severe atherosclerosis, with most deaths historically occurring around age 12 yr, with the causes of death largely by myocardial infarctions, strokes, or congestive heart failure. Although loss of hair, loss of subcutaneous fat and cardiovascular disease are common in the normal aging population, other common feature of aging, including tumors, cataracts, diabetes, and hyperlipidemia are not usually present in HGPS (Merideth et al. 2008). The paucity of at least some of these features (notably cancer) may simply be related to the short life spans of these patients.

Two major challenges to the identification of the genetic basis for HGPS were (1) its rarity, with a prevalence of about one in eight million, and (2) the fact that it nearly always occurs as a sporadic disorder, therefore creating difficulties for traditional genetic methodologies such as positional cloning. Cytogenetic abnormalities, however, localized the candidate region to chromosome 1q, and the cause of HGPS was found to be a de novo recurrent silent substitution in exon 11 of lamin A/C (*LMNA*), which activates a cryptic splice site, resulting in the production of a protein with a deletion of ∼50 amino acids at the carboxyl terminus (Eriksson et al. 2003). This truncated protein (termed progerin) causes a dramatic change in the shape of the nuclear envelope, thickening of the nuclear lamina, and loss of peripheral heterochromatin. Progerin also accumulates in fibroblasts and coronary arteries in an age-dependent fashion in the normal population (McClintock et al. 2007; Olive et al. 2010).

In addition to the alterations of the nucleus, there is strong evidence that the cellular phenotype of HGPS includes genomic instability. In HGPS fibroblasts, there is a delayed checkpoint response, and defective response to DNA damage (Liu et al. 2005, 2006). Furthermore, A-type lamins bind to mammalian telomeres in vivo, and play a role in telomere maintaining telomere length, structure, and function. $LMNA^{-/-}$ fibroblasts show aneuploidy, increased frequency of chromosome breaks, and defects in nonhomologous end-joining of dysfunctional telomeres (Andres and Gonzalez 2009).

This raises the question, then, of why cancers (a classic feature of many chromosomal instability syndromes) are not part of the usual picture of HGPS. Osteosarcoma at age 9 yr was reported in one girl with a missense variant in $LMNA$ that causes a later onset of HGPS, and is compatible with survival into the fifth decade. It may be, as the investigators suggest, that in most patients with HGPS the severe atherosclerotic disease with death in the second decade masks an increased cancer risk, which would be evident at an older age (Shalev et al. 2007).

HGPS is a good example of the productive synergy between families affected by rare disorders and scientists interested in investigating such disorders. The family of an affected child founded the Progeria Research Foundation with a mission to discover treatments and, potentially, cures of HGPS and related disorders of aging (www.progeriaresearch.org). Despite HGPS being one of the more recent gene mutations discovered to be the basis of a segmental progeroid syndrome, promising results of a clinical treatment trial of farnesylation inhibitors were published a mere 11 yr later (Gordon et al. 2014).

Ataxia Telangiectasia

Ataxia telangiectasia (AT) is a member of the group of rare, human diseases that are charac-

terized by marked chromosomal instability, increased cancer risk and progeroid features. AT patients are normal at birth, but typically develop difficulty walking in the second year of life. They gradually develop neurological impairments such as ataxia, ocular apraxia, dysarthric speech, and choreoathetoid movements. In addition, they have humeral and cellular immune defects, with frequent sinus or ear infections. The lifetime cancer risk for AT patients reaches 30%–40%, with the most common childhood malignancies being acute lymphocytic leukemia, non-Hodgkin and Hodgkin lymphomas. The most common adult malignancies are T-cell leukemias and solid tumors (Woods and Taylor 1992). The progeroid features in AT homozygotes include the increased risk of cancer, telangiectasias, graying hair, skin atrophy, and pigmentation changes. AT individuals have reduced fertility from abrupt apoptosis of germ cells in response to a defect in meiotic recombination. Heterozygous female ATM carriers are at increased risk of breast cancer (Thompson et al. 2005).

Chromosomal instability is evident in lymphoid and nonlymphoid cells from AT patients. Their frequent, spontaneous in vivo chromosomal aberrations include DSBs and telomere abnormalities (Meyn 1993). Cultured cells from AT patients are exquisitely sensitive to the cytotoxic effects of ionizing radiation.

The ATM gene was identified by positional cloning in 1995. The majority (85%) of mutations consists of nonsense and splice site mutations that produce prematurely truncated proteins (Savitsky et al. 1995). AT deficient cells show defects in multiple cell cycle checkpoints and in the signaling pathways that activate a network of responses to DNA DSBs.

Cockayne Syndrome

Two major forms of arteriosclerosis (atherosclerosis and arteriolosclerosis), hypertension, age-related renal pathology, presbycusis, cognitive decline, and the loss of subcutaneous adipose tissue are among the diverse phenotypes that permit the characterization of the Cockayne syndrome as a segmental progeroid syndrome

(Laugel 2013). Mutations in at least five loci have been associated with this syndrome, *CSA*, *CSB*, *XPB*, *XPD*, and *XPG*, thus documenting pathogenetic overlaps with Xeroderma pigmentosa (Jaarsma et al. 2013). The *XPB* locus has been characterized as a $3' \rightarrow 5'$ DNA helicase, a member of the "super family 2" (SF2) group of helicases and a subunit of the transcription factor complex TFIIH, which functions both in transcription and DNA repair (Fan and DuPrez 2015).

About two-thirds of Cockayne patients have mutations at *CSB* and approximately one-third at the *CSA* locus (Laugel 2013). The underlying pathogenesis involves defects in transcription-coupled excision repair of DNA (Marteijn et al. 2014). Oxidative stress and mitochondrial dysfunction have also been associated with the Cockayne syndrome and have been shown to be related to depletion of the catalytic subunit of DNA polymerase γ, the enzyme responsible for replicating mitochondrial DNA. That depletion was associated with the accumulation of a serine protease; of great potential therapeutic significance, the phenotype could be reversed by a serine protease inhibitor (Chatre et al. 2015).

Fanconi Anemia

An interest in Fanconi anemia as a segmental progeroid syndrome is immediately suggested by its earlier designation as a "pancytopenia" (Neveling et al. 2009). In addition to multiple and variably expressed developmental phenotypes, one observes many phenotypes that can be interpreted as evidence of premature aging, including endocrine abnormalities, gonadal failure, sarcopenia, osteoporosis, increased susceptibility to infectious agents, and progeroid features of skin (Neveling et al. 2009). Additional relevant observations have included nuclear aberrations and diminished replication of cultured fibroblasts (Willingale-Theune et al. 1989), hypersensitivity to oxidative stress (Liebetrau et al. 1997), mitochondrial damage (Pagano et al. 2014), and preneoplastic lesions such as leukoplakias of the oral cavity (Grein Cavalcanti et al. 2015) and early myelodysplasias

(Quentin et al. 2011). It has also been described as "a highly penetrant cancer susceptibility syndrome" (Auerbach 2009). Were it not for early deaths from bone marrow failure, infectious diseases ad highly malignant neoplasms, one would likely observe a much larger burden of cancers commonly seen within the general aging population.

Current evidence points to some 16 distinct genetic loci, mutations of which are associated with the clinical diagnoses of Fanconi anemia (Walden and Deans 2014). Both autosomal recessive and X-linked modes of inheritance have been implicated (Oostra et al. 2012). These loci all appear to be involved in a complex hierarchy of gene actions that focus on the repair of DNA cross-links, a ubiquitin ligase complex for two key substrates and downstream proteins involved in nuclease-dependent and recombinational functions (Walden and Deans 2014). DNA cross-linking agents have been associated with endogenous and exogenous sources of aldehydes. These notably include derivations from the peroxidation of lipids, one of a number of postulated major elements of the oxidative damage theory of aging (Schottker et al. 2015).

Myotonic Dystrophy

Although the name "myotonic dystrophy" (MD) suggests a unimodal progeroid syndrome—specifically, as a model for age-related sarcopenia (Malatesta et al. 2014), there is a much broader range of phenotypes. An expert on MD, Peter S. Harper, has stated that ". . . anything that can go wrong does go wrong in myotonic dystrophy" (Harper et al. 2001)! This is the case for both type 1 MD (also known as Steinart's disease) and a much milder form known as type 2 MD. These phenotypes (Meola and Cardani 2015) include myocardial conduction defects, insulin resistance, ocular cataracts, male hypogonadism and frontal baldness, respiratory failure, and a range of neurological and behavioral disorders, including sleep disorders (Axford and Pearson 2013; Laberge et al. 2013). Both types of MD are inherited as autosomal dominants. Although different genetic

Cite this article as *Cold Spring Harb Perspect Med* doi: 10.1101/cshperspect.a025882

loci are involved, they may share common pathogenetic mechanisms, particularly RNA toxicity (Meola and Cardani 2015). DM1 is thought to be caused by triplet repeat expansions (CTG) in *DMPK* (dystrophia myotonica-protein kinase), whereas DM2 is thought to be the result of repeat expansions in *CNBP* (CCHC-type zinc finger, nucleic acid binding protein) and, perhaps, other loci (Meola and Cardani 2015). When transcribed into CUG-containing RNA, mutant transcripts aggregate as nuclear foci that sequester RNA-binding proteins, resulting in a spliceopathy of downstream effector genes and accounting for the pleiotropic features. A variety of pathogenetic mechanisms have been suggested (Meola and Cardani 2015), but there is some concern that these might be downstream events (Bachinski et al. 2014). Once the fundamental pathogenetic mechanisms have been fully clarified, it will be important to seek evidence of their roles in normative aging.

Berardinelli–Seip Congenital Lipodystrophy

Given the often profound metabolic disturbances associated with generalized lipodystrophies, such disorders have been of interest to geroscientists at least since a 1978 review that included the Seip syndrome in a compilation of human genetic disorders of potential relevance to the pathobiology of aging (Martin 1978). Insulin resistance, fatty liver, hypertriglyceridemia, type 2 diabetes mellitus, and other progeroid features noted below warrant its consideration as a segmental progeroid syndrome.

A valuable review of the pathophysiologies of a wide range of both genetic and acquired lipodystrophies has recently been published, including the several genetic variants of the Seip syndrome (Nolis 2014). All currently recognized forms of the latter are autosomal recessive in nature. The type 1 disorder is caused by mutations at *AGPAT2*, which codes for 1-acylglycerol-3-phosphate *O*-acyltransferase-2, an enzyme involved in de novo phospholipid biosynthesis. Type 2 is caused by mutations at *BSCL2* (Berardinelli–Seip congenital lipodystrophy 2, also known as Seipin), which codes for a transmembrane protein that, like *AGPAT2*,

is localized to the endoplasmic reticulum and participates in the control of lipid droplet formation and adipocyte differentiation. The type 3 disorder involves mutations at *CAV1*, or caveolin 1, a plasma membrane scaffolding protein and oncogene. Finally, the type 4 disorder is associated with mutations at *PTRF*, coding for the polymerase I and transcript-release factor, which is required for dissociation of a transcription complex and is also involved in the organization of the caveolae of plasma membranes. These mutations result in variable expressions of striking losses of normal adipose tissue and abnormal accumulations of lipids in various viscera, including skeletal muscle, liver, and heart. In addition to the regional atrophy of subcutaneous tissues that is so common in normative aging, one also observes type 2 diabetes mellitus (Lawson 2009), cardiovascular lesions (Nelson et al. 2013) often associated with lipid abnormalities, sometimes with multiple xanthomas (Machado et al. 2013), psychomotor abnormalities (Wei et al. 2013), and what some regard as secondary abnormalities of mitochondrial oxidative phosphorylation (Jeninga et al. 2012). Gastrointestinal polyps, a common benign feature of normative aging, have also been observed (Agrawala et al. 2014).

Like all segmental progeroid syndromes, there are of course discordances with what one observes in normative aging, the most dramatic of which is the striking muscular hypertrophy associated with Berardinelli–Seip congenital lipodystrophy.

EXAMPLES OF UNIMODAL PROGEROID SYNDROMES

Dementias of the Alzheimer Type

We use here the nomenclature of "dementias of the Alzheimer type" as a heuristic device to emphasize the need to consider the hypothesis that there may be several age-related independent pathogenetic pathways converging on a final common pathway that involves the processing of the β amyloid precursor protein (APP) for the synthesis of what are currently considered to be the major neurotoxic moieties—oligomers and fibrils of amyloid β peptides, particularly those derived from Aβ 1–42 (Ow and Dunstan 2014). Aggregates of the hyperphosphorylated microtubule-associated protein tau (Nisbet et al. 2015) are often considered to be downstream pathological events, as is microglia-mediated neuroinflammation (Tang and Le 2015). Given the striking exponential increases in age-specific incidence and prevalence of the common sporadic forms of DATs (Larson et al. 1992), it is surprising that there has been comparatively little attention given to the underlying intrinsic processes of aging that "set the stage" for the emergence of this disorder(s).

Although aging is the major risk factor for the emergence of DATs, by far the most striking genetic risk factor for the common sporadic late onset forms of the disease is the epsilon 4 allele of the gene that codes for the synthesis of apolipoprotein E (*APOE4*) (Corder et al. 1993). There are also rare familial, earlier onset forms of the disease caused by autosomal dominant mutations at three loci involved in the synthesis of amyloid β: the β amyloid precursor gene (*APP*), presenilin 1 (*PS1*), and presenilin 2 (*PS2*) (Chouraki and Seshadri 2014).

Although the most frequent *APOE* allele in European and North American populations is the E3 allele, the E4 allele is thought to be the ancestral allele, as it is found in nonhuman primates and in fossil DNA sequences of two Denisovans, members of an ancient species of hominids (McIntosh et al. 2012). An attractive hypothesis for why E4 may have evolved in nonhuman primates and early hominids and why the allele may persist today is that it may protect us from lipotropic pathogens (Martin 1999). There is, in fact, some evidence that the allele protects against malaria (Fujioka et al. 2013).

A range of potential mechanisms that underlie the vulnerability of E4 carriers to DATs has been recently reviewed (Kim et al. 2014). Although a differential effect of APOE alleles on the clearance of amyloid β has been widely considered, there are several other possible mechanisms, certainly including those that may be related to an intrinsic deficiency in the delivery of lipids to maintain synaptic plasticity. Of special interest in this context is evidence that E4 carriers are deficient in repairing a variety of injuries, including those involved in the recovery from strokes and myocardial infarctions (Corder et al. 2000).

DATs have been considered by some investigators to be part of a group of neurodegenerative disorders resulting from deficiencies in DNA repair, with emphasis on the impacts on mitochondrial dysfunction for the case of DATs (Jeppesen et al. 2011). Impacts on cell cycle regulation and genomic integrity have also been related to APP functions (Chen et al. 2014). Thus, given further research, at least some forms of DATs may yet prove to be another example of a disorder of aging related to genomic instability.

Parkinson's Disease

Parkinson's disease (PD) affects >one million people in the United States, and >four million people worldwide. PD is a neurodegenerative disease characterized by bradykinesia, resting tremor, rigidity, postural instability, and responsiveness to dopaminergic therapy. Additional features include rapid eye movement (REM) behavior sleep disorder, depression, visual hallucinations, and dementia (Berg et al. 2014). Idiopathic PD occurs at older ages, is usually sporadic, and at autopsy, is characterized by neuronal loss (particularly of the pigmented neurons of the substantia nigra), astrocytic gliosis, and characteristic inclusions (Lewy bodies and dystrophic neurites). "Parkinsonism" is an umbrella term that includes monogenic and environmental (toxic) disorders with Parkinsonian features.

Great progress has been made in the identification of monogenic causes of PD, which can be clinically indistinguishable from idiopathic PD, especially those resulting in juvenile Parkinsonism (onset before age 20 yr), and early-onset PD (onset before age 50 yr). However, only ∼15% of PD is early-onset (a higher proportion of which are genetically determined); 85% of PD cases are of the late-onset variety. Mutated genes that cause early-onset PD include

Cite this article as *Cold Spring Harb Perspect Med* doi: 10.1101/cshperspect.a025882

those inherited as autosomal recessive traits (*PARKIN*, *PINK1*, *DJ1*) and those inherited as autosomal dominant traits (*SCNA*, which encodes α-synuclein, a key component of the intraneuronal Lewy body inclusions, and *LRRK2*) (Klein and Schlossmacher 2006). Identification of mutations in multiple genes sufficient to cause PD has highlighted the critical roles of sensing redox equilibrium and of mitochondrial dysfunction, including the importance of maintaining mitochondrial integrity in aging neurons. Although mutations in the genes causing early onset PD are rare in idiopathic forms of PD, mutations in the *LRRK2* gene encoding a leucine rich kinase cause both late-onset familial PD as well as 1%–3% of sporadic late-onset cases of PD (Farrer et al. 2005).

Idiopathic, sporadic, late-onset PD (the most common type) is considered a complex trait. Its onset is modified by aging and environmental factors that include exposures to neurotoxins, caffeine intake, and cigarette smoking. The latter poses the paradox that it protects against PD despite the characterization of cigarette smoke as a "gerontogen"—an agent that accelerates multiple features of aging (Martin 1987; Bernhard et al. 2007). Both epigenetic and genetic factors are of course also considered to contribute to susceptibilities to this complex sporadic, late-onset disorder. Polymorphic variants in multiple loci have been implicated, including some bearing mutations involved in early onset familial forms. Examples include *UCHL1* (*PARK5*), which encodes a ubiquitin carboxy-terminal hydrolase, and a locus coding for glucocerebrosidase β (*GBA*) (Nichols et al. 2009; Miyake et al. 2012).

In summary, genetic studies of both familial and sporadic forms of PD (both early-onset and late onset forms of PD) have yielded a new basis for understanding its pathogenesis and a wealth of targets for translational research and therapeutic interventions.

Xeroderma Pigmentosa (XP)

Given that virtually all human beings have variable degrees of exposure to ultraviolet light, it is not surprising that there is overlap in the age-related phenotypes seen in patients with this set of rare autosomal recessive UV hypersensitive disorders and the sun-exposed tissues of most of us as we age, including skin atrophy, actinic keratosis, hypo and hyperpigmentations, and a range of common geriatric neoplasms of the skin, including basal cell carcinomas, squamous carcinomas and melanomas. Concordances with how individuals usually age are also seen among some of the many eye pathologies in these patients, including ocular cataracts, drusen, keratitis, and pterygiums. Less intuitive overlaps involve neurological disorders such as hearing loss, cortical atrophy, ventricular dilatation, and impaired cognitive functions (Karass et al. 2014). Given this broader array of phenotypes, some might conclude that XP might be better classified under segmental progeroid syndromes. There are substantial variations in the extent to which such ancillary features are expressed, however, whereas the skin lesions are universally present and very severe. Moreover, although it seems reasonable to expect genomic instability to contribute to normative central nervous system (CNS) aging and its associated degenerative disorders, there is as yet no compelling evidence that this is a major contributor. Current research is more focused on alterations in proteostasis (Powers et al. 2009; Ben-Gedalya and Cohen 2012).

Initially described by a Hungarian physician in 1874, it was not until 1968 when James Cleaver made the seminal observation that the disease was caused by a defect in DNA repair (DiGiovanna and Kraemer 2012). We now know that there are eight different genetic loci at which bi-allelic mutations are involved in such defective DNA repair. Seven of them (XPA through XPG) are the result of aberrations in nucleotide excision repair, while the eighth form is the result of mutations in DNA polymerase η, which is involved in translesion DNA synthesis (Karass et al. 2014).

Transcription coupled repair is an area of interest in XP as well as in the Cockayne syndrome (see the discussion above). TFIIH, a nine-subunit general transcription factor, plays a key role in that modality of DNA repair (O'Gorman et al. 2005).

Attenuated Familial Polyposis

There is a range of gene mutations associated with highly penetrant precancerous and frankly malignant neoplasms of the gastrointestinal tract, notably of the colon (Jasperson 2012). These provide potentially informative clues for the role of biological aging in the genesis of a major type of human cancer, one whose age-specific incidence continues to rise exponentially well into the eighth decade of life. Mathematical modeling of such data has given support for the parallel process hypothesis, namely that multiple routes exist to the formation of cancer (Brody 2009). The complex genetic picture alluded to above is certainly consistent with that hypothesis. In keeping with the theme of this review, it is not surprising that there is strong evidence for genomic instability as a dominant factor in the pathogenesis of colon cancer (Rao and Yamada 2013). Relevant mutations include those involving mismatch repair of DNA (Durno et al. 2015) and germ line mutations in PolE and PolD (Valle et al. 2014).

We have here chosen one example of the multiple pathways to neoplasia—an autosomal dominant gene mutation that illustrates a type of gene action associated with a late onset form of colon cancer, one that would therefore more closely fit with what is seen in usual aging. Although this example, attenuated familial adenomatous polyposis, does not directly involve aberrations in DNA repair, it is of interest because its mutations are in unusual domains (3' and 5') of a gene (APC, adenomatous polyposis coli) (Ibrahim et al. 2014). More common mutations elsewhere in this gene result in much more virulent, earlier onset forms of the disorder, which involves the loss of a tumor suppressor mechanism involving WNT signaling (www.genecards.org/cgi-bin/carddisp.pl?gene=APC).

Age-Related Macular Degeneration

Age-related macular degeneration (AMD) is the commonest form of blindness among geriatric patients and is clearly coupled to processes of aging, although environmental factors, notably cigarette smoking, are important variables (Ratnapriya and Chew 2013). Allelic variations at some 20 different genetic loci have so far been described as being pathogenetically significant for the several classifications of the disorder, which may require "multiple hits" for its full expression (Fritsche et al. 2014). Since the seminal 2005 discovery of the importance of mutations at the complement factor H locus (Klein et al. 2005), a number of studies have highlighted the importance of oxidative stress, inflammation and the immune system in its pathogenesis (Gao et al. 2015), a view that fits with an emerging theme in geroscience referred to as "inflammaging" (Franceschi et al. 2007).

Osteoporosis

As both females and males age, especially females, there is a gradual loss of bone mineral density (BMD) such that affected individuals are prone to fractures of bone. Before the era of modern surgical techniques and effective antibiotic management, a sadly common clinical geriatric scenario was death from pneumonia of an aging female with some degree of sarcopenia and problems with balance who, as a result of loss of BMD, suffers a serious fracture of the hip followed by prolonged immobility. Useful clinical assays of BMD, including dual energy X-ray absorptiometry (DXA) (Nayak et al. 2015), have permitted large-scale epidemiological studies of osteoporosis. Using statistical criteria for the loss of BMD as compared with healthy young adults, estimates suggest that perhaps 8% of men and 30% of women over the age of 50 are osteoporotic (Kwan 2015).

Although most investigators emphasize a pathogenesis based on an imbalance between the rates of generation of new bone via the proliferation of osteoblasts and the rates of bone destruction via the proliferation and activities of osteoclasts, additional factors involving the endocrine and neuromuscular systems as well as nutrition are thought to be of importance (Kwan 2015). More than 60 genetic loci have been alleged to be associated with osteoporosis and some 15 have been associated with susceptibility to fractures, but large confirmational

Cite this article as Cold Spring Harb Perspect Med doi: 10.1101/cshperspect.a025882 studies will be required, especially given that only ∼5% of the heritability is so far explained by current genetic studies (Clark and Duncan 2015). Given strong evidence of the involvement of at least three biochemical pathways, however, (Clark and Duncan 2015), it seems likely that advances in precision medicine will indeed be capable of identifying those individuals at substantial risk and, moreover, clinically useful translational interventions tailored to affected individuals, some of which will hopefully be guided by studies of individuals who are exceptionally resistant to the age-related loss of BMD.

Atherosclerosis

Atherosclerosis is a major form of arteriosclerosis ("hardening of the arteries") (Fishbein and Fishbein 2009). The literature on atherogenesis, its pathology, and its relationships to the biology of aging, to the metabolism of lipids, to inflammation, and to genetics is vast. Only a few major points can be made in the context of this review. The first point to emphasize is that cardiovascular diseases, predominately ischemic heart disease and strokes caused by atherosclerosis, remain the major causes of death in our society (www.cdc.gov/nchs/fastats/leading-causes-of-death.htm). Second, the detection of high frequencies of calcified arterial lesions among the remains of humans who populated ancient civilizations is consistent with the hypothesis that it has always been an important disease of our species (Thomas et al. 2014). Third, aging is quite clearly a major risk factor in the emergence of the disease (Wang and Bennett 2012). DNA damage and repair, oxidative stress, telomere shortening, replicative senescence, apoptosis, inflammation, and mitochondrial dysfunctions are among the many processes of aging that have been associated with the disease (Wang and Bennett 2012; Sobenin et al. 2014; Wang et al. 2014).

Considering only coronary artery atherosclerosis, some 153 genome-wide associations have so far been identified, of which 56 have been validated, yet these ∼200 loci, even if fully validated, can explain only ∼10% of the genetic variance for the disorder (Bjorkegren et al. 2015). Older classical studies had already identified very significant gene mutations, most notably those involving the low-density lipoprotein receptor (Goldstein et al. 1975). Moreover, elsewhere in this review, we have noted the important contributions of mutations at the *WRN* and *LMNA* loci.

To return to the earlier point regarding the ancient origins of the disease, an interesting hypothesis suggests that atherosclerosis may be a price we have paid for the role of scavenger receptors of macrophages in the defense against infectious agents (Martin 1998). That hypothesis was suggested by an experiment in which deletion of a macrophage scavenger receptor from a mouse model of atherosclerosis resulted in the anticipated amelioration of the atherosclerosis (because macrophages bearing oxidized lipoproteins are key components of atherogenesis), but marked vulnerability of these mice to infectious agents (Suzuki et al. 1997).

EXAMPLES OF UNIMODAL ANTIGEROID ALLELES

Dementias of the Alzheimer Type

The fact that the *APOE4* allele is by far the major genetic risk factor for the sporadic, late-onset forms of DATs has been discussed above. The observation that the *APOE2* allele is protective, however (Corder et al. 1994), has had comparatively much less attention, as evidenced by searches of PubMed for "APOE4 and Alzheimer's disease" versus "APOE2 and Alzheimer's disease." As of this writing, those numbers are, respectively, 3201 versus 432. This, plus the fact that it is so difficult to find other well-established examples of unimodal antigeroid alleles, supports a major theme of this review, namely a plea for the need of much more research on this topic. For the present example, the importance of such research is not only relevant to the disease entities in question, DATs, it is also relevant to the broader issue of gene actions related to the heritability of longevity. The *APOE2* allele is among those that contribute to this heritability, which has been estimated to be of the order of 25%–33% (Drenos and Kirkwood 2010).

A recent review of the role of the *APOE2* allele in DATs, with the apt title beginning with "The forgotten APOE allele" considers various potential mechanisms of gene actions for this forgotten allele (Suri et al. 2013). These investigators consider a very wide range of potential mechanisms together with lines of supportive and conflicting lines of evidence. These include resistance of the apolipoprotein to denaturation, more favorable protein–protein interactions, a role in resisting the transformation of amyloid β monomers to cytotoxic oligomers, enhanced degradation and clearance of depositions of amyloid β, enhanced protection of neurons from apoptotic death, enhancement of anti-inflammatory and antioxidant functions, a role in decreasing the levels of neurofibrillary tangles, and promotion of synaptic integrity. These investigators properly conclude that much more research is required, including research that differentiates the above properties from that of the more prevalent E3 allele. Support for gene actions related to enhanced morphologies of dendrites and antioxidant functions is emphasized.

Atherosclerosis

Given the above reference to evidence for a role of the *APOE2* alleles in the enhancement of longevity, readers will perhaps be puzzled by the fact that homozygosity for that allele is associated with dysbetalipoproteinemia (type III hyperlipoproteinemia), a cause of "accelerated" atherogenesis (Phillips 2014). Thus, one could argue that the wild-type allele in most developed societies, *APOE3*, could be interpreted as being protective. Given the complex pathogenesis of this disorder, one can imagine a large number of allelic variants that are likely to provide even greater protection. An interesting effort to provide support for that scenario comes from recent in vivo gene expression studies of aortic cells from strains of pigeons that, under comparable conditions of husbandry and with comparable levels of serum cholesterol, are ei-

ther highly susceptible (White Carneau) or highly resistant (Show Racer) to the spontaneous development of atherosclerosis (Anderson et al. 2013). Although there is evidence of a major autosomal recessive gene responsible for these contrasting phenotypes, the study revealed some 48 differentially expressed loci that fell into multiple biochemical genetic pathways, including striking divergences in cytoskeletal remodeling, proteasome activity, cellular respiration, and the immune response, thereby revealing families of loci for which human polymorphic variants may be of relevance to different pathways toward relative resistance to the disorder. An independent complementary study on atherosclerosis-resistant versus atherosclerosis-susceptible Japanese quails revealed seven loci supporting the pathogenetic importance of cholesterol metabolism (Li et al. 2012). This has, of course, been well established for human atherogenesis, but deserves much more investigation regarding alleles providing resistance in human subjects.

There are many potentially fruitful areas of research regarding gene actions that may protect human subjects from atherosclerosis. One example might involve the differential regulation of the clonal senescence of cells of the vascular wall, as there is evidence of diminished clonal proliferation of vascular somatic cells from the abdominal aorta, which shows substantially more atherosclerosis than the thoracic aorta (Martin and Sprague 1972, 1973). Of interest in this context is the observation that significantly greater cloning efficiencies could be obtained from the aortic vascular wall cells from a long lived murine species (*Peromyscus leukopus*) as compared the laboratory mouse and that the clonal efficiencies from both systematically declined during aging (Martin et al. 1983). Variations in telomere function and structure could be one of the relevant underlying molecular variables (Chang and Harley 1995; Okuda et al. 2000).

The most recent actionable example of a unimodal antigeroid syndrome related to atherogenesis derives from observations of exceedingly low levels of LDL cholesterol associated with loss-of-function mutations at *PCSK9*, including haploinsufficiency (Awan et al. 2014; Roberts 2015). Gain-of-function mutations at *PCSK9* (proprotein convertase subtilisin/kexin type 9) were discovered as the third cause of familial hypercholesterolemia (Awan et al. 2014). Drug companies have now introduced monoclonal antibodies against that gene product in an effort to emulate the effects of loss-of-function mutations that appear to protect subjects from atherosclerosis and ischemic heart disease (www.nytimes.com/aponline/2015/06/09/us/ap-us-cholesterol-drug-fda.html).

Genetic Resistance to Environmental Carcinogens

There is no doubt that one of the secrets to the avoidance of cancer (especially lung cancer) and to increasing one's chances of living to the tenth and eleventh decades of life is to avoid cigarette smoke (Wilhelmsen et al. 2015). But why do some heavy cigarette smokers live well into their tenth and eleventh decades free of lung cancer (Rajpathak et al. 2011)? There is a very large literature on candidate polymorphic variants that can provide protection against some of the large numbers of carcinogenic compounds in cigarette smoke, a review of which is beyond the scope of this review. Polygenic models (Dragani et al. 1996; Galvan et al. 2008) are likely to be the most satisfactory approaches to uncovering various patterns of resistance and susceptibility as we approach the era of whole gene sequencing and precision medicine (Esplin et al. 2014)

CONCLUSIONS AND FUTURE DIRECTIONS

We have seen how skilled phenotypic characterizations, when combined with careful family histories and the modern tools of genomic analysis, has led to the definition of a subset of heritable disorders that show many, but not all, features consistent with accelerated rates of aging together with increased vulnerabilities to a range of geriatric disorders. These have become known as "segmental progeroid syndromes" (Martin 1978). Remarkably, biochemical genetic studies of these mutations have provided strong support for a fundamental mechanism

of aging—genomic instability (Fig. 1C). This mechanism of aging appears to have deep evolutionary roots, as exemplified by research on the limited replicative life spans of budding yeast mother cells (Xie et al. 2015).

We have also given a few examples of genetic disorders that predominately impact on a single major geriatric disorder—"unimodal progeroid disorders" (Martin 1982). As such, they can elucidate biochemical genetic mechanisms, thus opening the door to translational efforts in prevention and treatment. An exemplary example has been the discovery that all three autosomal dominant mutations responsible for early onset familial forms of dementias of the Alzheimer type point to the pivotal role of the differential processing of the β APP (Nhan et al. 2015).

With an eye toward the future, we suggest new efforts by geneticists and others to begin to define allelic variants that provide unusually robust resistance to fundamental mechanisms of aging or to specific physiological domains that typically decline during aging. Although productive research along these lines is being performed by colleagues who focus on the genetic basis of unusually long life spans (see Milman and Barzilai 2015), we argue for two additional avenues of research. The first could involve a somatic cell genetic approach that uses various types of immortalized human somatic cells. Examples could include searches for mutations or copy number variations that result in unusual resistance to agents that produce genomic instability, such as DNA cross-linking agents (Ogburn et al. 1997), epigenetic drifts in gene expression (Martin 2009), mitochondrial dysfunction (Meyer et al. 2013), or altered proteostasis (Morimoto and Cuervo 2014), each of which have been proposed as mechanisms of aging. Of greater and timelier clinical significance, however, would be research that uncovers unusual alleles that could explain the mechanisms of exceptionally robust resistance to declines in such phenotypes as cognition, sarcopenia, cardiovascular function, etc. This might be achieved via longitudinal studies performed during middle ages, when the force of natural selection has waned and when processes of aging begin to translate into varying degrees of pathophysiology, typically in the absence of significant comorbidity (Martin 2002). Another approach could involve an initiative comparable to Seattle's Sage Bionetworks, which searches for individuals who have genetic changes expected to cause severe illness but who remain perfectly healthy (en.wikipedia.org/wiki/Sage_Bionetworks). Why is this important for the longevity dividend initiative, which argues for a fundamental slowing of the intrinsic rate of aging as a pathway toward the postponement of "all" forms of age-related pathophysiology? The answer, we believe, is given by two of the classical evolutionary biological theories of aging—antagonistic pleiotropy (Williams 1957) and mutation accumulation (Medawar 1952). Both of these mechanisms predict what might be called "private susceptibilities" to very particular domains of aging. These theories predict idiosyncratic expressions of senescent phenotypes because of the uniqueness of individual genomes, expososomes, somatic mutations, and epigenetic drifts of gene expression. Thus, although major interventions in basic mechanisms of aging may well provide the robustness and resilience required for extended life spans, they may not guarantee freedom from individual "Achilles heels" that result in departures from robust health spans.

ACKNOWLEDGMENTS

We thank Professor M. Stephen Meyn, a medical geneticist at the University of Toronto, for his helpful review of this manuscript. We thank the National Cancer Institute and the National Institute on Aging for their long term support of the University of Washington's International Registry of Werner Syndrome (5R24AG042328-03,-S1). The Photo of Werner syndrome patient, registry#ROMA1010, was provided by Dr. Giovanni Neri, Universit Cattolica, Roma, Italy.

REFERENCES

*Reference is also in this collection.

Agrawala RK, Choudhury AK, Mohanty BK, Baliarsinha AK. 2014. Berardinelli–Seip congenital lipodystrophy: An autosomal recessive disorder with rare association of duodenocolonic polyps. *J Pediatr Endocrinol Metab* **27**: 989–991.

Anderson JL, Ashwell CM, Smith SC, Shine R, Smith EC, Taylor RL Jr. 2013. Atherosclerosis-susceptible and atherosclerosis-resistant pigeon aortic cells express different genes in vivo. *Poult Sci* **92**: 2668–2680.

Andres V, Gonzalez JM. 2009. Role of A-type lamins in signaling, transcription, and chromatin organization. *J Cell Biol* **187**: 945–957.

Auerbach AD. 2009. Fanconi anemia and its diagnosis. *Mutat Res* **668**: 4–10.

Awan Z, Baass A, Genest J. 2014. Proprotein convertase subtilisin/kexin type 9 (PCSK9): Lessons learned from patients with hypercholesterolemia. *Clin Chem* **60**: 1380–1389.

Axford MM, Pearson CE. 2013. Illuminating CNS and cognitive issues in myotonic dystrophy: Workshop report. *Neuromuscul Disord* **23**: 370–374.

Bachinski LL, Baggerly KA, Neubauer VL, Nixon TJ, Raheem O, Sirito M, Unruh AK, Zhang J, Nagarajan L, Timchenko LT, et al. 2014. Most expression and splicing changes in myotonic dystrophy type 1 and type 2 skeletal muscle are shared with other muscular dystrophies. *Neuromuscul Disord* **24**: 227–240.

Baidas S, Chen TJ, Kolev V, Wong LJ, Imholte J, Qin N, Meck J. 2004. Constitutional trisomy 8 mosaicism due to meiosis II non-disjunction in a phenotypically normal woman with hematologic abnormalities. *Am J Med Genet A* **124A**: 383–387.

Ben-Gedalya T, Cohen E. 2012. Quality control compartments coming of age. *Traffic* **13**: 635–642.

Berg D, Postuma RB, Bloem B, Chan P, Dubois B, Gasser T, Goetz CG, Halliday GM, Hardy J, Lang AE, et al. 2014. Time to redefine PD? Introductory statement of the MDS Task Force on the definition of Parkinson's disease. *Mov Disord* **29**: 454–462.

Bernhard D, Moser C, Backovic A, Wick G. 2007. Cigarette smoke—An aging accelerator? *Exp Gerontol* **42**: 160–165.

Bjorkegren JL, Kovacic JC, Dudley JT, Schadt EE. 2015. Genome-wide significant loci: How important are they?: Systems genetics to understand heritability of coronary artery disease and other common complex disorders. *J Am Col Cardiol* **65**: 830–845.

Blank A, Bobola MS, Gold B, Varadarajan S, Kolstoe DD, Meade EH, Rabinovitch PS, Loeb LA, Silber JR. 2004. The Werner syndrome protein confers resistance to the DNA lesions N3-methyladenine and O^6-methylguanine: Implications for WRN function. *DNA Repair (Amst)* **3**: 629–638.

Bloom D. 1954. Congenital telangiectatic erythema resembling lupus erythematosus in dwarfs; probably a syndrome entity. *AMA Am J Dis Child* **88**: 754–758.

Brody JP. 2009. Parallel routes of human carcinoma development: Implications of the age-specific incidence data. *PLoS ONE* **4**: e7053.

Capell BC, Collins FS. 2006. Human laminopathies: Nuclei gone genetically awry. *Nat Rev Genet* **7**: 940–952.

Centore RC, Yazinski SA, Tse A, Zou L. 2012. Spartan/C1orf124, a reader of PCNA ubiquitylation and a regulator of UV-induced DNA damage response. *Mol Cell* **46**: 625–635.

Chaganti RS, Schonberg S, German J. 1974. A manyfold increase in sister chromatid exchanges in Bloom's syndrome lymphocytes. *Proc Natl Acad Sci* **71**: 4508–4512.

Chang E, Harley CB. 1995. Telomere length and replicative aging in human vascular tissues. *Proc Natl Acad Sci* **92**: 11190–11194.

Chatre L, Biard DS, Sarasin A, Ricchetti M. 2015. Reversal of mitochondrial defects with CSB-dependent serine protease inhibitors in patient cells of the progeroid Cockayne syndrome. *Proc Natl Acad Sci* **112**: E2910–2919.

Chen L, Lee L, Kudlow BA, Dos Santos HG, Sletvold O, Shafeghati Y, Botha EG, Garg A, Hanson NB, Martin GM, et al. 2003. LMNA mutations in atypical Werner's syndrome. *Lancet* **362**: 440–445.

Chen Y, Neve R, Zheng H, Griffin W, Barger S, Mrak R. 2014. Cycle on Wheels: Is APP key to the AppBp1 pathway? *Austin Alzheimers Parkinsons Dis* **1**: id1008.

Cheung HH, Liu X, Canterel-Thouennon L, Li L, Edmonson C, Rennert OM. 2014. Telomerase protects Werner syndrome lineage-specific stem cells from premature aging. *Stem Cell Rep* **2**: 534–546.

Chouraki V, Seshadri S. 2014. Genetics of Alzheimer's disease. *Adv Genet* **87**: 245–294.

Clark GR, Duncan EL. 2015. The genetics of osteoporosis. *Br Med Bull* **113**: 73–81.

Clifford R, Louis T, Robbe P, Ackroyd S, Burns A, Timbs AT, Colopy GW, Dreau H, Sigaux F, Judde JG, et al. 2013. *SAMHD1* is mutated recurrently in chronic lymphocytic leukemia and is involved in response to DNA damage. *Blood* **123**: 1021–1031.

Corder EH, Saunders AM, Strittmatter WJ, Schmechel DE, Gaskell PC, Small GW, Roses AD, Haines JL, Pericak-Vance MA. 1993. Gene dose of apolipoprotein E type 4 allele and the risk of Alzheimer's disease in late onset families. *Science* **261**: 921–923.

Corder EH, Saunders AM, Risch NJ, Strittmatter WJ, Schmechel DE, Gaskell PC Jr, Rimmler JB, Locke PA, Conneally PM, Schmader KE, et al. 1994. Protective effect of apolipoprotein E type 2 allele for late onset Alzheimer disease. *Nat Genet* **7**: 180–184.

Corder EH, Basun H, Fratiglioni L, Guo Z, Lannfelt L, Viitanen M, Corder LS, Manton KG, Winblad B. 2000. Inherited frailty. ApoE alleles determine survival after a diagnosis of heart disease or stroke at ages 85+. *Ann NY Acad Sci* **908**: 295–298.

Croteau DL, Popuri V, Opresko PL, Bohr VA. 2014. Human RecQ helicases in DNA repair, recombination, and replication. *Annu Rev Biochem* **83**: 519–552.

Dale RC, Gornall H, Singh-Grewal D, Alcausin M, Rice GI, Crow YJ. 2010. Familial Aicardi–Goutieres syndrome due to *SAMHD1* mutations is associated with chronic arthropathy and contractures. *Am J Med Genet A* **152A**: 938–942.

De S, Kumari J, Mudgal R, Modi P, Gupta S, Futami K, Goto H, Lindor NM, Furuichi Y, Mohanty D, et al. 2012. RECQL4 is essential for the transport of p53 to mitochondria in normal human cells in the absence of exogenous stress. *J Cell Sci* **125**: 2509–2522.

DiGiovanna JJ, Kraemer KH. 2012. Shining a light on xeroderma pigmentosum. *J invest dermatol* **132**: 785–796.

Dragani TA, Canzian F, Pierotti MA. 1996. A polygenic model of inherited predisposition to cancer. *FASEB J* **10**: 865–870.

Drenos F, Kirkwood TB. 2010. Selection on alleles affecting human longevity and late-life disease: The example of apolipoprotein E. *PLoS ONE* **5**: e10022.

Drouin CA, Mongrain E, Sasseville D, Bouchard HL, Drouin M. 1993. Rothmund–Thomson syndrome with osteosarcoma. *J Am Acad Dermatol* **28**: 301–305.

Durno CA, Sherman PM, Aronson M, Malkin D, Hawkins C, Bakry D, Bouffet E, Gallinger S, Pollett A, Campbell B, et al. 2015. Phenotypic and genotypic characterisation of biallelic mismatch repair deficiency (BMMR-D) syndrome. *Eur J Cancer* **51**: 977–983.

Eriksson M, Brown WT, Gordon LB, Glynn MW, Singer J, Scott L, Erdos MR, Robbins CM, Moses TY, Berglund P, et al. 2003. Recurrent de novo point mutations in lamin A cause Hutchinson–Gilford progeria syndrome. *Nature* **423**: 293–298.

Esplin ED, Oei L, Snyder MP. 2014. Personalized sequencing and the future of medicine: Discovery, diagnosis and defeat of disease. *Pharmacogenomics* **15**: 1771–1790.

Fan L, DuPrez KT. 2015. XPB: An unconventional SF2 DNA helicase. *Prog Biophys Mol Biol* **117**: 174–181.

Farrer M, Stone J, Mata IF, Lincoln S, Kachergus J, Hulihan M, Strain KJ, Maraganore DM. 2005. LRRK2 mutations in Parkinson disease. *Neurology* **65**: 738–740.

Finch C, Kirkwood TBL. 2000. *Chance, development, and aging*. Oxford University Press, New York.

Fineman RM, Ablow RC, Howard RO, Albright J, Breg WR. 1975. Trisomy 8 mosaicism syndrome. *Pediatrics* **56**: 762–767.

Fishbein GA, Fishbein MC. 2009. Arteriosclerosis: Rethinking the current classification. *Arch Pathol Lab Med* **133**: 1309–1316.

Franceschi C, Capri M, Monti D, Giunta S, Olivieri F, Sevini F, Panourgia MP, Invidia L, Celani L, Scurti M, et al. 2007. Inflammaging and anti-inflammaging: A systemic perspective on aging and longevity emerged from studies in humans. *Mech Ageing Dev* **128**: 92–105.

Friedrich K, Lee L, Leistritz DF, Nurnberg G, Saha B, Hisama FM, Eyman DK, Lessel D, Nurnberg P, Li C, et al. 2010. *WRN* mutations in Werner syndrome patients: Genomic rearrangements, unusual intronic mutations and ethnic-specific alterations. *Hum Genet* **128**: 103–111.

Fritsche LG, Fariss RN, Stambolian D, Abecasis GR, Curcio CA, Swaroop A. 2014. Age-related macular degeneration: Genetics and biology coming together. *Annu Rev Genomics Hum Genet* **15**: 151–171.

Fujioka H, Phelix CF, Friedland RP, Zhu X, Perry EA, Castellani RJ, Perry G. 2013. Apolipoprotein E4 prevents growth of malaria at the intraerythrocyte stage: Implications for differences in racial susceptibility to Alzheimer's disease. *J Health Care Poor Underserved* **24**: 70–78.

Galvan A, Falvella FS, Spinola M, Frullanti E, Leoni VP, Noci S, Alonso MR, Zolin A, Spada E, Milani S, et al. 2008. A polygenic model with common variants may predict lung adenocarcinoma risk in humans. *Int J Can* **123**: 2327–2330.

Gao J, Liu RT, Cao S, Cui JZ, Wang A, To E, Matsubara JA. 2015. NLRP3 inflammasome: Activation and regulation in age-related macular degeneration. *Mediators Inflamm* **2015**: 690243.

German J, Bloom D, Passarge E. 1984. Bloom's syndrome XI. Progress report for 1983. *Clin Genet* **25**: 166–174.

Goldstein JL, Dana SE, Brunschede GY, Brown MS. 1975. Genetic heterogeneity in familial hypercholesterolemia: Evidence for two different mutations affecting functions of low-density lipoprotein receptor. *Proc Natl Acad Sci* **72**: 1092–1096.

Gordon LB, Brown WT, Collins FS. 1993. Hutchinson–Gilford progeria syndrome. *GeneReviews*, University of Washington, Seattle.

Gordon LB, Massaro J, D'Agostino RB Sr, Campbell SE, Brazier J, Brown WT, Kleinman ME, Kieran MW, Progeria Clinical Trials C. 2014. Impact of farnesylation inhibitors on survival in Hutchinson–Gilford progeria syndrome. *Circulation* **130**: 27–34.

Goto M, Miller RW, Ishikawa Y, Sugano H. 1996. Excess of rare cancers in Werner syndrome (adult progeria). *Cancer Epidemiol Biomarkers Prev* **5**: 239–246.

Grein Cavalcanti L, Fatima Lyko K, Lins Fuentes Araujo R, Miguel Amenabar J, Bonfim C, Carvalho Torres-Pereira C. 2015. Oral leukoplakia in patients with Fanconi anaemia without hematopoietic stem cell transplantation. *Pediatr Blood Cancer* **62**: 1024–1026.

Gupta S, De S, Srivastava V, Hussain M, Kumari J, Muniyappa K, Sengupta S. 2014. RECQL4 and p53 potentiate the activity of polymerase γ and maintain the integrity of the human mitochondrial genome. *Carcinogenesis* **35**: 34–45.

Harper PS, Brook JD, Newman E. 2001. *Myotonic dystrophy*. WB Saunders, New York.

Hisama FM, Lessel D, Leistritz D, Friedrich K, McBride KL, Pastore MT, Gottesman GS, Saha B, Martin GM, Kubisch C, et al. 2011. Coronary artery disease in a Werner syndrome-like form of progeria characterized by low levels of progerin, a splice variant of lamin A. *Am J Med Genet A* **155A**: 3002–3006.

Hoki Y, Araki R, Fujimori A, Ohhata T, Koseki H, Fukumura R, Nakamura M, Takahashi H, Noda Y, Kito S, et al. 2003. Growth retardation and skin abnormalities of the *Recql4*-deficient mouse. *Hum Mol Genet* **12**: 2293–2299.

Ibrahim A, Barnes DR, Dunlop J, Barrowdale D, Antoniou AC, Berg JN. 2014. Attenuated familial adenomatous polyposis manifests as autosomal dominant late-onset colorectal cancer. *Eur J Hum Genet* **22**: 1330–1333.

Jaarsma D, van der Pluijm I, van der Horst GT, Hoeijmakers JH. 2013. Cockayne syndrome pathogenesis: Lessons from mouse models. *Mech Ageing Dev* **134**: 180–195.

Jasperson KW. 2012. Genetic testing by cancer site: Colon (polyposis syndromes). *Cancer J* **18**: 328–333.

Jeninga EH, de Vroede M, Hamers N, Breur JM, Verhoeven-Duif NM, Berger R, Kalkhoven E. 2012. A patient with congenital generalized lipodystrophy due to a novel mutation in *BSCL2*: Indications for secondary mitochondrial dysfunction. *JIMD Reps* **4**: 47–54.

Jeppesen DK, Bohr VA, Stevnsner T. 2011. DNA repair deficiency in neurodegeneration. *Prog Neurobiol* **94**: 166–200.

Kamath-Loeb AS, Shen JC, Schmitt MW, Loeb LA. 2012. The Werner syndrome exonuclease facilitates DNA degradation and high fidelity DNA polymerization by human DNA polymerase δ. *J Biol Chem* **287**: 12480–12490.

Kaneko H, Kondo N. 2004. Clinical features of Bloom syndrome and function of the causative gene, BLM helicase. *Expert Rev Mol Diagn* **4**: 393–401.

Karass M, Naguib MM, Elawabdeh N, Cundiff CA, Thomason J, Steelman CK, Cone R, Schwenkter A, Jordan C, Shehata BM. 2014. Xeroderma pigmentosa: Three new cases with an in depth review of the genetic and clinical characteristics of the disease. *Fetal Pediatr Pathol* **34**: 120–127.

Kim J, Yoon H, Basak J, Kim J. 2014. Apolipoprotein E in synaptic plasticity and Alzheimer's disease: Potential cellular and molecular mechanisms. *Mol Cells* **37**: 767–776.

Kitao S, Shimamoto A, Goto M, Miller RW, Smithson WA, Lindor NM, Furuichi Y. 1999. Mutations in *RECQL4* cause a subset of cases of Rothmund–Thomson syndrome. *Nat Genet* **22**: 82–84.

Klein C, Schlossmacher MG. 2006. The genetics of Parkinson disease: Implications for neurological care. *Nat Clin Pract Neurol* **2**: 136–146.

Klein RJ, Zeiss C, Chew EY, Tsai JY, Sackler RS, Haynes C, Henning AK, SanGiovanni JP, Mane SM, Mayne ST, et al. 2005. Complement factor H polymorphism in age-related macular degeneration. *Science* **308**: 385–389.

Kwan P. 2015. Osteoporosis: From osteoscience to neuroscience and beyond. *Mech Ageing Dev* **145C**: 26–38.

Laberge L, Gagnon C, Dauvilliers Y. 2013. Daytime sleepiness and myotonic dystrophy. *Curr Neurol Neurosci Rep* **13**: 340.

Larson EB, Kukull WA, Katzman RL. 1992. Cognitive impairment: Dementia and Alzheimer's disease. *Annu Rev Public Health* **13**: 431–449.

Laugel V. 2013. Cockayne syndrome: The expanding clinical and mutational spectrum. *Mech Ageing Dev* **134**: 161–170.

Lawson MA. 2009. Lipoatrophic diabetes: A case report with a brief review of the literature. *J Adolesc Health* **44**: 94–95.

Lessel D, Saha B, Hisama F, Kaymakamzade B, Nurlu G, Gursoy-Ozdemir Y, Thiele H, Nurnberg P, Martin GM, Kubisch C, et al. 2014a. Atypical Aicardi–Goutieres syndrome: is the *WRN* locus a modifier? *Am J Med Genet A* **164A**: 2510–2513.

Lessel D, Vaz B, Halder S, Lockhart PJ, Marinovic-Terzic I, Lopez-Mosqueda J, Philipp M, Sim JC, Smith KR, Oehler J, et al. 2014b. Mutations in *SPRTN* cause early onset hepatocellular carcinoma, genomic instability and progeroid features. *Nat Genet* **46**: 1239–1244.

Li X, Schulte P, Godin DV, Cheng KM. 2012. Differential mRNA expression of seven genes involved in cholesterol metabolism and transport in the liver of atherosclerosis-susceptible and -resistant Japanese quail strains. *Genet Sel Evol* **44**: 20.

Liebetrau W, Runge TM, Baumer A, Henning C, Gross O, Schindler D, Poot M, Hoehn H. 1997. Exploring the role of oxygen in Fanconi's anemia. *Recent Results Cancer Res* **143**: 353–367.

Liu B, Wang J, Chan KM, Tjia WM, Deng W, Guan X, Huang JD, Li KM, Chau PY, Chen DJ, et al. 2005. Genomic instability in laminopathy-based premature aging. *Nat Med* **11:** 780–785.

Liu Y, Rusinol A, Sinensky M, Wang Y, Zou Y. 2006. DNA damage responses in progeroid syndromes arise from defective maturation of prelamin A. *J Cell Sci* **119:** 4644–4649.

Lord J, Lu AJ, Cruchaga C. 2014. Identification of rare variants in Alzheimer's disease. *Front Genet* **5:** 369.

Machado PV, Daxbacher EL, Obadia DL, Cunha EF, Alves Mde F, Mann D. 2013. Do you know this syndrome? Berardinelli–Seip syndrome. *An Bras Dermatol* **88:** 1011–1013.

Machwe A, Karale R, Xu X, Liu Y, Orren DK. 2011. The Werner and Bloom syndrome proteins help resolve replication blockage by converting (regressed) Holliday junctions to functional replication forks. *Biochemistry* **50:** 6774–6788.

Malatesta M, Cardani R, Pellicciari C, Meola G. 2014. RNA transcription and maturation in skeletal muscle cells are similarly impaired in myotonic dystrophy and sarcopenia: The ultrastructural evidence. *Front Aging Neurosci* **6:** 196.

Marteijn JA, Lans H, Vermeulen W, Hoeijmakers JH. 2014. Understanding nucleotide excision repair and its roles in cancer and ageing. *Nat Rev Mol Cell Biol* **15:** 465–481.

Martin GM. 1978. Genetic syndromes in man with potential relevance to the pathobiology of aging. *Birth Defects Orig Art Ser* **14:** 5–39.

Martin GM. 1982. Syndromes of accelerated aging. *Natl Cancer Inst Monogr* **60:** 241–247.

Martin GM. 1987. Interactions of aging and environmental agents: The gerontological perspective. *Prog Clin Biol Res* **228:** 25–80.

Martin GM. 1998. Atherosclerosis is the leading cause of death in the developed societies. *Am J Pathol* **153:** 1319–1320.

Martin GM. 1999. APOE alleles and lipophylic pathogens. *Neurobiol Aging* **20:** 441–443.

Martin GM. 2002. Help wanted: Physiologists for research on aging. *Sci Aging Knowledge Environ* **2002:** vp2.

Martin GM. 2005. Genetic modulation of senescent phenotypes in *Homo sapiens*. *Cell* **120:** 523–532.

Martin GM. 2009. Epigenetic gambling and epigenetic drift as an antagonistic pleiotropic mechanism of aging. *Aging Cell* **8:** 761–764.

Martin GM. 2012. Stochastic modulations of the pace and patterns of ageing: Impacts on quasi-stochastic distributions of multiple geriatric pathologies. *Mech Ageing Dev* **133:** 107–111.

Martin GM, Oshima J. 2000. Lessons from human progeroid syndromes. *Nature* **408:** 263–266.

Martin GM, Sprague CA. 1972. Clonal senescence and atherosclerosis. *Lancet* **2:** 1370–1371.

Martin GM, Sprague CA. 1973. Symposium on in vitro studies related to atherogenesis. Life histories of hyperplastoid cell lines from aorta and skin. *Exp Mol Pathol* **18:** 125–141.

Martin GM, Ogburn CE, Wight TN. 1983. Comparative rates of decline in the primary cloning efficiencies of smooth muscle cells from the aging thoracic aorta of two murine species of contrasting maximum life span potentials. *Am J Pathol* **110:** 236–245.

McClintock D, Ratner D, Lokuge M, Owens DM, Gordon LB, Collins FS, Djabali K. 2007. The mutant form of lamin A that causes Hutchinson–Gilford progeria is a biomarker of cellular aging in human skin. *PLoS ONE* **2:** e1269.

McIntosh AM, Bennett C, Dickson D, Anestis SF, Watts DP, Webster TH, Fontenot MB, Bradley BJ. 2012. The apolipoprotein E (*APOE*) gene appears functionally monomorphic in chimpanzees (*Pan troglodytes*). *PLoS ONE* **7:** e47760.

Medawar P. 1952. *An unsolved problem in biology*. H.K. Lewis, London.

Meola G, Cardani R. 2015. Myotonic dystrophies: An update on clinical aspects, genetic, pathology, and molecular pathomechanisms. *Biochim Biophys Acta* **1852:** 594–606.

Merideth MA, Gordon LB, Clauss S, Sachdev V, Smith AC, Perry MB, Brewer CC, Zalewski C, Kim HJ, Solomon B, et al. 2008. Phenotype and course of Hutchinson–Gilford progeria syndrome. *N Engl J Med* **358:** 592–604.

Meyer JN, Leung MC, Rooney JP, Sendoel A, Hengartner MO, Kisby GE, Bess AS. 2013. Mitochondria as a target of environmental toxicants. *Toxicol Sci* **134:** 1–17.

Meyn MS. 1993. High spontaneous intrachromosomal recombination rates in ataxia-telangiectasia. *Science* **260:** 1327–1330.

* Milman S, Barzilai N. 2015. Dissecting the mechanisms underlying unusually successful human health span and life span. *Cold Spring Harb Persect Med* doi: 10.1101/cshperspect.a025098.

Miyake Y, Tanaka K, Fukushima W, Kiyohara C, Sasaki S, Tsuboi Y, Yamada T, Oeda T, Shimada H, Kawamura N, et al. 2012. *UCHL1* S18Y variant is a risk factor for Parkinson's disease in Japan. *BMC Neurol* **12:** 62.

Morimoto RI, Cuervo AM. 2014. Proteostasis and the aging proteome in health and disease. *J Gerontol A Biol Sci Med Sci* **69** (Suppl. 1): S33–S38.

Nayak S, Edwards DL, Saleh AA, Greenspan SL. 2015. Systematic review and meta-analysis of the performance of clinical risk assessment instruments for screening for osteoporosis or low bone density. *Osteoporos Int* **26:** 1543–1554.

Nelson MD, Victor RG, Szczepaniak EW, Simha V, Garg A, Szczepaniak LS. 2013. Cardiac steatosis and left ventricular hypertrophy in patients with generalized lipodystrophy as determined by magnetic resonance spectroscopy and imaging. *Am J Cardiol* **112:** 1019–1024.

Neveling K, Endt D, Hoehn H, Schindler D. 2009. Genotype-phenotype correlations in Fanconi anemia. *Mutat Res-Fund Mol M* **668:** 73–91.

Nhan HS, Chiang K, Koo EH. 2015. The multifaceted nature of amyloid precursor protein and its proteolytic fragments: friends and foes. *Acta Neuropathol* **129:** 1–19.

Nichols WC, Pankratz N, Marek DK, Pauciulo MW, Elsaesser VE, Halter CA, Rudolph A, Wojcieszek J, Pfeiffer RF, Foroud T, et al. 2009. Mutations in GBA are associated with familial Parkinson disease susceptibility and age at onset. *Neurology* **72:** 310–316.

Nisbet RM, Polanco JC, Ittner LM, Gotz J. 2015. Tau aggregation and its interplay with amyloid-β. *Acta Neuropathol* **129:** 207–220.

Nolis T. 2014. Exploring the pathophysiology behind the more common genetic and acquired lipodystrophies. *J Hum Genet* **59:** 16–23.

Ogburn CE, Oshima J, Poot M, Chen R, Hunt KE, Gollahon KA, Rabinovitch PS, Martin GM. 1997. An apoptosis-inducing genotoxin differentiates heterozygotic carriers for Werner helicase mutations from wild-type and homozygous mutants. *Hum Genet* **101:** 121–125.

O'Gorman W, Thomas B, Kwek KY, Furger A, Akoulitchev A. 2005. Analysis of U1 small nuclear RNA interaction with cyclin H. *J Biol Chem* **280:** 36920–36925.

Okuda K, Khan MY, Skurnick J, Kimura M, Aviv H, Aviv A. 2000. Telomere attrition of the human abdominal aorta: relationships with age and atherosclerosis. *Atherosclerosis* **152:** 391–398.

Olive M, Harten I, Mitchell R, Beers JK, Djabali K, Cao K, Erdos MR, Blair C, Funke B, Smoot L, et al. 2010. Cardiovascular pathology in Hutchinson–Gilford progeria: Correlation with the vascular pathology of aging. *Arterioscler Thromb Vasc Biol* **30:** 2301–2309.

Oostra AB, Nieuwint AW, Joenje H, de Winter JP. 2012. Diagnosis of Fanconi anemia: Chromosomal breakage analysis. *Anemia* **2012:** 238731.

Oshima J, Hisama FM. 2014. Search and insights into novel genetic alterations leading to classical and atypical Werner syndrome. *Gerontology* **60:** 239–246.

Oshima J, Martin GM, Hisama FM. 2014. Werner syndrome. In *GeneReviews(R)* (ed. Pagon RA, Adam MP, Bird TD, Dolan CR, Fong CT, Smith RJH, Stephens K). University of Washington, Seattle, WA.

Ow SY, Dunstan DE. 2014. A brief overview of amyloids and Alzheimer's disease. *Protein Sci* **23:** 1315–1331.

Pagano G, Shyamsunder P, Verma RS, Lyakhovich A. 2014. Damaged mitochondria in Fanconi anemia—An isolated event or a general phenomenon? *Oncoscience* **1:** 287–295.

Palles C, Cazier JB, Howarth KM, Domingo E, Jones AM, Broderick P, Kemp Z, Spain SL, Guarino E, Salguero I, et al. 2013. Germline mutations affecting the proofreading domains of POLE and POLD1 predispose to colorectal adenomas and carcinomas. *Nat Genet* **45:** 136–144.

Pelosini C, Martinelli S, Ceccarini G, Magno S, Barone I, Basolo A, Fierabracci P, Vitti P, Maffei M, Santini F. 2014. Identification of a novel mutation in the polymerase delta 1 (*POLD1*) gene in a lipodystrophic patient affected by mandibular hypoplasia, deafness, progeroid features (MDPL) syndrome. *Metabolism* **63:** 1385–1389.

Phillips MC. 2014. Apolipoprotein E isoforms and lipoprotein metabolism. *IUBMB Life* **66:** 616–623.

Powers ET, Morimoto RI, Dillin A, Kelly JW, Balch WE. 2009. Biological and chemical approaches to diseases of proteostasis deficiency. *Annu Rev Biochem* **78:** 959–991.

Quentin S, Cuccini W, Ceccaldi R, Nibourel O, Pondarre C, Pages MP, Vasquez N, d'Enghien CD, Larghero J, de Latour RP, et al. 2011. Myelodysplasia and leukemia of Fanconi anemia are associated with a specific pattern of genomic abnormalities that includes cryptic *RUNX1/AML1* lesions. *Blood* **117:** E161–E170.

Rajpathak SN, Liu Y, Ben-David O, Reddy S, Atzmon G, Crandall J, Barzilai N. 2011. Lifestyle factors of people with exceptional longevity. *J Am Geriatr Soc* **59:** 1509–1512.

Rao CV, Yamada HY. 2013. Genomic instability and colon carcinogenesis: From the perspective of genes. *Front Oncol* **3:** 130.

Ratnapriya R, Chew EY. 2013. Age-related macular degeneration—Clinical review and genetics update. *Clin Genet* **84:** 160–166.

Rice GI, Bond J, Asipu A, Brunette RL, Manfield IW, Carr IM, Fuller JC, Jackson RM, Lamb T, Briggs TA, et al. 2009. Mutations involved in Aicardi–Goutieres syndrome implicate *SAMHD1* as regulator of the innate immune response. *Nat Genet* **41:** 829–832.

Roberts R. 2015. A genetic basis for coronary artery disease. *Trends Cardiovasc Med* **25:** 171–178.

Ruijs MW, van Andel RN, Oshima J, Madan K, Nieuwint AW, Aalfs CM. 2003. Atypical progeroid syndrome: An unknown helicase gene defect? *Am J Med Genet A* **116A:** 295–299.

Saha B, Cypro A, Martin GM, Oshima J. 2014. Rapamycin decreases DNA damage accumulation and enhances cell growth of *WRN*-deficient human fibroblasts. *Aging Cell* **13:** 573–575.

Savitsky K, Bar-Shira A, Gilad S, Rotman G, Ziv Y, Vanagaite L, Tagle DA, Smith S, Uziel T, Sfez S, et al. 1995. A single ataxia telangiectasia gene with a product similar to PI-3 kinase. *Science* **268:** 1749–1753.

Schottker B, Saum KU, Jansen EH, Boffetta P, Trichopoulou A, Holleczek B, Dieffenbach AK, Brenner H. 2015. Oxidative stress markers and all-cause mortality at older age: A population-based cohort study. *J Gerontol A Biol Sci Med Sci* **70:** 518–524.

Sebastiani P, Solovieff N, Dewan AT, Walsh KM, Puca A, Hartley SW, Melista E, Andersen S, Dworkis DA, Wilk JB, et al. 2012. Genetic signatures of exceptional longevity in humans. *PLoS ONE* **7:** e29848.

Shalev SA, De Sandre-Giovannoli A, Shani AA, Levy N. 2007. An association of Hutchinson–Gilford progeria and malignancy. *Am J Med Genet A* **143A:** 1821–1826.

Shimamoto A, Kagawa H, Zensho K, Sera Y, Kazuki Y, Osaki M, Oshimura M, Ishigaki Y, Hamasaki K, Kodama Y, et al. 2014. Reprogramming suppresses premature senescence phenotypes of Werner syndrome cells and maintains chromosomal stability over long-term culture. *PLoS ONE* **9:** e112900.

Siitonen HA, Kopra O, Kaariainen H, Haravuori H, Winter RM, Saamanen AM, Peltonen L, Kestila M. 2003. Molecular defect of RAPADILINO syndrome expands the phenotype spectrum of *RECQL* diseases. *Hum Mol Genet* **12:** 2837–2844.

Sobenin IA, Zhelankin AV, Sinyov VV, Bobryshev YV, Orekhov AN. 2014. Mitochondrial aging: Focus on mitochondrial DNA damage in atherosclerosis—A mini-review. *Gerontology* **61:** 343–349.

Stephenson JB. 2008. Aicardi–Goutieres syndrome (AGS). *Eur J Paediatr Neurol* **12:** 355–358.

Cite this article as *Cold Spring Harb Perspect Med* doi: 10.1101/cshperspect.a025882

Suri S, Heise V, Trachtenberg AJ, Mackay CE. 2013. The forgotten APOE allele: A review of the evidence and suggested mechanisms for the protective effect of APOE varepsilon2. *Neurosci Biobehav Rev* **37:** 2878–2886.

Suzuki H, Kurihara Y, Takeya M, Kamada N, Kataoka M, Jishage K, Ueda O, Sakaguchi H, Higashi T, Suzuki T, et al. 1997. A role for macrophage scavenger receptors in atherosclerosis and susceptibility to infection. *Nature* **386:** 292–296.

Takemoto M, Yamaga M, Furuichi Y, Yokote K. 2015. Astaxantin improved non-alcoholic fatty liver disease in Werner syndrome with diabetes. *J Am Geristr Soc* **63:** 1271–1273.

Talaei F, van Praag VM, Henning RH. 2013. Hydrogen sulfide restores a normal morphological phenotype in Werner syndrome fibroblasts, attenuates oxidative damage and modulates mTOR pathway. *Pharmacol Res* **74:** 34–44.

Tang Y, Le W. 2015. Differential Roles of M1 and M2 Microglia in neurodegenerative diseases. *Mol Neurobiol* doi: 10.1007/s12035-014-9070-5.

Thomas GS, Wann LS, Allam AH, Thompson RC, Michalik DE, Sutherland ML, Sutherland JD, Lombardi GP, Watson L, Cox SL, et al. 2014. Why did ancient people have atherosclerosis?: From autopsies to computed tomography to potential causes. *Global Heart* **9:** 229–237.

Thompson D, Duedal S, Kirner J, McGuffog L, Last J, Reiman A, Byrd P, Taylor M, Easton DF. 2005. Cancer risks and mortality in heterozygous ATM mutation carriers. *J Natl Cancer Inst* **97:** 813–822.

Valle L, Hernandez-Illan E, Bellido F, Aiza G, Castillejo A, Castillejo MI, Navarro M, Segui N, Vargas G, Guarinos C, et al. 2014. New insights into POLE and POLD1 germline mutations in familial colorectal cancer and polyposis. *Hum Mol Genet* **23:** 3506–3512.

Van Maldergem L, Siitonen HA, Jalkh N, Chouery E, De Roy M, Delague V, Muenke M, Jabs EW, Cai J, Wang LL, et al. 2006. Revisiting the craniosynostosis–radial ray hypoplasia association: Baller–Gerold syndrome caused by mutations in the *RECQL4* gene. *J Med Genet* **43:** 148–152.

Vennos EM, James WD. 1995. Rothmund–Thomson syndrome. *Dermatol Clin* **13:** 143–150.

Walden H, Deans AJ. 2014. The Fanconi anemia DNA repair pathway: Structural and functional insights into a complex disorder. *Annu Rev Biophys* **43:** 257–278.

Wang JC, Bennett M. 2012. Aging and atherosclerosis: Mechanisms, functional consequences, and potential therapeutics for cellular senescence. *Circ Res* **111:** 245–259.

Wang LL, Levy ML, Lewis RA, Chintagumpala MM, Lev D, Rogers M, Plon SE. 2001. Clinical manifestations in a cohort of 41 Rothmund–Thomson syndrome patients. *Am J Med Genet* **102:** 11–17.

Wang M, Jiang L, Monticone RE, Lakatta EG. 2014. Proinflammation: The key to arterial aging. *Trends Endocrinol Metab* **25:** 72–79.

Weedon MN, Ellard S, Prindle MJ, Caswell R, Lango Allen H, Oram R, Godbole K, Yajnik CS, Sbraccia P, Novelli G, et al. 2013. An in-frame deletion at the polymerase active site of POLD1 causes a multisystem disorder with lipodystrophy. *Nat Genet* **45:** 947–950.

Wei S, Soh SL, Qiu W, Yang W, Seah CJ, Guo J, Ong WY, Pang ZP, Han W. 2013. Seipin regulates excitatory synaptic transmission in cortical neurons. *J Neurochem* **124:** 478–489.

Wilhelmsen L, Dellborg M, Welin L, Svardsudd K. 2015. Men born in 1913 followed to age 100 years. *Scand Cardiovasc J* **49:** 45–48.

Williams GC. 1957. Pleiotropy, natural selection, and the evolution of senescence. *Evolution* **11:** 398–411.

Willingale-Theune J, Schweiger M, Hirsch-Kauffmann M, Meek AE, Paulin-Levasseur M, Traub P. 1989. Ultrastructure of Fanconi anemia fibroblasts. *J Cell Sci* **93:** 651–665.

Woods CG, Taylor AM. 1992. Ataxia telangiectasia in the British Isles: The clinical and laboratory features of 70 affected individuals. *Q J Med* **82:** 169–179.

Wu L, Davies SL, Levitt NC, Hickson ID. 2001. Potential role for the BLM helicase in recombinational repair via a conserved interaction with RAD51. *J Biol Chem* **276:** 19375–19381.

Xie Z, Jay KA, Smith DL, Zhang Y, Liu Z, Zheng J, Tian R, Li H, Blackburn EH. 2015. Early telomerase inactivation accelerates aging independently of telomere length. *Cell* **160:** 928–939.

Dissecting the Mechanisms Underlying Unusually Successful Human Health Span and Life Span

Sofiya Milman[1,2] and Nir Barzilai[1,2,3]

[1]Department of Medicine, Division of Endocrinology, Albert Einstein College of Medicine, New York, New York 10461

[2]Institute for Aging Research, Albert Einstein College of Medicine, New York, New York 10461

[3]Department of Genetics, Albert Einstein College of Medicine, Bronx, New York 10461

Correspondence: nir.barzilai@einstein.yu.edu

Humans age at different rates, and families with exceptional survival provide the opportunity to understand why some people age slower than others. Unique features exhibited by centenarians include a family history of longevity, compression of morbidity with resultant extension of health span, and biomarkers such as low-circulating insulin-like growth factor 1 (IGF-1) and elevated high-density lipoprotein (HDL) cholesterol levels. Given the rarity of the centenarian phenotype, it has not been surprising that the use of discovery methods that relied on common population single nucleotide polymorphisms (SNPs) to unlock the genetic determinants of exceptional longevity have not yielded significant results. Conversely, gene sequencing has resulted in discoveries of functional gene variants that support several of the centenarian phenotypes. These discoveries have led to the strategic developments of drugs that may delay aging and prolong health span.

THE RATIONALE FOR STUDYING HUMAN EXCEPTIONAL LONGEVITY

The United States government annually publishes a report on the rate of death from individual diseases, stratified by age groups. What is striking about these reports is that the rate of death increases logarithmically with advancing age for all diseases associated with aging, including heart disease, cancer, stroke, diabetes mellitus type 2 (T2DM), and Alzheimer's disease (AD) (Fig. 1). This shows that aging is a major risk factor that all of these age-related diseases have in common. To put these statistics in perspective, elevated low-density lipoprotein (LDL) cholesterol level, which is one of the best known and aggressively treated risk factors for heart disease, the most common cause of death among older adults, is associated with a three-fold increase in the risk for heart disease. However, advancing the age from 30 years to 80 years raises the rate of death from each of the age-related diseases by as much as 100- to 1000-fold. If we accept the notion that aging is the common and major risk factor for all age-related diseases, then we conclude that unless aging

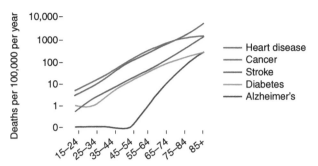

Figure 1. Rate of death per 100,000 people per year for age-related diseases, stratified by age groups (from Health, United States, 2014; www.cdc.gov/nchs/hus.htm).

itself is delayed our best attempts at preventing each disease individually will result in exchanging one disease for another. The return for curing individual diseases is small. For example, statistical models project that delaying cancer would result in an increase of only 0.8% in the population of older adults over a 50-year period, whereas delaying aging would lead to an increase of ~7% in the population, with most of these individuals being free of disability (Goldman et al. 2013). Furthermore, this delay in aging would yield ~$7.1 trillion in social benefit to the population (Goldman et al. 2013).

When thinking about aging, it is important to recognize that chronological and biological age are not the same. It is well recognized by all that some individuals appear younger than their chronological age, whereas others appear older. This observation highlights an opportunity for scientific discovery that until recently has been missed, that is, to try to understand the biology of why some age faster while others age slower. At one extreme of the spectrum of the rate of aging are rare segmental progeroid syndromes that are thought to accelerate various aging phenotypes. The responsible gene mutations have suggested genomic instability as an important mechanism of aging (Martin 2005). More recently, scientists have become interested in studying people with exceptional longevity, which are located at the other end of the rate of aging spectrum, in an effort to discover the genetic and biological determinants of delayed aging.

Centenarians are a unique group of individuals that constitute an example of delayed aging.

This delay in aging can only be accomplished if it results in the extension of disease-free survival and, indeed, this appears to be the case in many centenarians. Analyses from the New England Centenarian Study (Andersen et al. 2012), the Long Life Family Study (LLFS) (Sebastiani et al. 2013), and the Longevity Genes Program (Ismail 2014) have provided evidence that individuals with exceptional longevity manifest compression of morbidity, meaning that they spend a smaller percentage of their life being ill, and, as a result, their health span approximates their life span. These studies revealed significant delay in the ages of onset for most age-related diseases among individuals with exceptional longevity, including hypertension, cardiovascular disease (CVD), cancer, T2DM, stroke, osteoporosis, and AD. Thus, not only do centenarians live longer, they live healthier. Although a large proportion of centenarians delay or escape from age-related diseases altogether (Evans et al. 2014), a number of individuals achieve exceptional longevity despite having developed one or several of these diseases (Andersen et al. 2012; Ailshire et al. 2015). This suggests that these people likely possess protective factors that allow them to be resilient and survive despite health ailments.

The inherent differences between chronological and biological age, and between the diverse rates of aging, offer scientists opportunities to study the variations in the biology and genetics among these different groups. As exemplified in the literature, several mechanisms have already been identified that can delay aging in a variety of animal models. Investigating whether these same mechanisms apply to humans with exceptional longevity serves to validate these discoveries as important for human aging. Furthermore, studies are underway for discovery of age-delaying mechanisms that are specific to humans by using centenarian populations. The rationale for studying centenarians is that they are the "poster children" for what we are ultimately trying to achieve—extension of health span and not merely life span.

THE EVIDENCE THAT LONGEVITY IS INHERITED

Demographers and epidemiologists have attributed ∼15%–30% of the variation in life span to heritable factors. Several studies have found positive correlations between the life spans of the parents and their biological children (Atzmon et al. 2004; Schoenmaker et al. 2006; Westendorp et al. 2009). However, the advances of modern medicine that include preventive measures and treatments, have extended the life spans of the newer generations beyond what would have been predicted based on their inheritance. Thus, offspring whose parents died from CVD resulting from hereditary hyperlipidemia, can now enjoy an extension of their life span through treatment with cholesterol-lowering medications and interventions such as coronary artery bypass graft surgery or revascularization of coronary arteries with angioplasty. Despite these significant medical advances, achievement of exceptional longevity remains a rare occurrence. Yet, exceptional longevity clusters in families point to a strong relationship between genetics and longevity.

Data suggests that the offspring of parents who achieved a life span of at least 70 years have a much greater probability of living longer compared with the offspring of parents with shorter life spans, with this association becoming stronger as the parental life span lengthens (Gavrilov et al. 2001). This relationship is even more pronounced in families with exceptional longevity. Siblings of centenarians have been shown to be ∼4–5 times more likely to achieve longevity, with male siblings being 17 times

more likely to become centenarians themselves (Perls et al. 1998, 2002). The parents of centenarians were found to be seven times more likely to have survived to age 90 and beyond, compared with parents of those with the usual life span (Atzmon et al. 2004). Even if genetics account for smaller differences observed in the rate of aging, identification of these genes is important for planning strategies that can delay the aging process. Furthermore, because exceptional longevity is heritable, studying the families of centenarians to identify genetic determinants of exceptional longevity offers great promise for discovery.

Familial longevity is likely mediated through protection from age-related diseases, which is inherited by the offspring from their parents. Centenarians and their offspring have a lower prevalence and later age of onset of heart disease, stroke, hypertension, T2DM, AD, and cancer (Anderson et al. 1991; Atzmon et al. 2004; Adams et al. 2008; Lipton et al. 2010; Altmann-Schneider et al. 2012). This heritable protection from disease has also been shown in several large studies. A prospective population-based study found that the incidence of AD was 43% lower in offspring of parents with exceptional longevity compared with offspring of parents with more usual life spans over a 23-year follow-up (Lipton et al. 2010). A similar association was also found in a study conducted in a population whose parents achieved more modest longevity. In a secondary analysis of the Diabetes Prevention Program (DPP), a large clinical trial designed to compare strategies for T2DM prevention in individuals at high risk for T2DM, parental longevity was associated with a delay in the incidence of T2DM in the offspring, with the children of parents with longest life spans experiencing the greatest delay in disease onset (Florez et al. 2011). The effect of parental life span on diabetes prevention was found to be just as strong as the effect of metformin, an antidiabetic drug used in this study (Florez et al. 2011). These results show that extended parental life span is strongly associated with better health outcomes in the offspring, even in populations who achieve less extreme degrees of longevity.

Although environmental influences may have a significant effect on health and life span in the general population, this does not seem to be the case in centenarians. A study that compared individuals with exceptional longevity to their contemporaries who did not achieve longevity found that centenarians were as likely as their shorter-lived peers to have been overweight or obese (Rajpathak et al. 2011). Furthermore, the proportion of centenarians who smoked, consumed alcohol daily, had not participated in regular physical activity, or had not followed a low-calorie diet throughout their middle age was similar to that among their peers from the same birth cohort. In fact, as many as 60% of male and 30% of female centenarians had been smokers (Rajpathak et al. 2011). Thus, the centenarians had not engaged in a healthier lifestyle compared with their peers. This supports the notion that people with exceptional longevity possess genomic factors that protect them from the environmental influences that may be detrimental to health.

GENETICS OF EXCEPTIONAL LONGEVITY

For more than a decade, centenarian populations of diverse Americans, as well as ethnically homogeneous populations of Mormons, Ashkenazi Jews (AJs), Icelandics, Okinawan Japanese, Italians, Irish, and Dutch, among others, have served as cohorts for studies to identify longevity genes or longevity-associated biological pathways. These studies relied on candidate genes and genome-wide association studies (GWAS) that included genotyping of large populations. One of the strengths of GWAS compared with the candidate gene approach is that these studies are unbiased. Their results may provide insights into novel mechanisms of longevity. Several research groups have conducted GWAS for longevity (Beekman et al. 2010; Sebastiani et al. 2012), yet none yielded significant results after appropriate statistical corrections for multiple comparisons were applied. One exception was the finding of the *APOE2* genotype, although its identification may have been the result of ascertainment bias, because individuals with the *APOE4* allele, who are at higher risk for developing Alzheimer's dementia, are less likely to be recruited into population studies (Nebel et al. 2011).

There are several explanations for these disappointing results. First, relying on common genetic variants that occur at frequencies from 5% to 49% in the population to study such a rare event as exceptional longevity (one that occurs at a rate of 1/6000–1/10,000 in the general population) may result in missing the rarer longevity-associated genotypes. This also underscores the need for exon or whole-genome sequencing to discover rare mutations. Second, applying GWAS to genetically diverse populations requires a very large study cohort to account for genomic diversity and to identify relatively rare genetic variants. Thus, most studies have lacked sufficient power for such discoveries.

Following this logic, it is not surprising that many important genetic discoveries were made in populations that show comparatively small levels of genetic diversity. One such example is the Icelandic population, which originated from a small number of founders and expanded to ~500,000 people. Others include the Amish and AJs, a larger population (Barzilai et al. 2003; Atzmon et al. 2008, 2009b, 2010; Suh et al. 2008). The advantage of studying a genetically homogeneous population was exemplified by a recent study, which showed that the addition of each AJ subject contributed 20 times more genetic variability to the cohort as compared with adding a European subject to a cohort of European origin of identical sample size (Carmi et al. 2014).

There are several ways in which genetically similar populations can contribute to genetic and biological discovery. One is if the population has a higher frequency of carriers of a particular genotype and its associated phenotype caused by the founder effect, as is the case with breast cancer caused by mutations in the *BRCA* genes among AJ women. Another is that single nucleotide polymorphisms (SNPs) that are novel or rare in the general population will occur at higher frequencies in a homogenous population. This will result in the associated rare phenotype, such as longevity, to be more amenable to withstand the rigorous statistical analysis that is performed on genetic data.

Third, many SNPs that are statistically significant, but below the threshold for GWAS, may still be relevant. Last, it is possible that numerous SNPs contribute in combination to the phenotype. Indeed, Sebastiani et al. (2012) have identified 281 SNPs that can distinguish centenarians from controls.

Although discovery of longevity-associated genes has been met with several challenges, many genes have been identified that are associated with risk for CVD, AD, T2DM, and other age-related diseases. One attractive hypothesis has been that centenarians lack these disease-associated genes, thus being protected by a more "perfect genome." However, it has become clear from GWAS that centenarians harbor as many disease-associated genotypes as controls. Furthermore, a whole-genome sequence analysis of 44 centenarians revealed that this group carried a total of 227 autosomal and 7 X-chromosome coding single nucleotide variants (SNVs) that are likely to cause disease according to the ClinVar database (Freudenberg-Hua et al. 2014). Among these are variants associated with Parkinson's disease, AD, neurodegenerative diseases, neoplastic, and cardiac diseases. Despite >95 years of exposure to these risky genotypes, none of the centenarians showed any of the diseases for which they were genetic carriers. These observations led to the conclusion that there are longevity-associated protective genotypes in centenarians that delay aging or specifically protect against the manifestation of age-related diseases.

Although the GWAS approach did not prove to be particularly helpful in identifying longevity genes, some success stories have emerged through the application of the candidate gene approach. Several genes were selected for investigation because they were previously implicated in aging, and SNPs within these genes were suggested to be linked with longevity. These included PON1 (Bonafe et al. 2002; Rea et al. 2004; Franceschi et al. 2005; Marchegiani et al. 2006; Tan et al. 2006), insulin-like growth factor 1 (IGF-1) (Bonafe et al. 2003; Kojima et al. 2004; van Heemst et al. 2005), PAPR-1, cytokine genes, genes that code for enzymatic antioxidants such as superoxide dismutases (Andersen et al. 1998;

Mecocci et al. 2000), and components of lipid metabolism (Barzilai et al. 2006; Vergani et al. 2006). Other genes that have been implicated in human aging, and not only longevity, are updated on the Aging Gene Database (see genomics .senescence.info/genes).

However, not all discoveries resulted in improved understanding of the biology of aging. One of the most notable discoveries of a longevity-associated gene, which has been validated by numerous research groups, is the FOXO3a genotype. As summarized by Kahn (2014), the FOXO3a genotypes are rather common, the identified SNPs within the gene localize to intronic or noncoding regions, and despite sequencing of the whole gene by several groups, no functional mutations have thus far been identified in the regions of the gene that would predict altered protein function. Furthermore, assays of cells with the FOXO3a genotype variants also have not been, thus far, associated with functional changes. Finally, no identifiable phenotype has yet been linked with these FOXO3a genotypes and they have not been related to risk or protection from disease. In fact, a panel of experts did not agree on whether a drug that displaces FOXO3a from the nucleus to the cytoplasm would induce longevity or shorten the life span (Monsalve and Olmos 2011). The example of FOXO3a shows that even a validated genotype does not always translate into better understanding of the biology of longevity.

There are also other challenges that researchers face studying longevity. In addition to the usual problems and pitfalls of association studies, particularly in the new age of "big data" brought on by whole-genome sequencing (Lawrence et al. 2005), there is another problem that is particular to longevity studies—that of identifying appropriate controls for a cohort of exceptionally long-lived individuals. This has been a challenge because the ideal controls, individuals of the same birth cohort as the centenarians but who have not achieved exceptional longevity, are all deceased. One approach to overcome this challenge has been to rely on the innovative experimental design in which the progeny of centenarians, who have inherited about half of their genome from the centenarian parent, are compared with their spouses who do not have a parental history of longevity and thus can serve as matched controls (Barzilai et al. 2001).

GENOMIC DISCOVERIES AND MECHANISMS FOR EXCEPTIONAL LONGEVITY

The Longevity Genes Project (LGP) and LonGenity are studies that include families of AJs with exceptional longevity. Because longevity carries a substantial genetic component, these studies conduct genomic and detailed phenotype analyses in the families with exceptional longevity in an effort to determine the functions of genes of interest. Using the candidate gene approach in this AJ cohort, several favorable homozygous genotypes were identified in multiple genes, which were associated with unique biological phenotypes.

The cholesterol ester transfer protein (*CETP*) gene codon 405 isoleucine to valine variant was associated with low levels of plasma CETP, high levels of high-density lipoprotein (HDL) cholesterol, and large lipoprotein particle size. This genotype was also shown to be protective against cognitive decline and AD in an independent diverse population (Sanders et al. 2010). This same genotype was validated by another research group in an Italian population (Vergani et al. 2006). Three other genotypes in the *CETP* gene were also found to be significantly associated with longevity in the LLFS study. Although none of the other studies have confirmed these findings, it is important to keep in mind that a particular SNP may not show a similar phenotype in all populations. Therefore, the biological phenotype itself should be tested for association with longevity rather than a particular SNP that may have differential expression in varying populations. Further complicating matters is the possibility that the gene with the significant action may be in linkage disequilibrium with the SNP and that there may be genetic variations at that associated locus.

Another lipid-related genotype, homozygosity for the apolipoprotein C-3 (*APOC-3*)−641 C allele was also associated with exceptional longevity in AJs (Atzmon et al. 2006). It too showed a unique lipid phenotype and low levels of plasma APOC-3 (Atzmon et al. 2006). In a striking example of validation, carriers of a different *APOC3* genotype in a homogenous Pennsylvania Amish population also showed low APOC-3 levels, a favorable lipid phenotype, better arterial health score, and enhanced longevity (Pollin et al. 2008). These findings show the power of discovery in selected genetically homogeneous populations. The *APOC-3* genotype was also identified to be related to exceptional longevity in the LLFS, but the phenotype associated with this SNP has not yet been revealed.

ADIPOQ is another longevity-associated genotype. Adiponectin is a fat-derived peptide with powerful effects on lipids and metabolism. A deletion at +2019 in the adiponectin (*ADIPOQ*) gene was associated in the AJ cohorts with longevity, which was also related to a phenotype of high adiponectin levels, independent of fat mass (Atzmon et al. 2008).

A longevity-associated genotype whose discovery has already made an impact on clinical practice is that of the thyroid stimulating hormone receptor (*TSHR*) (Atzmon et al. 2009a,b). The metabolic rate theory of aging suggests that, in nature, there exists an inverse relationship between basal metabolic rate and aging, with several hypothyroid mammalian models showing longer life span. Centenarians have higher plasma thyroid stimulating hormone (TSH) levels, although they are not hypothyroid, and their offspring also show this phenotype with significant heritability (Atzmon et al. 2009a; Rozing et al. 2010). These clinical features have been supported by a National Health and Nutrition Examination Survey (NHANES III) conducted across the United States and led to the recommendation to not supplement older adults with mild elevations in TSH with thyroid hormone (Tabatabaie and Surks 2013).

In nature, disruption of the growth hormone (GH)/IGF-1 action has led to extension of life span. Spontaneous and experimentally induced partial disruptions of the GH/IGF-1 pathway, including genetic alterations, are associated with a small body size (dwarfism) across species (Brown-Borg et al. 1996). Thus, small dogs have longer life spans than large dogs (Samaras and Elrick 2002). Models of IGF-1 deficiency show numerous indices of delayed aging, including enhanced stress resistance and a major increase in life span (Kenyon et al. 1993; Brown-Borg et al. 1996). On the other hand, reduced levels of IGF-1 in humans, while protective against cancer, have been linked with higher risk for CVD and diabetes (Sandhu et al. 2002; Burgers et al. 2011), suggesting a more complex physiological role for IGF-1 in humans. Several SNPs in genes within the insulin/IGF-1-signaling pathway have been associated with and validated in exceptional longevity, but, for the most part, no specific phenotype related to these SNPs has been identified (Pawlikowska et al. 2009). An exception to this has been the identification of a functional IGF-1 receptor (*IGF-1R*) gene mutation discovered after sequencing the *IGF-1* and *IGF-1R* genes of centenarians (Suh et al. 2008). Heterozygous mutations in the *IGF-1R* gene have been overrepresented among centenarians compared with the controls without familial longevity and have been associated with high-serum IGF-1 levels in the setting of reduced activity of the *IGF-1R*, as measured in transformed lymphocytes (Tazearslan et al. 2011). Partial IGF-1 resistance conferred by these longevity-associated *IGF-1R* genotypes was confirmed in a study conducted on wild-type cells transformed with the mutant genes (Tazearslan et al. 2011). A particular *IGF-1R* genotype was also associated with longevity in the LLFS; however, its associated phenotype has not yet been defined.

Another example that highlights the importance of GH/IGF-1 signaling in extended health span comes from a population of Laron Dwarfs, who are carriers of a rare mutation in the GH receptor (*GHR*) gene that results in GHR deficiency. A group with this genotype was studied in Ecuador and appears to have a negligible prevalence of type 2 diabetes mellitus and cancer (Guevara-Aguirre et al. 2011). Although they did not live long, clearly they have been protected from major age-related diseases.

Finally, among females with exceptional longevity, those with IGF-1 levels below the median exhibited significantly longer survival compared with those with levels above the median (Fig. 2) (Milman et al. 2014). However, this relationship between IGF-1 levels and survival was not observed in males with exceptional longevity. On the other hand, among males and females who achieved longevity and had a history of cancer, lower IGF-1 levels predicted longer survival (Milman et al. 2014). Thus, low IGF-1 levels predict life expectancy in exceptionally long-lived individuals, supporting the role of the GH/IGF-1 pathway in exceptional longevity.

Interest in telomeres and their association with aging led to significant research efforts aimed at identifying the role of telomere length in exceptional longevity. Telomere length or mass assessment showed that centenarians have longer telomeres, that this length is inherited in their offspring, and is associated with decreased incidence of the metabolic syndrome (MS), T2DM, and cognitive decline (Atzmon et al. 2010). This longevity-associated telomere phenotype has also been related to a genetic "fingerprint" in the telomerase genes in centenarians (Atzmon et al. 2010).

Other genomic mechanisms, no doubt, also contribute to aging, including epigenomic variations. Sirtuins, resveratrol, and other specific activators have been used to induce histone deacetylation and activation of the *SIRT1* gene, thereby resulting in longevity in a variety of animal models and in high-fat fed mice. However, no significant association between *SIRT1* genotypes and longevity has been reported in humans thus far (Han et al. 2014). Methylation patterns have been noted to change with aging and may affect the transcribed DNA. Initial studies have shown significant differences in methylation patterns between centenarians and younger controls, with several groups currently pursuing this line of research. Finally, longevity-associated microRNAs have been identified, but their effects still need to be determined (Gombar et al. 2012).

EXCEPTIONAL LONGEVITY LEADING TO AGE-DELAYING DRUGS

The goal of longevity research is to identify pathways that are relevant to human aging and

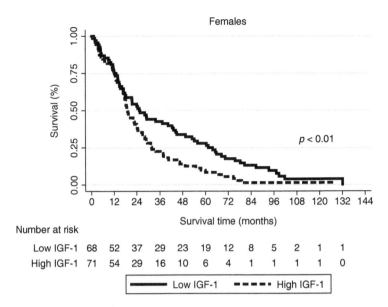

Figure 2. Kaplan–Meier survival curves for females with IGF-1 levels above and below the median. (From Milman et al. 2014; adapted, with permission.)

to develop drugs that will delay aging by targeting these pathways. Longevity and extension of healthy life span have been achieved in models via a variety of genetic manipulations, drugs, and environmental influences, thereby providing the preclinical foundation needed to proceed to drug development. The main obstacle facing the development of drugs for the treatment of aging is the fact that the U.S. Food and Drug Administration (FDA) does not consider aging as a preventable condition. Even if there would be a popular demand for drugs that delay aging, the pharmaceutical industry would not develop drugs that will not be reimbursed by health insurance companies. The same was true for hypertension, until studies showed that lowering blood pressure prevented CVD, including strokes.

The pharmaceutical industry has relied on genetic discoveries made in longevity studies, as well as other studies, to identify individuals who have naturally occurring genetic variants or mutations that confer desirable phenotypes. The goals for pharmaceutical development is to create drugs whose actions would mimic those of the favorable genetic variants. Observing the carriers of these genetic variants for any

detrimental health effects informs drug makers of any potential side effects that may arise from a drug that targets the desired pathway. For example, the observation that centenarians are enriched with a unique *CETP* genotype that exposes them to a lifetime of lower CETP levels that is also associated with high HDL level and large lipoprotein particle size, suggests that decreased CETP function is safe (Barzilai et al. 2003). In fact, a CETP inhibitor is currently being tested in a phase 3 trial by a leading pharmaceutical company (Cannon et al. 2010). Similar observations were made about the APOC-3 protein, and an APOC-3 inhibitor is also being tested in a phase 3 trial by another pharmaceutical company (Graham et al. 2013; Lee et al. 2013).

Another class of agents whose actions on aging may be predicted through longevity research are monoclonal antibodies directed against the IGF-1 receptor. These were initially developed by several pharmaceutical industries as antineoplastic therapies; however, they were not successful at treating cancer because of a significant degree of mutagenesis within cancer cells that eventually made them resistant to these drugs. Nonetheless, these compounds are available for preclinical testing in aging research. Similarly, the GH/IGF-1 pathway, which may be important for human aging, can be targeted by the GHR antagonist that is currently in clinical use for the treatment of acromegaly, a condition of GH excess (Kopchick 2003). Although the above-mentioned therapeutics are not presently being developed for longevity, these drugs may be tested in the future for the indication of delaying aging and age-associated diseases.

Other drugs may target aging more specifically, although they are in clinical use for other indications. One example is a class of drugs that inhibit the mammalian target of rapamycin (mTOR) enzyme. These drugs are primarily used as immune modulators post organ transplantation, but recently also have been shown to increase the immune response to vaccinations in the elderly (Mannick et al. 2014), thereby demonstrating their potential utility in the treatment of health conditions associated with aging.

Another drug of interest is metformin, the first line drug treatment for T2DM. Several research groups tested the effect of metformin on aging and showed that it caused extension in life span and health span in many rodent models (Anisimov et al. 2008, 2010, 2011; Smith et al. 2010; Martin-Montalvo et al. 2013). Metformin also extended the life span of nematodes (Cabreiro et al. 2013), suggesting that its action is mediated via an evolutionary conserved mechanism. Numerous investigators looked at the potential antiaging effects of this drug in populations treated with metformin for T2DM. The large United Kingdom Prospective Diabetes Study (UKPDS) convincingly showed that metformin reduced the incidence of CVD (Holman et al. 2008; Anfossi et al. 2010). This finding has been validated and reproduced by other studies and meta-analysis (Johnson et al. 2005; Lamanna et al. 2011; Roumie et al. 2012; Hong et al. 2013; Whittington et al. 2013). In addition, a number of studies suggested that metformin use is associated with a decreased incidence of cancer (Libby et al. 2009; Landman et al. 2010; Lee et al. 2011; Monami et al. 2011; Tseng 2012), with many animal and cell models demonstrating the inhibitory effects of metformin on tumorigenesis (Seibel et al. 2008;

Tosca et al. 2010; Liu et al. 2011; Salani et al. 2012; Anisimov and Bartke 2013; Karnevi et al. 2013; Quinn et al. 2013). The proposed mechanisms of action for metformin's effect on inhibiting tumorigenesis include decrease in insulin production and its action, decrease in IGF-1 signaling, and AMP-activated protein kinase (AMPK) activation.

In the future, other compounds discovered to be important for longevity may be developed into drugs. For example, the level of humanin, a mitochondrial-derived peptide, decreases with aging but has been shown to increase up to threefold in the offspring of centenarians (Muzumdar et al. 2009), thus making it an attractive candidate for drug development.

CONCLUDING REMARKS

This article shows that, via the use of biologic and genetic experimental methods, scientists can determine why some people age more slowly or more rapidly than others. Such discoveries in humans, as opposed to those in other animal models, have the advantage of being directly relevant to human longevity and can be relied on by pharmaceutical developers looking to establish the safety of drugs whose actions mimic the function of the genetic variants found in centenarians. Thus it follows that if functional mutations or SNPs that are more common in centenarians are also deemed safe in that population, then drugs that mimic the desired actions are worth developing. This kind of drug development should result in unique drugs that target not only specific diseases but also aging. The barrier for development of drugs that target aging is that, at present, aging is not an indication for treatment by the FDA. There is an urgent need to change this paradigm to accelerate drug development and realize the longevity dividend.

ACKNOWLEDGMENTS

S.M. is supported by the National Center for Advancing Translational Sciences of the National Institutes of Health under Award Number KL2TR001071. N.B. is supported by Grants from the National Institutes of Health (NIH) (P01AG021654), The Nathan Shock Center of Excellence for the Biology of Aging (P30AG038072), the Glenn Center for the Biology of Human Aging (Paul Glenn Foundation for Medical Research), NIH R37 AG18381 (Barzilai Merit Award), and NIH/NIA 1 R01AG044829. The content is solely the responsibility of the authors and does not necessary represent the official views of the NIH.

REFERENCES

Adams ER, Nolan VG, Andersen SL, Perls TT, Terry DF. 2008. Centenarian offspring: Start healthier and stay healthier. *J Am Geriatr Soc* **56:** 2089–2092.

Ailshire JA, Beltran-Sanchez H, Crimmins EM. 2015. Becoming centenarians: Disease and functioning trajectories of older US Adults as they survive to 100. *J Gerontol A Biol Sci Med Sci* **70:** 193–201.

Altmann-Schneider I, van der Grond J, Slagboom PE, Westendorp RG, Maier AB, van Buchem MA, de Craen AJ. 2012. Lower susceptibility to cerebral small vessel disease in human familial longevity: The Leiden Longevity Study. *Stroke* **44:** 9–14.

Andersen HR, Jeune B, Nybo H, Nielsen JB, Andersen-Ranberg K, Grandjean P. 1998. Low activity of superoxide dismutase and high activity of glutathione reductase in erythrocytes from centenarians. *Age Ageing* **27:** 643–648.

Andersen SL, Sebastiani P, Dworkis DA, Feldman L, Perls TT. 2012. Health span approximates life span among many supercentenarians: Compression of morbidity at the approximate limit of life span. *J Gerontol A Biol Sci Med Sci* **67:** 395–405.

Anderson KM, Odell PM, Wilson PW, Kannel WB. 1991. Cardiovascular disease risk profiles. *Am Heart J* **121:** 293–298.

Anfossi G, Russo I, Bonomo K, Trovati M. 2010. The cardiovascular effects of metformin: Further reasons to consider an old drug as a cornerstone in the therapy of type 2 diabetes mellitus. *Curr Vasc Pharmacol* **8:** 327–337.

Anisimov VN, Bartke A. 2013. The key role of growth hormone-insulin-IGF-1 signaling in aging and cancer. *Crit Rev Oncol Hematol* **87:** 201–223.

Anisimov VN, Berstein LM, Egormin PA, Piskunova TS, Popovich IG, Zabezhinski MA, Tyndyk ML, Yurova MV, Kovalenko IG, Poroshina TE, et al. 2008. Metformin slows down aging and extends life span of female SHR mice. *Cell Cycle* **7:** 2769–2773.

Anisimov VN, Egormin PA, Piskunova TS, Popovich IG, Tyndyk ML, Yurova MN, Zabezhinski MA, Anikin IV, Karkach AS, Romanyukha AA. 2010. Metformin extends life span of HER-2/neu transgenic mice and in combination with melatonin inhibits growth of transplantable tumors in vivo. *Cell Cycle* **9:** 188–197.

Anisimov VN, Berstein LM, Popovich IG, Zabezhinski MA, Egormin PA, Piskunova TS, Semenchenko AV, Tyndyk ML, Yurova MN, Kovalenko IG, et al. 2011. If started early in life, metformin treatment increases life span and postpones tumors in female SHR mice. *Aging (Albany NY)* **3:** 148–157.

Atzmon G, Schechter C, Greiner W, Davidson D, Rennert G, Barzilai N. 2004. Clinical phenotype of families with longevity. *J Am Geriatr Soc* **52:** 274–277.

Atzmon G, Rincon M, Schechter CB, Shuldiner AR, Lipton RB, Bergman A, Barzilai N. 2006. Lipoprotein genotype and conserved pathway for exceptional longevity in humans. *PLoS Biol* **4:** e113.

Atzmon G, Pollin TI, Crandall J, Tanner K, Schechter CB, Scherer PE, Rincon M, Siegel G, Katz M, Lipton RB, et al. 2008. Adiponectin levels and genotype: A potential regulator of life span in humans. *J Gerontol A Biol Sci Med Sci* **63:** 447–453.

Atzmon G, Barzilai N, Hollowell JG, Surks MI, Gabriely I. 2009a. Extreme longevity is associated with increased serum thyrotropin. *J Clin Endocrinol Metab* **94:** 1251–1254.

Atzmon G, Barzilai N, Surks MI, Gabriely I. 2009b. Genetic predisposition to elevated serum thyrotropin is associated with exceptional longevity. *J Clin Endocrinol Metab* **94:** 4768–4775.

Atzmon G, Cho M, Cawthon RM, Budagov T, Katz M, Yang X, Siegel G, Bergman A, Huffman DM, Schechter CB, et al. 2010. Evolution in health and medicine Sackler colloquium: Genetic variation in human telomerase is associated with telomere length in Ashkenazi centenarians. *Proc Natl Acad Sci* **107:** 1710–1717.

Barzilai N, Gabriely I, Gabriely M, Iankowitz N, Sorkin JD. 2001. Offspring of centenarians have a favorable lipid profile. *J Am Geriatr Soc* **49:** 76–79.

Barzilai N, Atzmon G, Schechter C, Schaefer EJ, Cupples AL, Lipton R, Cheng S, Shuldiner AR. 2003. Unique lipoprotein phenotype and genotype associated with exceptional longevity. *JAMA* **290:** 2030–2040.

Barzilai N, Atzmon G, Derby CA, Bauman JM, Lipton RB. 2006. A genotype of exceptional longevity is associated with preservation of cognitive function. *Neurology* **67:** 2170–2175.

Beekman M, Nederstigt C, Suchiman HE, Kremer D, van der Breggen R, Lakenberg N, Alemayehu WG, de Craen AJ, Westendorp RG, Boomsma DI, et al. 2010. Genome-wide association study (GWAS)–identified disease risk alleles do not compromise human longevity. *Proc Natl Acad Sci* **107:** 18046–18049.

Bonafe M, Marchegiani F, Cardelli M, Olivieri F, Cavallone L, Giovagnetti S, Pieri C, Marra M, Antonicelli R, Troiano L, et al. 2002. Genetic analysis of paraoxonase (PON1) locus reveals an increased frequency of Arg192 allele in centenarians. *Eur J Hum Genet* **10:** 292–296.

Bonafe M, Barbieri M, Marchegiani F, Olivieri F, Ragno E, Giampieri C, Mugianesi E, Centurelli M, Franceschi C, Paolisso G. 2003. Polymorphic variants of insulin-like growth factor I (IGF-I) receptor and phosphoinositide 3-kinase genes affect IGF-I plasma levels and human longevity: Cues for an evolutionarily conserved mechanism of life span control. *J Clin Endocrinol Metab* **88:** 3299–3304.

Brown-Borg HM, Borg KE, Meliska CJ, Bartke A. 1996. Dwarf mice and the ageing process. *Nature* **384:** 33.

Burgers AM, Biermasz NR, Schoones JW, Pereira AM, Renehan AG, Zwahlen M, Egger M, Dekkers OM. 2011. Meta-analysis and dose-response metaregression: Circulating insulin-like growth factor I (IGF-I) and mortality. *J Clin Endocrinol Metab* **96:** 2912–2920.

Cabreiro F, Au C, Leung KY, Vergara-Irigaray N, Cocheme HM, Noori T, Weinkove D, Schuster E, Greene ND, Gems D. 2013. Metformin retards aging in *C. elegans* by altering microbial folate and methionine metabolism. *Cell* **153:** 228–239.

Cannon CP, Shah S, Dansky HM, Davidson M, Brinton EA, Gotto AM, Stepanavage M, Liu SX, Gibbons P, Ashraf TB, et al. 2010. Safety of anacetrapib in patients with or at high risk for coronary heart disease. *N Engl J Med* **363:** 2406–2415.

Carmi S, Hui KY, Kochav E, Liu X, Xue J, Grady F, Guha S, Upadhyay K, Ben-Avraham D, Mukherjee S, et al. 2014. Sequencing an Ashkenazi reference panel supports population-targeted personal genomics and illuminates Jewish and European origins. *Nat Commun* **5:** 4835.

Evans CJ, Ho Y, Daveson BA, Hall S, Higginson IJ, Gao W. 2014. Place and cause of death in centenarians: A population-based observational study in England, 2001 to 2010. *PLoS Med* **11:** e1001653.

Florez H, Ma Y, Crandall JP, Perreault L, Marcovina SM, Bray GA, Saudek CD, Barrett-Connor E, Knowler WC. 2011. Parental longevity and diabetes risk in the Diabetes Prevention Program. *J Gerontol A Biol Sci Med Sci* **66:** 1211–1217.

Franceschi C, Olivieri F, Marchegiani F, Cardelli M, Cavallone L, Capri M, Salvioli S, Valensin S, De Benedictis G, Di Iorio A, et al. 2005. Genes involved in immune response/inflammation, IGF1/insulin pathway and response to oxidative stress play a major role in the genetics of human longevity: The lesson of centenarians. *Mech Ageing Dev* **126:** 351–361.

Freudenberg-Hua Y, Freudenberg J, Vacic V, Abhyankar A, Emde AK, Ben-Avraham D, Barzilai N, Oschwald D, Christen E, Koppel J, et al. 2014. Disease variants in genomes of 44 centenarians. *Mol Genet Genomic Med* **2:** 438–450.

Gavrilov L, Gavrilova N, Semyonova V, Evdokushkina G. 2001. Parental age effects on human longevity. In *Annual Meeting of the Gerontological Society of America*, Chicago, November 15–18.

Goldman DP, Cutler D, Rowe JW, Michaud PC, Sullivan J, Peneva D, Olshansky SJ. 2013. Substantial health and economic returns from delayed aging may warrant a new focus for medical research. *Health Aff (Millwood)* **32:** 1698–1705.

Gombar S, Jung HJ, Dong F, Calder B, Atzmon G, Barzilai N, Tian XL, Pothof J, Hoeijmakers JH, Campisi J, et al. 2012. Comprehensive microRNA profiling in B-cells of human centenarians by massively parallel sequencing. *BMC Genomics* **13:** 353.

Graham MJ, Lee RG, Bell TA III, Fu W, Mullick AE, Alexander VJ, Singleton W, Viney N, Geary R, Su J, et al. 2013. Antisense oligonucleotide inhibition of apolipoprotein C-III reduces plasma triglycerides in rodents, nonhuman primates, and humans. *Circ Res* **112:** 1479–1490.

Guevara-Aguirre J, Balasubramanian P, Guevara-Aguirre M, Wei M, Madia F, Cheng CW, Hwang D, Martin-Montalvo

A, Saavedra J, Ingles S, et al. 2011. Growth hormone receptor deficiency is associated with a major reduction in pro-aging signaling, cancer, and diabetes in humans. *Sci Transl Med* **3:** 70ra13.

Han J, Atzmon G, Barzilai N, Suh Y. 2014. Genetic variation in Sirtuin 1 (SIRT1) is associated with lipid profiles but not with longevity in Ashkenazi Jews. *Transl Res* **165:** 480–481.

Holman RR, Paul SK, Bethel MA, Matthews DR, Neil HA. 2008. 10-year follow-up of intensive glucose control in type 2 diabetes. *N Engl J Med* **359:** 1577–1589.

Hong J, Zhang Y, Lai S, Lv A, Su Q, Dong Y, Zhou Z, Tang W, Zhao J, Cui L, et al. 2013. Effects of metformin versus glipizide on cardiovascular outcomes in patients with type 2 diabetes and coronary artery disease. *Diabetes Care* **36:** 1304–1311.

Ismail KNL, Sebastiani P, Milman S, Perls TT, Barzilai N. 2014. Individuals with exceptional longevity exhibit increased health span and delayed onset of age-related diseases. In *Presidential Poster Session of the American Geriatrics Society Annual Meeting*, Orlando, FL, May 15–18.

Johnson JA, Simpson SH, Toth EL, Majumdar SR. 2005. Reduced cardiovascular morbidity and mortality associated with metformin use in subjects with type 2 diabetes. *Diabet Med* **22:** 497–502.

Kahn AJ. 2014. FOXO3 and related transcription factors in development, aging, and exceptional longevity. *J Gerontol A Biol Sci Med Sci* **70:** 421–425.

Karnevi E, Said K, Andersson R, Rosendahl AH. 2013. Metformin-mediated growth inhibition involves suppression of the IGF-I receptor signalling pathway in human pancreatic cancer cells. *BMC Cancer* **13:** 235.

Kenyon C, Chang J, Gensch E, Rudner A, Tabtiang R. 1993. A *C. elegans* mutant that lives twice as long as wild type. *Nature* **366:** 461–464.

Kojima T, Kamei H, Aizu T, Arai Y, Takayama M, Nakazawa S, Ebihara Y, Inagaki H, Masui Y, Gondo Y, et al. 2004. Association analysis between longevity in the Japanese population and polymorphic variants of genes involved in insulin and insulin-like growth factor 1 signaling pathways. *Exp Gerontol* **39:** 1595–1598.

Kopchick JJ. 2003. Discovery and mechanism of action of pegvisomant. *Eur J Endocrinol* **148:** S21–25.

Lamanna C, Monami M, Marchionni N, Mannucci E. 2011. Effect of metformin on cardiovascular events and mortality: A meta-analysis of randomized clinical trials. *Diabetes Obes Metab* **13:** 221–228.

Landman GW, Kleefstra N, van Hateren KJ, Groenier KH, Gans RO, Bilo HJ. 2010. Metformin associated with lower cancer mortality in type 2 diabetes: ZODIAC-16. *Diabetes Care* **33:** 322–326.

Lawrence RW, Evans DM, Cardon LR. 2005. Prospects and pitfalls in whole genome association studies. *Philos Trans R Soc Lond B Biol Sci* **360:** 1589–1595.

Lee MS, Hsu CC, Wahlqvist ML, Tsai HN, Chang YH, Huang YC. 2011. Type 2 diabetes increases and metformin reduces total, colorectal, liver and pancreatic cancer incidences in Taiwanese: A representative population prospective cohort study of 800,000 individuals. *BMC Cancer* **11:** 20.

Lee RG, Crosby J, Baker BF, Graham MJ, Crooke RM. 2013. Antisense technology: An emerging platform for cardiovascular disease therapeutics. *J Cardiovasc Transl Res* **6**: 969–980.

Libby G, Donnelly LA, Donnan PT, Alessi DR, Morris AD, Evans JM. 2009. New users of metformin are at low risk of incident cancer: A cohort study among people with type 2 diabetes. *Diabetes Care* **32**: 1620–1625.

Lipton RB, Hirsch J, Katz MJ, Wang C, Sanders AE, Verghese J, Barzilai N, Derby CA. 2010. Exceptional parental longevity associated with lower risk of Alzheimer's disease and memory decline. *J Am Geriatr Soc* **58**: 1043–1049.

Liu B, Fan Z, Edgerton SM, Yang X, Lind SE, Thor AD. 2011. Potent anti-proliferative effects of metformin on trastuzumab-resistant breast cancer cells via inhibition of erbB2/IGF-1 receptor interactions. *Cell Cycle* **10**: 2959–2966.

Mannick JB, Del Giudice G, Lattanzi M, Valiante NM, Praestgaard J, Huang B, Lonetto MA, Maecker HT, Kovarik J, Carson S, et al. 2014. mTOR inhibition improves immune function in the elderly. *Sci Transl Med* **6**: 268ra179.

Marchegiani F, Marra M, Spazzafumo L, James RW, Boemi M, Olivieri F, Cardelli M, Cavallone L, Bonfigli AR, Franceschi C. 2006. Paraoxonase activity and genotype predispose to successful aging. *J Gerontol A Biol Sci Med Sci* **61**: 541–546.

Martin GM. 2005. Genetic modulation of senescent phenotypes in *Homo sapiens*. *Cell* **120**: 523–532.

Martin-Montalvo A, Mercken EM, Mitchell SJ, Palacios HH, Mote PL, Scheibye-Knudsen M, Gomes AP, Ward TM, Minor RK, Blouin MJ, et al. 2013. Metformin improves healthspan and lifespan in mice. *Nat Commun* **4**: 2192.

Mecocci P, Polidori MC, Troiano L, Cherubini A, Cecchetti R, Pini G, Straatman M, Monti D, Stahl W, Sies H, et al. 2000. Plasma antioxidants and longevity: A study on healthy centenarians. *Free Radic Biol Med* **28**: 1243–1248.

Milman S, Atzmon G, Huffman DM, Wan J, Crandall JP, Cohen P, Barzilai N. 2014. Low insulin-like growth factor-1 level predicts survival in humans with exceptional longevity. *Aging Cell* **13**: 769–771.

Monami M, Colombi C, Balzi D, Dicembrini I, Giannini S, Melani C, Vitale V, Romano D, Barchielli A, Marchionni N, et al. 2011. Metformin and cancer occurrence in insulin-treated type 2 diabetic patients. *Diabetes Care* **34**: 129–131.

Monsalve M, Olmos Y. 2011. The complex biology of FOXO. *Curr Drug Targets* **12**: 1322–1350.

Muzumdar RH, Huffman DM, Atzmon G, Buettner C, Cobb LJ, Fishman S, Budagov T, Cui L, Einstein FH, Poduval A, et al. 2009. Humanin: A novel central regulator of peripheral insulin action. *PLoS ONE* **4**: e6334.

Nebel A, Kleindorp R, Caliebe A, Nothnagel M, Blanche H, Junge O, Wittig M, Ellinghaus D, Flachsbart F, Wichmann HE, et al. 2011. A genome-wide association study confirms *APOE* as the major gene influencing survival in long-lived individuals. *Mech Ageing Dev* **132**: 324–330.

Pawlikowska L, Hu D, Huntsman S, Sung A, Chu C, Chen J, Joyner AH, Schork NJ, Hsueh WC, Reiner AP, et al. 2009. Association of common genetic variation in the insulin/IGF1 signaling pathway with human longevity. *Aging Cell* **8**: 460–472.

Perls TT, Bubrick E, Wager CG, Vijg J, Kruglyak L. 1998. Siblings of centenarians live longer. *Lancet* **351**: 1560.

Perls TT, Wilmoth J, Levenson R, Drinkwater M, Cohen M, Bogan H, Joyce E, Brewster S, Kunkel L, Puca A. 2002. Life-long sustained mortality advantage of siblings of centenarians. *Proc Natl Acad Sci* **99**: 8442–8447.

Pollin TI, Damcott CM, Shen H, Ott SH, Shelton J, Horenstein RB, Post W, McLenithan JC, Bielak LF, Peyser PA, et al. 2008. A null mutation in human *APOC3* confers a favorable plasma lipid profile and apparent cardioprotection. *Science* **322**: 1702–1705.

Quinn BJ, Dallos M, Kitagawa H, Kunnumakkara AB, Memmott RM, Hollander MC, Gills JJ, Dennis PA. 2013. Inhibition of lung tumorigenesis by metformin is associated with decreased plasma IGF-I and diminished receptor tyrosine kinase signaling. *Cancer Prev Res (Phila)* **6**: 801–810.

Rajpathak SN, Liu Y, Ben-David O, Reddy S, Atzmon G, Crandall J, Barzilai N. 2011. Lifestyle factors of people with exceptional longevity. *J Am Geriatr Soc* **59**: 1509–1512.

Rea IM, McKeown PP, McMaster D, Young IS, Patterson C, Savage MJ, Belton C, Marchegiani F, Olivieri F, Bonafe M, et al. 2004. Paraoxonase polymorphisms PON1 192 and 55 and longevity in Italian centenarians and Irish nonagenarians. A pooled analysis. *Exp Gerontol* **39**: 629–635.

Roumie CL, Hung AM, Greevy RA, Grijalva CG, Liu X, Murff HJ, Elasy TA, Griffin MR. 2012. Comparative effectiveness of sulfonylurea and metformin monotherapy on cardiovascular events in type 2 diabetes mellitus: A cohort study. *Ann Intern Med* **157**: 601–610.

Rozing MP, Houwing-Duistermaat JJ, Slagboom PE, Beekman M, Frolich M, de Craen AJ, Westendorp RG, van Heemst D. 2010. Familial longevity is associated with decreased thyroid function. *J Clin Endocrinol Metab* **95**: 4979–4984.

Salani B, Maffioli S, Hamoudane M, Parodi A, Ravera S, Passalacqua M, Alama A, Nhiri M, Cordera R, Maggi D. 2012. Caveolin-1 is essential for metformin inhibitory effect on IGF1 action in non-small-cell lung cancer cells. *FASEB J* **26**: 788–798.

Samaras TT, Elrick H. 2002. Height, body size, and longevity: Is smaller better for the human body? *West J Med* **176**: 206–208.

Sanders AE, Wang C, Katz M, Derby CA, Barzilai N, Ozelius L, Lipton RB. 2010. Association of a functional polymorphism in the cholesteryl ester transfer protein (CETP) gene with memory decline and incidence of dementia. *JAMA* **303**: 150–158.

Sandhu MS, Heald AH, Gibson JM, Cruickshank JK, Dunger DB, Wareham NJ. 2002. Circulating concentrations of insulin-like growth factor-I and development of glucose intolerance: A prospective observational study. *Lancet* **359**: 1740–1745.

Schoenmaker M, de Craen AJ, de Meijer PH, Beekman M, Blauw GJ, Slagboom PE, Westendorp RG. 2006. Evidence of genetic enrichment for exceptional survival using a family approach: The Leiden Longevity Study. *Eur J Hum Genet* **14**: 79–84.

Sebastiani P, Solovieff N, Dewan AT, Walsh KM, Puca A, Hartley SW, Melista E, Andersen S, Dworkis DA, Wilk

JB, et al. 2012. Genetic signatures of exceptional longevity in humans. *PLoS ONE* **7**: e29848.

Sebastiani P, Sun FX, Andersen SL, Lee JH, Wojczynski MK, Sanders JL, Yashin A, Newman AB, Perls TT. 2013. Families enriched for exceptional longevity also have increased health-span: Findings from the Long Life Family Study. *Front Public Health* **1**: 38.

Seibel SA, Chou KH, Capp E, Spritzer PM, von Eye Corleta H. 2008. Effect of metformin on IGF-1 and IGFBP-1 levels in obese patients with polycystic ovary syndrome. *Eur J Obstet Gynecol Reprod Biol* **138**: 122–124.

Smith DL Jr, Elam CF Jr., Mattison JA, Lane MA, Roth GS, Ingram DK, Allison DB. 2010. Metformin supplementation and life span in Fischer-344 rats. *J Gerontol A Biol Sci Med Sci* **65**: 468–474.

Suh Y, Atzmon G, Cho MO, Hwang D, Liu B, Leahy DJ, Barzilai N, Cohen P. 2008. Functionally significant insulin-like growth factor I receptor mutations in centenarians. *Proc Natl Acad Sci* **105**: 3438–3442.

Tabatabaie V, Surks MI. 2013. The aging thyroid. *Curr Opin Endocrinol Diabetes Obes* **20**: 455–459.

Tan Q, Christiansen L, Bathum L, Li S, Kruse TA, Christensen K. 2006. Genetic association analysis of human longevity in cohort studies of elderly subjects: An example of the *PON1* gene in the Danish 1905 birth cohort. *Genetics* **172**: 1821–1828.

Tazearslan C, Huang J, Barzilai N, Suh Y. 2011. Impaired IGF1R signaling in cells expressing longevity-associated human *IGF1R* alleles. *Aging Cell* **10**: 551–554.

Tosca L, Rame C, Chabrolle C, Tesseraud S, Dupont J. 2010. Metformin decreases IGF1-induced cell proliferation and protein synthesis through AMP-activated protein kinase in cultured bovine granulosa cells. *Reproduction* **139**: 409–418.

Tseng CH. 2012. Diabetes, metformin use, and colon cancer: A population-based cohort study in Taiwan. *Eur J Endocrinol* **167**: 409–416.

Van Heemst D, Beekman M, Mooijaart SP, Heijmans BT, Brandt BW, Zwaan BJ, Slagboom PE, Westendorp RG. 2005. Reduced insulin/IGF-1 signalling and human longevity. *Aging Cell* **4**: 79–85.

Vergani C, Lucchi T, Caloni M, Ceconi I, Calabresi C, Scurati S, Arosio B. 2006. I405V polymorphism of the cholesteryl ester transfer protein (CETP) gene in young and very old people. *Arch Gerontol Geriatr* **43**: 213–221.

Westendorp RG, van Heemst D, Rozing MP, Frolich M, Mooijaart SP, Blauw GJ, Beekman M, Heijmans BT, de Craen AJ, Slagboom PE. 2009. Nonagenarian siblings and their offspring display lower risk of mortality and morbidity than sporadic nonagenarians: The Leiden longevity study. *J Am Geriatr Soc* **57**: 1634–1637.

Whittington HJ, Hall AR, McLaughlin CP, Hausenloy DJ, Yellon DM, Mocanu MM. 2013. Chronic metformin associated cardioprotection against infarction: Not just a glucose lowering phenomenon. *Cardiovasc Drugs Ther* **27**: 5–16.

The Companion Dog as a Model for the Longevity Dividend

Kate E. Creevy[1], Steven N. Austad[2], Jessica M. Hoffman[3], Dan G. O'Neill[4], and Daniel E.L. Promislow[5]

[1]Department of Small Animal Medicine and Surgery, College of Veterinary Medicine, University of Georgia, Athens, Georgia 30602

[2]Department of Biology, University of Alabama at Birmingham, Birmingham, Alabama 35294

[3]Department of Genetics, University of Georgia, Athens, Georgia 30602

[4]Veterinary Epidemiology, Economics and Public Health, The Royal Veterinary College, Hatfield, Herts AL9 7TA, United Kingdom

[5]Departments of Pathology and Biology, University of Washington, Seattle, Washington 98195

Correspondence: promislo@uw.edu

The companion dog is the most phenotypically diverse species on the planet. This enormous variability between breeds extends not only to morphology and behavior but also to longevity and the disorders that affect dogs. There are remarkable overlaps and similarities between the human and canine species. Dogs closely share our human environment, including its many risk factors, and the veterinary infrastructure to manage health in dogs is second only to the medical infrastructure for humans. Distinct breed-based health profiles, along with their well-developed health record system and high overlap with the human environment, make the companion dog an exceptional model to improve understanding of the physiological, social, and economic impacts of the longevity dividend (LD). In this review, we describe what is already known about age-specific patterns of morbidity and mortality in companion dogs, and then explore whether this existing evidence supports the LD. We also discuss some potential limitations to using dogs as models of aging, including the fact that many dogs are euthanized before they have lived out their natural life span. Overall, we conclude that the companion dog offers high potential as a model system that will enable deeper research into the LD than is otherwise possible.

In a 2006 paper in *The Scientist*, Olshansky and colleagues introduced the notion of the longevity dividend (LD), which is defined as the social, economic, and health bonuses for both individuals and populations that accrue as a result of medical interventions to slow the rate of human aging (Olshansky et al. 2006). These investigators posit that, although permanently curing one or two chronic age-related diseases (e.g., heart disease, cancer) might lengthen both life span and health span a bit, a much more potent effect would result from slowing the underlying aging processes, which would reduce the risk and/or delay the age of onset for most or all age-related maladies. These maladies include fatal and disabling diseases but also include nonfatal conditions, such as joint pain, hearing loss, dementia, or muscle weakness, which can degrade the quality of life.

The LD hypothesis rests on two key assumptions, one pathophysiological and the other demographic. First, the LD hypothesis assumes that if we ameliorate the underlying biological processes that drive aging, then necessarily we will reduce the frequency or delay the onset of most or all age-related disorders. Second, if we succeed in retarding the rate of aging, the period of debility toward the end of life will be compressed or at least maintained rather than lengthened. That is, we will extend the healthy phase of life without also extending the unhealthy phase.

Empirical evidence supporting both of these assumptions is mixed. In model laboratory species, interventions that increase longevity have been shown to retard some age-related functional declines, but to exacerbate others. For instance, with dietary restrictions (DRs) in laboratory mice, the poster child for extended health span studies, many aspects of health are in fact extended (Weindruch and Walford 1988). However, there are also some health downsides, such as increased susceptibility to some infectious diseases (Gardner 2005; Goldberg et al. 2015) and slowed wound healing throughout life (Reiser et al. 1995), not just near its end. Also, many genetic and environmental alterations that extend life span are associated with decreased fertility (Austad 2014). Although compromised fertility might not be directly related to mortality, it is certainly an important measure of the functional limitations of aging that Fries (1980) discussed in his classic paper on the compression of morbidity. A recent study in the laboratory nematode, *Caenorhabditis elegans*, was really the first to attempt to define worm health comprehensively and found while some—but by no means all—of four common worm longevity mutations increased the period of healthy life, all of these mutations also increased the duration of unhealthy life (Bansal et al. 2015).

Even what we know about humans, whose life spans have nearly doubled in the past 180 years (Oeppen and Vaupel 2002), is complicated. Mortality in general, and especially among the elderly, has continued to decline in virtually all developed countries, including the United States. This decline has been accompanied by prolonged health and morbidity compression in a number of countries (Thatcher et al. 2010). However, some indicators of later-life health, such as functional mobility, appear to be have worsened since the last years of the 20th century (Crimmins and Beltran-Sanchez 2011), and this is a particularly worrisome trend among cohorts of the "near elderly" (Freedman et al. 2013). Additionally, in global surveys over the past 20 years, life expectancy at birth appears to be increasing more rapidly than healthy life expectancy, suggesting that more unhealthy years of life may be in our future (Salomon et al. 2012).

These debates make clear that to fully appreciate the consequences of extended longevity, both beneficial and deleterious, we need to develop a comprehensive understanding of the underlying causes of mortality, of how risks associated with these causes change with age, and of the impact on health span of reducing or eliminating cause-specific mortality as contrasted with an overall slowing of the rate of aging.

These challenges are addressed throughout this collection, including discussion of the potential to study the LD in common laboratory model species, such as the nematode worm *C. elegans*, the fruit fly *Drosophila melanogaster*, and the house mouse *Mus musculus*. Although these model organisms have obvious strengths for providing insight into the dynamics of life span versus health span, they also suffer from various limitations. Especially in worms and flies, we have a limited understanding of the underlying causes of death and our measures of health in these species are also not well developed, tending to focus on traits such as stress resistance, locomotion, mating behavior, or feeding rates (Burger and Promislow 2006; Bansal et al. 2015). In mice, the limitations are somewhat different. Although mouse postmortem histopathology is quite sophisticated, determination of health status in living animals is crude and still being developed. Currently, mouse health assays mainly use tools that were developed to assess genetically induced diseases. Mice that are commonly used in research also die of a restricted range of causes that do not reflect common causes of human deaths. For instance, mice do not spontaneously develop atherosclerosis and seldom develop neurodegenerative diseases. Even the common cancers that spontaneously develop in mice differ considerably from the common cancers in people. Mouse cancers are primarily lymphomas and sarcomas rather than the carcinomas that dominate the cancer spectrum in aging humans (Chang 2005).

In this review, we explore the possibility that companion dogs might offer an excellent model to study the potential of the LD to increase human health span. Dogs offer a number of strengths relative to existing animal model systems. First, our understanding of age-specific morbidity and mortality in companion dogs, and the medical infrastructure to treat their diseases, are second only to that of humans. Second, variation between breeds in their frequency of different age-specific diseases points to a strong genetic component. Third, the complex nature of dog behavior and physiology, both healthy and pathological, and the fact that these companion animals share our homes and our daily routines, suggests that we can develop meaningful measures of health span in companion dogs that will also be relevant in humans. Fourth, the shorter average life spans of dogs, ~12 years (O'Neill et al. 2013) compared with humans means that we can obtain results much faster, enabling speedier recommendations for improved human health. And, finally, there is value to understanding aging in dogs not only as a model for humans, but also because dogs are enormously valued as companions and owners want their older dogs to be healthy.

DOGS, AGING, AND THE LONGEVITY DIVIDEND

Dogs have been living in association with humans for at least 15,000 years and likely were domesticated multiple times across Eurasia from groups of wolves that foraged on refuse from human encampments (Larson et al. 2012; Freedman et al. 2014). Millennia of co-evolution followed, such that humans became particularly sensitive to dogs' postural communication and vice versa (Miklosi and Soproni 2006; Kaminski et al. 2012), and dogs have been used in a vast array of human activities from hunting to herding to bomb, drug, and cadaver detection to assistance for the disabled to simple companionship.

Variability

Thanks to selective breeding for particular purposes and the aesthetics of breed development, dogs stand out as the most phenotypically variable mammal on earth. From even the most cursory observation of dogs playing in the park, the breadth of both behavioral and morphological diversity within this species is striking. Equally impressive is the demographic variation that we see between different dog breeds, with life expectancy varying from ~6 years to 16 years (Egenvall et al. 2005a; Adams et al. 2010; O'Neill et al. 2013).

This variation points to a strong genetic basis not only for visible phenotypes, but also for longevity and disease predisposition. Moreover, breed-specific risks of morbidity and mortality are strongly associated with breed-specific differences in size (Li et al. 1996; Greer et al. 2007; Kraus et al. 2013). Unlike the pattern across mammalian species, in which larger species tend to live longer (Austad and Fischer 1991; Promislow 1993), within the domesticated dog, the smaller breeds, such as Chihuahuas and Toy Poodles, typically live longer than the largest breeds, such as Great Danes and Irish Wolfhounds (Kraus et al. 2013). Indeed, in dogs, on average, an increase in size by 10 kg is associated with 6 months to a year of reduced life span (Greer et al. 2007; O'Neill et al. 2013). A similar within-species pattern of exceptionally small genotypes living longer than larger genotypes is found in mice, rats, and horses (Rollo 2002; Miller and Austad 2006).

Just why small breeds or genotypes within species tend to live longer than large breeds or genotypes is not yet fully understood. In mice, a broad swath of evidence implicates circulating levels of insulin-like growth factor 1 (IGF-1) as a major contributor (Tatar et al. 2003; Yuan et al. 2009). It turns out that much of the body size variation among dogs is explained by polymorphism in the IGF-1 gene (Sutter et al. 2007; Greer et al. 2011). This hints, but does not yet prove, that like in mice, IGF-1 signaling might also be a major contributor to longevity differences in dogs.

To fully understand the potential for an LD, we need to understand much more than just the genetics of longevity. We need to understand details of the relationship between morbidity and mortality as well as their underlying causes. The companion dog has the potential to contribute greatly in this area because of the wide breed-specific variation in both morbidity and mortality (Bonnett et al. 2005; Egenvall et al. 2005a; Fleming et al. 2011; Kraus et al. 2013).

Mortality

From a purely demographic perspective, the study of aging is the study of population patterns in age-specific mortality rates. The LD highlights the importance of measuring aging in terms of age-related changes in pathophysiological processes and comorbidities. But, at present, the state-of-the-art focuses on age-specific mortality rates, so let us start with a bit of background and mortality rate analysis. The LD hypothesis assumes that the rate of aging can be slowed. What, precisely, is the rate of aging? Biodemographers commonly focus on the "actuarial rate of aging," which is determined by age-specific mortality rates. Age-specific mortality rates can be intuitively thought of as crude measures of population health. If, say, 1% of 25-year-olds die in one population and 10% of 25-year-olds die in another, then the latter population appears to be substantially less healthy at age 25, assuming the two populations are in comparable environments. In almost every animal species studied, including humans, adult mortality rates increase with age in a pattern described effectively by the Gompertz model (Gompertz 1825; Promislow 1991; Finch 1990). The Gompertz model describes μ_x, the

mortality rate at age x, using two parameters: a "baseline" mortality rate α, usually the lowest mortality rate achieved in adulthood, which occurs around the time of sexual maturity, and an age-dependent parameter β:

$$\mu_x = \alpha e^{\beta x}. \tag{1}$$

If we plot the logarithm of mortality rate against age, the Gompertz curve is a straight line, where the actuarial rate of aging is given by β, which is the slope of this line. In contrast, α can be thought of as the "intercept" or "elevation" of this line. Intuitively, the Gompertz slope measures the rate of increasing susceptibility to death. A steep slope indicates organisms that deteriorate quickly. A slightly more detailed equation, known as the Siler model, includes a common pattern of initially high juvenile mortality that declines with age (Fig. 1).

From a Gompertzian perspective, longevity in a population can be increased by decreasing α, which lowers mortality by an equal proportion at all ages, or by decreasing β, which reduces the slope, or a combination of both. Biodemographers often focus on decreasing the slope as the only valid metric of a slowed aging rate, because the slope is a measure of how fast things deteriorate. Although this is a valid demographic point, from a health span perspective, it may be preferable to lower α. Assuming that the absolute mortality rate at any age is an indicator of population health at that age, then lowering α increases health at all ages by an equivalent amount, whereas decreasing β increases the health of the young by a small amount and of the old by a larger amount. Best of all, of course, would be to lower both.

Interestingly, whether genetic or environmental interventions that lengthen life do so by lowering α versus β might depend on the species in question. Such interventions in flies mostly lower α (Promislow et al. 1996), whereas in worms they mostly lower β (Chen et al. 2007). In mice, it depends on the intervention—some lower one parameter, some the other (de Magalhaes et al. 2005). In humans, most of the increase in life span seen over the past two centuries is caused by a dramatic re-

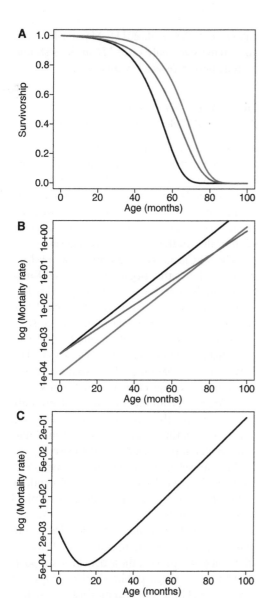

Figure 1. Survivorship curves (*A*) and age-specific mortality curves (*B*) for a population following a Gompertz mortality model (Gompertz 1825). According to the Gompertz mortality model, mortality increases exponentially over time, with $\mu_x = \alpha e^{\beta x}$, where μ_x is the mortality rate at age x, α is the "baseline" mortality rate, and β is the rate of aging. The black line shows standard mortality. Survivorship can be increased (or mortality lowered) either by reducing the baseline mortality (red) or by slowing the rate of aging (blue). Most real-world examples show a much greater reduction in α than in β. (*C*) The more realistic Siler mortality model (Siler 1979) is shown, where $\mu_x = \alpha_1 e^{-\beta_1 x} + \alpha_2 + \alpha_3 e^{\beta_3 x}$.

duction in early-age mortality, such that we see a large decrease in intercept but an increase in slope (Burger et al. 2012).

Where does the domestic dog fit into this pattern? Is the longevity advantage in small breeds compared with large ones primarily owing to a decrease in the Gompertz slope or intercept? A recent study by one of us suggests that variation in longevity among dog breeds is primarily a function of changing the slope (Kraus et al. 2013). The study relied on data from the Veterinary Medical Database (VMDB) in which exact age at death for the dogs is unknown. Ages at death in the VMDB are placed into bins of varying sizes. For older dogs, one bin comprises animals that died at any time between 10 and 15 years of age, and dogs older than that are simply listed as 15+. As other data sets with more detailed age-at-death information are compiled, such as VetCompass (O'Neill et al. 2013), we will be able to more accurately determine exactly how rates of aging vary between breeds, assuming that we can come to an agreement on a definition of aging.

Does it matter for the LD whether longevity is enhanced by lowering α versus lowering β? In its original formulation, a goal of the LD is "a modest deceleration in the rate of aging sufficient to delay all aging-related diseases and disorders by about seven years" (Olshansky et al. 2006, p. 32). This would seem to indicate that a lower slope was the goal. On the other hand, 7 years was chosen as a target, because the mortality rate in humans doubles approximately every 7 years. Thus, one could imagine that the investigators envisioned the LD occurring because all aging-related diseases and disorders were delayed by 7 years, halving the mortality rate at every age—reducing α, in other words, to what it had been 7 years earlier in life (Olshansky et al. 2006). Thus, 50-year-olds would, in principle, achieve the health that 43-year-olds formerly enjoyed, and 57 would be the new 50. This sounds a lot like retarded aging to many of us.

Morbidity

Any reasonable goal of health-related research should include improving and prolonging health. Let us admit that age-specific mortality is a very crude measure of population health. Its related concept, morbidity, describes the consequence of myriad diseases that can significantly affect frailty and health span. A full understanding of the potential for an LD requires that we understand age-specific risks of mortality and morbidity. Dog breeds vary not only in "when" they die, they also vary tremendously in "why" they die. For the common dog breeds, there is an extensive literature on common ailments and causes of death. For example, Doberman Pinschers have high rates of morbidity and mortality caused by cardiomyopathy, Dachshunds suffer from intervertebral disc disease, and Miniature Schnauzers develop diabetes more often than other common breeds (Hess et al. 2000; Wess et al. 2010). And even for rarer breeds, we can identify highly specific "morbidity profiles" for each breed from large health databases that provide sufficient study sample sizes (Fleming et al. 2011).

Health is not simply delaying the onset or slowing the progression of fatal or potentially fatal diseases. As Bellows et al. (2015) make clear in a recent comprehensive review of healthy aging, many nonfatal maladies of aging, such as joint pain, hearing and vision loss, and muscle weakness, are common in aging dogs. Of particular relevance to this review, we find that just as in humans there are many fatal and nonfatal maladies in dogs that increase with age. As Waters (2011) has noted, what we can measure in humans, we can measure in dogs. So identifying and exploring these maladies in dogs should be as successful as it is for humans. This illustrates the possibility to examine whether slowing aging in dogs might delay the onset, or reduce the frequency, of cancer, heart disease, and diabetes as well as reduce joint pain and maintain mobility and sensory function.

As we noted previously, body size accounts for much of the variation in the timing of death across dog breeds. The same is true for the onset of many age-related diseases or conditions. Cancer incidence increases dramatically with age in all breeds, but it occurs later in smaller breeds compared with large ones (Bonnett et al. 2005; Fleming et al. 2011). The same is true of cataracts (Urfer et al. 2011). Moving forward, comprehensive analyses of LD in dogs should include body size as a key covariate.

DOES WHAT WE KNOW ABOUT DOG AGING SUPPORT THE LD HYPOTHESIS?

To set the stage for LD, Olshansky et al. (2006) first noted that even if one were to eliminate the diseases that top the list as causes of mortality in human populations, one would see only a modest life span extension. To determine whether the same is true in dogs, we explored five causes of mortality in the VetCompass data set.

The VetCompass Programme is a novel research initiative that began in the United Kingdom and is now being extended worldwide. The philosophy is to capture the cumulative veterinary clinical experience via a large-scale collection of veterinary electronic clinical records that are made accessible for research. The VetCompass Programme (see www.rvc.ac.uk/vetcompass) was developed at the Royal Veterinary College in conjunction with the University of Sydney with the aim of sharing, analyzing, and disseminating veterinary clinical information to develop an evidence database on the health and welfare of companion animals. VetCompass began collecting data in 2009 and, by 2014, almost 500 veterinary clinics in the United Kingdom have already shared data on more than two million dogs, which offer detailed health experiences covering up to 5 years of life for individual dogs. To date, these data have supported 12 peer-reviewed publications on dogs that cover general longevity and morbidity as well as specific disorders.

The most common cause of death, cancer, kills one in every six dogs in the VetCompass database. And yet, even if we were to eliminate all deaths caused by cancer, the increase in life span is modest (Fig. 2A). In contrast, if we slow aging by 10%, a relatively small amount compared with what we can already accomplish in the laboratory in yeast, worms, flies, and mice, we see a similar increase in life span (Fig. 2B,C).

But companion dogs can also allow us to think more carefully about LD. Recall that the LD hypothesis assumes that ameliorating the

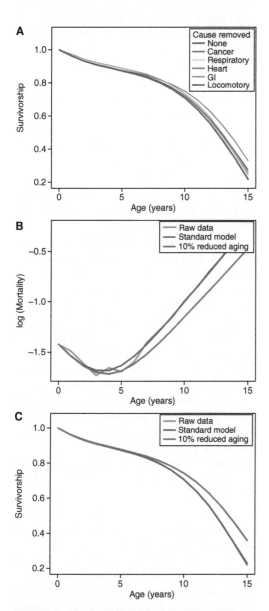

Figure 2. Longevity dividend (LD) calculations for dogs from the VetCompass database (O'Neill et al. 2013). (*A*) Survival curves for all dogs (blue line) and hypothetical survival curves for populations in which a single cause of mortality is omitted (see text for details). All-cause mortality curves (*B*) and survivorship curves (*C*) are shown for actual data (yellow), and based on fitted mortality curves using the Siler model (Siler 1979), $\mu_x = \alpha_1 e^{-\beta_1 x} + \alpha_2 + \alpha_3 e^{\beta_3 x}$, for all mortality (blue) and for reduced aging rate, where β_3 is the rate of aging and has been reduced by 10% (red).

underlying processes of aging will (1) delay the onset and/or slow the progression of aging-related diseases and conditions, and (2) compress the period of ill health at the end of life. We have already seen that many smaller dog breeds live substantially longer than the largest dog breeds, and these small breeds also develop many aging-related diseases at later ages than large breeds. This is clearly consistent with the LD hypothesis. However, we currently have no evidence about the relative length of illness or morbidity late in life among the various breeds. A study to determine whether there is morbidity compression among smaller breeds would be a valuable addition to this field of knowledge.

One complication in interpreting the LD perspective is that the prevalence of some diseases, such as infectious diseases, decreases with age in dogs (Fleming et al. 2011). Is it possible that retarding the rate of aging might prolong the period of susceptibility to diseases most likely to strike early in life? For instance, parvoviral enteritis is an often-fatal disease that preferentially strikes young dogs. There is now an effective vaccine, but before the widespread use of that vaccine, most cases of parvoviral enteritis struck dogs <6 months of age and very few cases occurred in dogs older than 2 years (Mason et al. 1987). This situation is complicated to interpret for several reasons. First, because the most susceptible dogs were not sexually mature and therefore not yet actually aging according to most interpretations, its relevance to the LD hypothesis is not obvious. Interventions thought to slow aging are unlikely to be imposed before sexual maturity. Second, this increased susceptibility may be because of an immature, prepubescent immune system, so this is also difficult to reconcile with the LD hypothesis. On the other hand, interventions that have successfully lengthened life in laboratory animals, such as DR or rapamycin, have been associated with some signs of immune suppression (Gardner 2005; Nikolich-Zugich and Messaoudi 2005; Ritz et al. 2008; Goldberg et al. 2015), potentially interpretable as a return to an immature immune system. Thus, slowing at least some aspects of aging and extending "potential" longevity could potentially increase some diseases early in life. In thinking about the LD then, it becomes imperative that we have a comprehensive understanding not only of the effects of age on different pathophysiological systems and processes, but also how specific interventions influence these systems and processes. A necessary step in the development of interventions that may lead to an LD must include assessment of the period in the life span that the intervention is likely to be most effective at reducing both early- and late-life diseases and debilities. As a case in point, which has been extensively studied in dogs, we now turn to the effects of sterilization on longevity and patterns of disease.

Reproduction, Sterility, and the LD

Companion animals in the United States are often surgically sterilized at the request of their owners to prevent unwanted reproduction or to limit unwanted behaviors. Surgical sterilization has been associated with longer life in a number of large studies on multibreed/mixed breed dogs (Bronson 1982; Michell 1999; Hoffman et al. 2013; O'Neill et al. 2013) as well as in human males (Hamilton and Mestler 1969; Min et al. 2012). Consequently, a large and accessible population of dogs is available to assess the health impact of this longevity-associated procedure.

The most comprehensive assessment of how surgical sterilization affects the causes of death in dogs is that by Hoffman et al. (2013). In that study of more than 40,000 companion dogs in the VMDB known to have died in veterinary teaching hospitals, sterilization was associated with a 19% increase in longevity (14% in males, 26% in females). The VMDB is limited to animals presented to tertiary care hospitals, which could lead to numerous biases (Fleming et al. 2011; Hoffman et al. 2013). However, a contemporaneous study in the United Kingdom using VetCompass data, based on dogs presented to primary care veterinary clinics, reported similar longevity benefits of neutering, at least in females (O'Neill et al. 2013, 2015). Hoffman et al. found—as have a number of previous studies—a higher prevalence of death by cancer

in sterilized compared with intact dogs and also a greater prevalence of deaths owing to immune-mediated diseases (see also Hart et al. 2014). On the other hand, sterilized dogs were less likely to die from infectious diseases, trauma, vascular, and degenerative diseases compared with intact dogs. Although we would expect longer-lived (sterilized) dogs to die from diseases that are more prevalent at late age (e.g., cancer), even after controlling for this potential confound statistically, these patterns were still evident. Interestingly, in a study of sterilized versus intact people, deaths from infectious diseases were also reduced in the sterilized group (Hamilton and Mestler 1969).

The "timing" of sterilization vis-à-vis an animal's life history can also be important in the development of diseases. As sterilization removes an entire organ system and its endocrine axis, it would stand to reason that whether that system is removed before adulthood, when the endocrine axis becomes particularly active, might be expected to have large effects throughout life. In the United States, dogs are traditionally sterilized around the time of sexual maturity or shortly thereafter, but debate exists over the relative health costs and benefits of pre-versus postmaturation sterilization. As a gross generalization, dogs are often considered to be sexually mature by 1 year of age. This generalization masks a great deal of breed and individual variation, however, as sexual maturity in dogs varies according to body size. Small dogs can enter puberty as early as 5 to 6 months and some large dogs as late as 2 years of age (Rice 2008).

A number of studies have examined the effects of the timing of sterilization on subsequent health and/or behavior. One of these found that in dogs adopted from animal shelters, animals categorized as having been sterilized at less than 6 months of age (median = 2.5 months) were more likely to contract parvoviral enteritis than dogs sterilized later (median = 1 year) (Howe et al. 2001). However, this result is confounded by the fact that parvoviral enteritis is more common in puppies compared with adults as noted before. Breed status of the animals was not identified in this study and the median length of follow-up was only 4 years after sterilization;

therefore, later-life health problems could not be ascertained. A much larger study—also of dogs adopted from shelters, also without breed identification, also with ~4 years of follow-up—found that early (<5.5 months) sterilization increased the prevalence of hip dysplasia and, in females only, was also associated with more incontinence and cystitis (Spain et al. 2004). It should be emphasized that these studies only examined the variable of age at sterilization and not the variable of sterilization itself; there were no intact animals for comparison.

The timing of sterilization and its comparison with nonsterilized dogs has also been examined in several studies. Specifically, a large study of Golden Retrievers, which are particularly predisposed to cancer (Fleming et al. 2011), found that sterilization itself as well as the timing of sterilization affected the type of cancers later developed (Torres de la Riva et al. 2013). Specifically, sterilization at any age increased the prevalence of mast cell tumors in females only, although late-neutered (>2 months) females had a higher prevalence of hemangiosarcoma than either intact or early-neutered females. Early neutering of males increased the rate of lymphosarcoma relative to either intact or late-neutered males. Early neutering also increased the rate of skeletal abnormalities, such as hip dysplasia and cranial cruciate ligament tears.

A study of Rottweilers has examined how the status and timing of sterilization affects longevity and cause of death (Waters et al. 2009). Like some other studies, this one found that female dogs lived longer than males (Egenvall et al. 2005b). Unlike a number of multibreed, multisex studies, however, this one found that sterilizing female Rottweilers before 4.5 years of age shortened life. Moreover, females in a group of animals chosen for the study because of their exceptional longevity (>30% greater than breed longevity average) were more likely than females of average life span to have retained their ovaries for more than 4 years. The longest-lived dogs were also at a considerably reduced risk of dying from cancers. A conclusion from this study then could be that exposure to ovarian hormones has a generally beneficial effect on longevity. As the results of this study

are at variance with multiple other longevity studies that did not include the timing of sterilization, it is worth considering some possible reasons for the difference. Most obviously, this result could be a particular feature of the Rottweiler breed. Second, this result could be a function of the selection of the study population. Among the usual longevity dogs in this group, 73% died of cancer. This is a substantially higher incidence of fatal cancers than reported for dogs as a whole (Bronson 1982) or for other populations of Rottweilers (Hoffman et al. 2013). Third, the timing of sterilization was very different in the usual longevity group (>40% sterilized within the first year of life) and the exceptional longevity group (40% sterilized by about 3 years of age) suggesting that the dogs may have been treated quite differently in terms of medical attention or weight maintenance or the quality and amount of their food. Note that exceptional longevity dogs were on average 5 kg lighter than usual longevity dogs, died considerably less often of cancer, including osteosarcoma to which Rottweilers are particularly prone, and had a much lower incidence of gastrointestinal deaths among noncancer deaths than usual longevity dogs. Provocatively, there were many more deaths from "frailty," in the exceptionally long-lived dogs. Frailty in this case was defined as a "combination of age-related disabilities, including deficits in mobility, cognition, hearing, eyesight, and inability to maintain body weight" (Waters et al. 2009, p. 753). This description does not seem particularly compatible with the LD. At present, it is not possible to choose among these various interpretations of the Rottweiler study.

Longevity is a composite variable of myriad events and susceptibilities. Although the net effect of sterilization is to increase overall life span in dogs, this longevity comes about through complex effects on causes of death in specific dog genotypes. Can health span become the tool by which we measure the overall costs and benefits of an intervention that decreases the risk of some causes of mortality at the expense of increasing the rate of other causes? Can we determine that the causes of death or length of debility that become more prevalent in longer-lived individuals detract from the value of those added months or years of life? The ability to evaluate dogs as individuals with respect to physiological and biochemical parameters, cognitive function, activity level, task performance, longevity, genetics, pharmacogenomics, and so on, suggests that dogs could be a powerful model for the creation of a standardized assessment of health span. Moreover, companion dogs offer the potential to identify interventions that could improve health span, providing an LD for both companion animals and their owners.

Delayers, Escapers, Survivors, and the LD

In a previous section, we highlighted age-related changes in various causes of death, including cancer, in a large population of dogs. As in humans, mice, rats, and virtually every other vertebrate species, dogs generally show a dramatic increase in risk of cancer with age (Finch 1990). But there is a subtler pattern in the cancer data that recapitulates another pattern seen in humans. In studies of human centenarians—those paragons of successful aging—researchers have suggested that three routes to exceptional longevity can be observed. There are people who live long despite being diagnosed with age-associated illnesses earlier in life, called Survivors; those who escape age-associated illnesses until late in life, called Delayers; and those who seem to escape most age-associated diseases to the very end of life, called Escapers (Evert et al. 2003). A study in Rottweilers suggests that they, too, might include Delayers or Escapers (Cooley et al. 2003). Within a cohort of 345 animals, among the oldest dogs, cancer rates were actually lower than in younger dogs. It is as though the oldest dogs made up a pool of high-quality Escapers that were comparatively immune to cancer. If a young cohort includes dogs of variable quality, these "Escapers" might simply reflect the highest quality individuals that were both likely to live the longest and were inherently resistant to cancer and other diseases.

To determine if this is true in other dog breeds, we examined age-specific patterns of cancer prevalence in the Veterinary Medical Database (see wvdb.org) for a subset of very

common breeds. Interestingly, we saw the same pattern in our data set for large-breed, relatively short-lived dogs. To our surprise, however, we did not observe the same pattern in smaller, longer-lived breeds (Fig. 3). This observation could be because of the fact that the highest two age classes in the VMDB (10–15 years and 15+ years) represent a large span of age collapsed into only two categories, so might mask underlying age-specific trends in the smaller breeds in which many individuals survive to these ages. Within the VetCompass data, which benefits from offering the precise age at death for all individual dogs, we observe a pattern of all breeds showing decreased rates of cancer in the later ages regardless of size (Fig. 3). However, at least for now, the VetCompass sample sizes for each age class are much smaller than the VMDB, such that late age data is relatively noisy.

From the perspective of the LD, we are left to wonder whether an increase in life expectancy is accompanied by an increase in the frequency of Escapers and/or Delayers, which would, indeed, pay an LD. Alternatively, it could simply mean more Survivor individuals living with, and suffering from, age-related morbidities.

LIMITATIONS OF THE DOG AS A MODEL FOR LD

The companion dog presents us with a powerful model to understand the genetic and environmental determinants of age-specific morbidity and mortality. However, as a translational model, dogs do have some important limitations. First, although there is extensive overlap between causes of death in humans and dogs (e.g., Waters 2011), there are also many differences. For example, the spectrum of cancers that is common in dogs includes some cancers that are rare in humans (e.g., hemangiosarcoma, mast cell tumor), although some are relatively common in children (e.g., osteosarcoma).

Second, euthanasia is a common end-of-life decision in dogs, in contrast with humans. A longevity study of dogs in the United Kingdom identified that euthanasia accounted for 86% of deaths, while only 14% of deaths were unassisted (O'Neill et al. 2013). This means

Cite this article as *Cold Spring Harb Perspect Med* doi: 10.1101/cshperspect.a026633

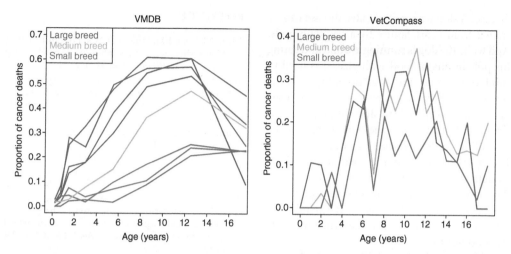

Figure 3. Proportion of dogs dying of cancer for specific breeds. (*Left*) Veterinary Medical Database (VMDB) data. (*Right*) VetCompass data. Blue lines indicate large breed dogs (VMDB: Labrador Retriever, Golden Retriever, Rottweiler, and Boxer; VetCompass: Labrador Retriever, Golden Retriever, Rottweiler, and German Shepherd). Orange line indicates medium breed dogs (VMDB: Beagle; VetCompass: Border Collie, American Staffordshire Terrier, English Springer Spaniel, Cocker Spaniel), and the red lines indicate small-breed dogs (VMDB: Miniature Poodle, Dachshund, and Yorkshire Terrier; VetCompass: Cavalier King Charles Spaniel, Jack Russell Terrier, West Highland White Terrier, and Yorkshire Terrier). Notice the trend of high rates of cancer in both data sets with a significant decline in the oldest ages.

that longevity in dogs is not strictly a measure of how long the dogs could live but of how long the dogs' caregivers (owners and veterinarians) believe that they should live. However, the benefit of this situation is that there is usually a defined reason stated to explain the decision to euthanize a dog; 87% of dogs had a cause of death ascribed (O'Neill et al. 2013). Moreover, although euthanasia tells us nothing directly about the underlying cause of disability, in most cases it likely occurs when morbidity dramatically inhibits the dog's activities of daily living and/or quality of life. As such, it may well be a reasonable measure of the end of "health span" for a dog.

Fries (1980) proposed that we measure functional restriction and disability in studies of morbidity and mortality. Fortunately, in many cases, this is straightforward for a dog: Can it walk, run, climb stairs, etc.? At the same time, other disabilities might be more subtle and harder to detect by both owner and vet, especially when relating to disorders about which dog do not visibly "complain." This might be particularly true of cognitive and psychological

problems, which form one of the most common causes of years lived with disability in human populations (Murray et al. 2012).

CONCLUSIONS

The question remains: If we ameliorate diseases of old age in dogs through extraordinary means, will we prolong the absolute and relative life period of frailty and disability? As is clear from this collection, the answers to this question are far from simple. What is meant by "extraordinary means" depends on the temporal, medical, and species context. A century ago, dogs and humans alike could die from sepsis resulting from a tooth abscess. Today, advanced dental care and antibiotics have dramatically reduced those risks of morbidity and mortality. However, although the lifesaving effects of the antibiotics might be dramatic, the treatment is unlikely to have a noticeable effect on subsequent frailty or disability.

Even when the treatment would seem to qualify as "extraordinary means," the effects on frailty might depend on the details of the disease. A dog with mitral valve disease can be spared from heart failure by a few daily pills. And while the dog is running around and eating his pills in bits of food with glee, it is hard to think of his condition as a "disability." On the other hand, if that dog were to progress to heart failure, then keeping him alive with intravenous drugs and an oxygen cage would seem to be extending frailty. Unless, of course, he goes back to running around and eating everything in sight the next day (which is not uncommon). The harder question to answer here is whether a period of poor health has been extended or whether the animal has been returned to a state of good health.

Of course, the dog might not go back to his normal activity or normal appetite ever again, instead teetering in a state of near failure for days or weeks until he finally succumbs or his owners elect for euthanasia. Here we might well say that medical intervention had extended a period of poor health. The dog lived longer, but his state of health in that time would meet no one's definition of health span.

In recent years, the companion dog has emerged as an excellent model to improve our understanding of the determinants of age-specific morbidity and mortality. For centuries, in the course of selecting dogs with specifically defined traits, breeders have unintentionally created lineages that vary considerably in life span and health. We are just now beginning to collect detailed data on age-specific morbidity and mortality across breeds and environments. As we do so, we will have an unprecedented opportunity to explore and better understand the limitations and potential of the LD. Especially in light of the fact that those same treatments (drugs, diagnostic tests, and surgical procedures) used to care for dogs are often also used in humans, there is a critical need to answer the questions posed here. Answers to these important questions will require a concerted and collaborative effort involving not only veterinarians and biodemographers, but also medical ethicists. We are confident that in the coming years, the well-loved canine species will allow us to answer many of the questions posed throughout the literature.

REFERENCES

Adams VJ, Evans KM, Sampson J, Wood JL. 2010. Methods and mortality results of a health survey of purebred dogs in the UK. *J Small Anim Pract* **51:** 512–24.

Austad SN. 2014. The evolutionary basis of aging. In *Molecular and cellular biology of aging* (ed. Vijg JC, Lithgow G). Gerontological Society of America, Washington, DC.

Austad SN, Fischer KE. 1991. Mammalian aging, metabolism, and ecology: Evidence from the bats and marsupials. *J Gerontol* **46:** B47–B53.

Bansal A, Zhu LJ, Yen K, Tissenbaum HA. 2015. Uncoupling lifespan and healthspan in *Caenorhabditis elegans* longevity mutants. *Proc Natl Acad Sci* **112:** E277–E286.

Bellows J, Colitz CM, Daristotle L, Ingram DK, Lepine A, Marks SL, Sanderson SL, Tomlinson J, Zhang J. 2015. Defining healthy aging in older dogs and differentiating healthy aging from disease. *J Am Vet Med Assoc* **246:** 77–89.

Bonnett BN, Egenvall A, Hedhammar A, Olson P. 2005. Mortality in over 350,000 insured Swedish dogs from 1995–2000. I: Breed-, gender-, age- and cause-specific rates. *Acta Vet Scand* **46:** 105–120.

Bronson RT. 1982. Variation in age at death of dogs of different sexes and breeds. *Am J Vet Res* **43:** 2057–2059.

Burger JM, Promislow DEL. 2006. Are functional and demographic senescence genetically independent? *Exp Gerontol* **41:** 1108–1116.

Burger O, Baudisch A, Vaupel JW. 2012. Human mortality improvement in evolutionary context. *Proc Natl Acad Sci* **109:** 18210–18214.

Chang S. 2005. Modeling aging and cancer in the telomerase knockout mouse. *Mutat Res* **576:** 39–53.

Chen J, Senturk D, Wang JL, Muller HG, Carey JR, Caswell H, Caswell-Chen EP. 2007. A demographic analysis of the fitness cost of extended longevity in *Caenorhabditis elegans*. *J Gerontol A Biol Sci Med Sci* **62:** 126–135.

Cooley DM, Schlittler DL, Glickman LT, Hayek M, Waters DJ. 2003. Exceptional longevity in pet dogs is accompanied by cancer resistance and delayed onset of major diseases. *J Gerontol A Biol Sci Med Sci* **58:** B1078–B1084.

Crimmins EM, Beltran-Sanchez H. 2011. Mortality and morbidity trends: Is there compression of morbidity? *J Gerontol B Psychol Sci Soc Sci* **66:** 75–86.

de Magalhaes JP, Cabral JA, Magalhaes D. 2005. The influence of genes on the aging process of mice: A statistical assessment of the genetics of aging. *Genetics* **169:** 265–274.

Egenvall A, Bonnett BN, Hedhammar A, Olson P. 2005a. Mortality in over 350,000 insured Swedish dogs from 1995–2000. II: Breed-specific age and survival patterns and relative risk for causes of death. *Acta Vet Scand* **46:** 121–136.

Egenvall A, Penell JC, Bonnett BN, Olson P, Pringle J. 2005b. Morbidity of Swedish horses insured for veterinary care between 1997 and 2000: Variations with age, sex, breed and location. *Vet Rec* **157:** 436–443.

Evert J, Lawler E, Bogan H, Perls T. 2003. Morbidity profiles of centenarians: Survivors, delayers, and escapers. *J Gerontol A Biol Sci Med Sci* **58:** 232–237.

Cite this article as *Cold Spring Harb Perspect Med* doi: 10.1101/cshperspect.a026633

Finch CE. 1990. *Longevity, senescence, and the genome.* University of Chicago Press, Chicago.

Fleming JM, Creevy KE, Promislow DE. 2011. Mortality in North American dogs from 1984 to 2004: An investigation into age-, size-, and breed-related causes of death. *J Vet Intern Med* **25:** 187–198.

Freedman VA, Spillman BC, Andreski PM, Cornman JC, Crimmins EM, Kramarow E, Lubitz J, Martin LG, Merkin SS, Schoeni RF, et al. 2013. Trends in late-life activity limitations in the United States: An update from five national surveys. *Demography* **50:** 661–671.

Freedman AH, Gronau I, Schweizer RM, Ortega-Del Vecchyo D, Han E, Silva PM, Galaverni M, Fan Z, Marx P, Lorente-Galdos B, et al. 2014. Genome sequencing highlights the dynamic early history of dogs. *PLoS Genet* **10:** e1004016.

Fries JF. 1980. Aging, natural death, and the compression of morbidity. *N Engl J Med* **303:** 130–135.

Gardner EM. 2005. Caloric restriction decreases survival of aged mice in response to primary influenza infection. *J Gerontol A Biol Sci Med Sci* **60:** 688–694.

Goldberg EL, Romero-Aleshire MJ, Renkema KR, Ventevogel MS, Chew WM, Uhrlaub JL, Smithey MJ, Limesand KH, Sempowski GD, Brooks HL, et al. 2015. Lifespan-extending caloric restriction or mTOR inhibition impair adaptive immunity of old mice by distinct mechanisms. *Aging Cell* **14:** 130–138.

Gompertz B. 1825. On the nature of the function expressive of the law of human mortality and on a new mode of determining life contingencies. *Phil Trans R Soc Lond* **1825:** 513–585.

Greer KA, Canterberry SC, Murphy KE. 2007. Statistical analysis regarding the effects of height and weight on life span of the domestic dog. *Res Vet Sci* **82:** 208–214.

Greer KA, Hughes LM, Masternak MM. 2011. Connecting serum IGF-1, body size, and age in the domestic dog. *Age (Dordr)* **33:** 475–483.

Hamilton JB, Mestler GE. 1969. Mortality and survival: Comparison of eunuchs with intact men in a mentally retarded population. *J Gerontol* **24:** 395–411.

Hart BL, Hart LA, Thigpen AP, Willits NH. 2014. Long-term health effects of neutering dogs: Comparison of Labrador Retrievers with Golden Retrievers. *PLoS ONE* **9:** e102241.

Hess RS, Saunders HM, Van Winkle TJ, Ward CR. 2000. Concurrent disorders in dogs with diabetes mellitus: 221 cases (1993–1998). *J Am Vet Med Assoc* **217:** 1166–1173.

Hoffman JM, Creevy KE, Promislow DEL. 2013. Reproductive capability is associated with lifespan and cause of death in companion dogs. *PLoS ONE* **8:** e61082.

Howe LM, Slater MR, Boothe HW, Hobson HP, Holcom JL, Spann AC. 2001. Long-term outcome of gonadectomy performed at an early age or traditional age in dogs. *J Am Vet Med Assoc* **218:** 217–221.

Kaminski J, Schulz L, Tomasello M. 2012. How dogs know when communication is intended for them. *Dev Sci* **15:** 222–32.

Kraus C, Pavard S, Promislow DEL. 2013. The size-life span trade-off decomposed: Why large dogs die young. *Am Nat* **181:** 492–505.

Larson G, Karlsson EK, Perri A, Webster MT, Ho SY, Peters J, Stahl PW, Piper PJ, Lingaas F, Fredholm M, et al. 2012. Rethinking dog domestication by integrating genetics, archeology, and biogeography. *Proc Natl Acad Sci* **109:** 8878–8883.

Li Y, Deeb B, Pendergrass W, Wolf N. 1996. Cellular proliferative capacity and life span in small and large dogs. *J Gerontol A Biol Sci Med Sci* **51:** B403–B408.

Mason MJ, Gillett NA, Muggenburg BA. 1987. Clinical, pathological, and epidemiologic aspects of canine parvoviral enteritis in an unvaccinated closed beagle colony—1978–1985. *J Am Anim Hosp Assoc* **23:** 183–192.

Michell AR. 1999. Longevity of British breeds of dog and its relationships with sex, size, cardiovascular variables and disease. *Vet Rec* **145:** 625–629.

Miklosi A, Soproni K. 2006. A comparative analysis of animals' understanding of the human pointing gesture. *Anim Cogn* **9:** 81–93.

Miller RA, Austad SN. 2006. Growth and aging: Why do big dogs die young? In *Handbook of the biology of aging*, 6th ed. (ed. Masoro JE, Austad SN). Academic, San Diego.

Min KJ, Lee CK, Park HN. 2012. The lifespan of Korean eunuchs. *Curr Biol* **22:** R792–R793.

Murray CJ, Vos T, Lozano R, Naghavi M, Flaxman AD, Michaud C, Ezzati M, Shibuya K, Salomon JA, Abdalla S, et al. 2012. Disability-adjusted life years (DALYs) for 291 diseases and injuries in 21 regions, 1990–2010: A systematic analysis for the Global Burden of Disease Study 2010. *Lancet* **380:** 2197–2223.

Nikolich-Zugich J, Messaoudi I. 2005. Mice and flies and monkeys too: Caloric restriction rejuvenates the aging immune system of non-human primates. *Exp Gerontol* **40:** 884–893.

Oeppen J, Vaupel JW. 2002. Demography. Broken limits to life expectancy. *Science* **296:** 1029–1031.

Olshansky SJ, Perry D, Miller RA, Butler RN. 2006. In pursuit of the longevity dividend. *The Scientist*, March 1, pp. 28–36.

O'Neill DG, Church DB, McGreevy PD, Thomson PC, Brodbelt DC. 2013. Longevity and mortality of owned dogs in England. *Vet J* **198:** 638–643.

O'Neill DG, Church DB, McGreevy PD, Thomson PC, Brodbelt DC. 2015. Longevity and mortality of cats attending primary care veterinary practices in England. *J Feline Med Surg* **17:** 125–133.

Promislow DEL. 1991. Senescence in natural populations of mammals: A comparative study. *Evolution* **45:** 1869–1887.

Promislow DEL. 1993. On size and survival: Progress and pitfalls in the allometry of life span. *J Gerontol* **48:** B115–B123.

Promislow DEL, Tatar M, Khazaeli AA, Curtsinger JW. 1996. Age-specific patterns of genetic variance in *Drosophila melanogaster*. I: Mortality. *Genetics* **143:** 839–848.

Reiser K, McGee C, Rucker R, Mcdonald R. 1995. Effects of aging and caloric restriction on extracellular matrix biosynthesis in a model of injury repair in rats. *J Gerontol A Biol Sci Med Sci* **50A:** B40–B47.

Rice D. 2008. *The complete book of dog breeding.* Barrons, Hauppauge, NY.

Ritz BW, Aktan I, Nogusa S, Gardner EM. 2008. Energy restriction impairs natural killer cell function and increases the severity of influenza infection in young adult male C57BL/6 mice. *J Nutr* **138:** 2269–2275.

Rollo CD. 2002. Growth negatively impacts the life span of mammals. *Evol Dev* **4:** 55–61.

Salomon JA, Wang H, Freeman MK, Vos T, Flaxman AD, Lopez AD, Murray CJ. 2012. Healthy life expectancy for 187 countries, 1990–2010: A systematic analysis for the Global Burden Disease Study 2010. *Lancet* **380:** 2144–2162.

Siler W. 1979. Competing-risk model for animal mortality. *Ecology* **60:** 750–757.

Spain CV, Scarlett JM, Houpt KA. 2004. Long-term risks and benefits of early-age gonadectomy in dogs. *J Am Vet Med Assoc* **224:** 380–387.

Sutter NB, Bustamante CD, Chase K, Gray MM, Zhao K, Zhu L, Padhukasahasram B, Karlins E, Davis S, Jones PG, et al. 2007. A single *IGF1* allele is a major determinant of small size in dogs. *Science* **316:** 112–115.

Tatar M, Bartke A, Antebi A. 2003. The endocrine regulation of aging by insulin-like signals. *Science* **299:** 1346–1351.

Thatcher AR, Cheung SL, Horiuchi S, Robine JM. 2010. The compression of deaths above the mode. *Demogr Res* **22:** 505–538.

Torres de la Riva G, Hart BL, Farver TB, Oberbauer AM, Messam LL, Willits N, Hart LA. 2013. Neutering dogs: Effects on joint disorders and cancers in golden retrievers. *PLoS ONE* **8:** e55937.

Urfer SR, Greer K, Wolf NS. 2011. Age-related cataract in dogs: A biomarker for life span and its relation to body size. *Age (Dordr)* **33:** 451–460.

Waters DJ. 2011. Aging research 2011: Exploring the pet dog paradigm. *ILAR J* **52:** 97–105.

Waters DJ, Kengeri SS, Clever B, Booth JA, Maras AH, Schlittler DL, Hayek MG. 2009. Exploring mechanisms of sex differences in longevity: Lifetime ovary exposure and exceptional longevity in dogs. *Aging Cell* **8:** 752–755.

Weindruch R, Walford RL. 1988. *The retardation of aging and disease by dietary restriction.* Charles C Thomas, Springfield, IL.

Wess G, Schulze A, Butz V, Simak J, Killich M, Keller LJ, Maeurer J, Hartmann K. 2010. Prevalence of dilated cardiomyopathy in Doberman Pinschers in various age groups. *J Vet Intern Med* **24:** 533–538.

Yuan R, Tsaih SW, Petkova SB, De Evsikova CM, Xing S, Marion MA, Bogue MA, Mills KD, Peters LL, Bult CJ, et al. 2009. Aging in inbred strains of mice: Study design and interim report on median lifespans and circulating IGF1 levels. *Aging Cell* **8:** 277–287.

Translating the Science of Aging into Therapeutic Interventions

James L. Kirkland

Robert and Arlene Kogod Center on Aging, Mayo Clinic, Rochester, Minnesota 55905

Correspondence: kirkland.james@mayo.edu

Life and health span have been extended in experimental animals using drugs that are potentially translatable into humans. Considerable effort is needed beyond the usual steps in drug development to devise the models, and realistic preclinical and clinical trial strategies are required to advance these agents into clinical application. It will be important to focus on subjects who already have symptoms or are at imminent risk of developing disorders related to fundamental aging processes, to use short-term, clinically relevant outcomes, as opposed to long-term outcomes, such as health span or life span, and to validate endpoint measures so they are acceptable to regulatory agencies. Funding is a roadblock, as is shortage of investigators with combined expertise in the basic biology of aging, clinical geriatrics, and investigational new drug clinical trials. Strategies for developing a path from the bench to the bedside are reviewed for interventions that target fundamental aging mechanisms.

Aging is the largest risk factor for many of the major chronic diseases that account for the bulk of morbidity, mortality, and health costs in developing and developed countries (Alliance for Aging Research 2012; Goldman et al. 2013; Kirkland 2013a,b). Indeed, for most of these conditions, chronological age leads all other known predictors combined. These age-related chronic disorders include atherosclerosis, type 2 diabetes, most cancers, dementias, Parkinson's disease, other neurodegenerative diseases, arthritis, renal dysfunction, blindness, frailty, and sarcopenia, among many others. Key pathogenic mechanisms that predispose to these disorders are shared with mechanisms associated with aging, including chronic, low-grade, nonmicrobial inflammation, cellular senescence, accumulation of damaged macromolecules (DNA,

proteins, carbohydrates, and lipids), and stem and progenitor cell dysfunction. These age-associated mechanisms are, in turn, targets of most of the genetic, environmental, and pharmacological interventions that appear to be effective in extending life span in lower mammals. Early indications suggest these interventions may also impact health span, resilience, and multiple age-related disorders in experimental animals.

Based on these observations, the "geroscience hypothesis" has been proposed. By targeting fundamental aging processes, it may be possible to delay, prevent, alleviate, or treat the major age-related chronic disorders as a group, instead of one at a time. Even if a single major chronic disease, such as atherosclerosis, were eradicated, as transformative as such an advance would be, it would only add 2 or 3 years to life expectancy (Olshansky et al. 1990; Fried et al. 2009). However, targeting the intersection between fundamental aging mechanisms and processes that lead to chronic diseases could alleviate multiple age-related disorders. For this to be achieved, a great deal of effort will be required to design and conduct the preclinical, clinical proof-of-concept, and formal clinical trials acceptable to regulatory agencies and the medical community. Here, we consider the process of translating drugs that target fundamental aging mechanisms into clinical application, potential indications for which agents that target fundamental aging processes may first be tested, personnel and resources needed to do so, and regulatory and intellectual property issues in developing feasible interventions. Although the path to clinical application of agents that target fundamental aging mechanisms will likely be difficult, the potential to transform health care as we know it is unparalleled.

HEALTH SPAN

Most members of the public appear to be more interested in an increased health span, the portion of life span during which function is sufficient to maintain independence, productivity, and well-being, rather than extended life span at all costs. Loss of autonomy and control predict mortality (Fry and Debats 2006). The elderly are not resigned to an old age of dysfunction and frailty (Fry 2000). Although study is needed about what the public hopes to gain from biomedical research, it seems there is widespread interest in supporting studies to enhance health span and compress the period of morbidity near the end of life. Increasing emphasis is being placed by the basic biology of the aging community on studying interventions that enhance health span in experimental animals, in addition to studying life span (Kirkland and Peterson 2009; Tatar 2009; Kirkland 2013a). Investigators have started to test if these interventions alleviate chronic diseases, frailty, and resilience in experimental animals.

Limits to health span include disability, frailty, chronic diseases, and, of course, life span (Kirkland 2013a). Disabilities are functional deficits that are consequences of diseases earlier in life, developmental disorders, or accidents. Frailty is a clinical syndrome that involves loss of resilience, with failure to recover from acute problems, such as pneumonia, stroke, influenza, heart attacks, dehydration, or fractures (Fried et al. 2001; Walston et al. 2002, 2006, 2009; Bandeen-Roche et al. 2006; Rockwood et al. 2006; Leng et al. 2007; Kanapuru and Ershler 2009; Qu et al. 2009). Frailty can be diagnosed using clinically validated scales that are moderately sensitive and specific. These scales include combinations of assessments of endurance, weakness, fatigue, weight loss, activity, confusion, and chronic disease and disability burden (Fried et al. 2001; Bandeen-Roche et al. 2006; Lucicesare et al. 2010; Rockwood and Mitnitski 2011). Loss of resilience, or ability to recover after acute insults, can precede overt clinical frailty in the absence of acute, superimposed stresses. Steps have begun to develop tests and markers of frailty for use in experimental animals.

Prevalence of loss of resilience and frailty increases with age (Fried et al. 2001; Walston et al. 2002, 2006, 2009; Bandeen-Roche et al. 2006, 2009; Rockwood et al. 2006; Leng et al. 2007; Kanapuru and Ershler 2009; Qu et al. 2009; Lucicesare et al. 2010; Rockwood and Mitnitski 2011). Frailty predisposes to chronic diseases, dependency, and increased mortality. It is linked to the "geriatric syndromes" of immobility, falling, sarcopenia, cachexia, depression, and confusion, as well as the low grade, nonmicrobial, chronic inflammation implicated in the initiation and progression of multiple chronic diseases. Age-related chronic diseases and frailty individually and in combination contribute to poor responses to treatments, such as chemotherapy, surgery, transplantation, immunization, and rehabilitation. They can initiate a downward spiral of dysfunction that progresses to dependency, institutionalization, and death. Intriguingly, low-grade, chronic inflammation is linked to frailty, dementias, depression, atherosclerosis, cancers, diabetes, and many other age-related diseases as well as advanced chronological age itself, cellular senescence, macromolecular dysregulation, and progenitor cell dysfunction. If and how chronic inflammation causally links these processes remains to be ascertained (Ferrucci et al. 1999; Harris et al. 1999; Brown et al. 2001; Pradhan et al. 2001; Walston et al. 2002; Bruunsgaard and Pedersen 2003; Bruunsgaard et al. 2003; Cesari et al. 2003; Spranger et al. 2003; Hu et al. 2004; Pai et al. 2004; Margolis et al. 2005; Tuomisto et al. 2006; Leng et al. 2007; Howren et al. 2009; Kanapuru and Ershler 2009; Srikrishna and Freeze 2009; O'Connor et al. 2010; Schetter et al. 2010; Kirkland 2013a; Tchkonia et al. 2013; Kirkland and Tchkonia 2014).

ARE THERE TRANSLATABLE INTERVENTIONS?

Aging has increasingly become recognized as a potentially modifiable risk factor for chronic diseases and frailty. Supporting this are the findings, reviewed elsewhere in this series, that (1) maximum life span is extended and age-related diseases can be delayed by several single-gene mutations (Bartke 2011). This suggests that pathways affected by these mutations may be good therapeutic targets. (2) People who live beyond 100, a trait that is sometimes partly heritable, often have delayed onset of age-related diseases and disabilities (Lipton et al. 2010), with compression of morbidity and enhanced health span. (3) Calorie restriction, which increases maximum life span, is linked to delayed onset of chronic diseases in animal models (Anderson and Weindruch 2012). (4) Factors produced by stem cells from young animals enhance cardiac, muscle, and brain repair capacity in the older animals in parabiotic, cross-circulated pairs of young and old mice (Conboy et al. 2005; Lavasani et al. 2012; Katsimpardi et al. 2014; Sinha et al. 2014). (5) Senescent cell accumulation is associated with chronic inflammation, which in turn promotes several age-related chronic diseases and frailty (Tchkonia et al. 2013: Kirkland and Tchkonia 2014). Importantly, removing senescent cells enhances health span in progeroid mice (Baker et al. 2011). Thus, there are multiple indications that fundamental aging mechanisms can be targeted. Indeed, drugs have been discovered that enhance life span and health span in rodents. These in-

clude rapamycin, drugs related to rapamycin (rapalogs), metformin, 17 α-estradiol, angiotensin converting enzyme inhibitors (and possibly angiotensin receptor blockers), flavonoids related to resveratrol, and aspirin and salsalate (Ferder et al. 1993; Linz et al. 1997, 2000; Basso et al. 2007; Anisimov et al. 2008, 2010, 2011; Strong et al. 2008; Harrison et al. 2009, 2014; Santos et al. 2009; Smith et al. 2010; Lamming et al. 2013; Mercken et al. 2014; Mitchell et al. 2014). Metformin may even enhance survival in older humans (Bannister et al. 2014). Several of these agents also appear to delay cancers, cognitive decline, cardiac dysfunction, and other age-related diseases and disabilities in several species, including humans (Chiasson et al. 2002; Zeymer et al. 2004; Basso et al. 2005; Dykens et al. 2005; Johnson et al. 2005; Crandall et al. 2006; Hanon et al. 2008; Seibel et al. 2008; Libby et al. 2009; Anfossi et al. 2010; Landman et al. 2010; Li et al. 2010; Tosca et al. 2010; Davies et al. 2011; Huang et al. 2011; Lamanna et al. 2011; Lee et al. 2011; Liu et al. 2011; Monami et al. 2011; Pasternak et al. 2011; Majumder et al. 2012; Roumie et al. 2012; Salani et al. 2012; Tseng 2012; Anisimov and Bartke 2013; Goldfine et al. 2013; Hong et al. 2013; Karnevi et al. 2013; Martin-Montalvo et al. 2013; Quinn et al. 2013; Whittington et al. 2013; Kennedy and Pennypacker 2014; Mannick et al. 2014; Richardson et al. 2014). A pipeline of even more interventions is developing that seem to enhance life span in experimental animals. Many of these interventions are promising, but not yet published. Because drugs that increase life span and health span in mammals now exist, it is plausible that, by targeting fundamental aging mechanisms, clinical interventions might be developed in enough time to influence the health of many people alive today, that could delay, prevent, alleviate, or possibly even partly reverse age-related diseases and disabilities as a group, instead of one at a time.

Targeting the intersection between aging and chronic disease predisposition could circumvent a problem encountered in elucidating the pathogenesis of many of chronic diseases in humans. Many major chronic age-related conditions, including Alzheimer's disease or atherosclerosis, only occur in humans or a few other species. Most age-related chronic diseases are only clinically evident by the time the disease is advanced at the molecular and cellular levels. This makes delineation of etiological mechanisms challenging, because it is obviously not feasible to obtain appropriate tissue samples for study sufficiently early during disease development from human subjects. By targeting the upstream, basic aging mechanisms that predispose to these diseases, these difficulties could be circumvented.

Because several potentially effective interventions that target aging processes have been shown to be effective in increasing life span or health span in experimental animals, there is an opportunity to select those interventions that are more likely to be translatable into clinical applications for humans. Lifestyle interventions, such as caloric restriction, would be particularly challenging, especially in the context of an obesity epidemic. For a pharmacological intervention to be translatable, it would need to have (1) low toxicity plus few side effects, (2) effectiveness by oral administration, (3) once daily, or less frequent dosing, (4) stability, (5) scalability and low manufacturing cost, (6) detectability in blood, and (7) effectiveness if administered in later life or once symptoms have already started to develop. Interventions that need to be started in childhood or early adulthood in asymptomatic subjects with the expectation of health benefit in later life would be difficult to translate into humans. To be acceptable to regulators, these interventions would need to be almost completely devoid of side effects. Furthermore, it would take decades to conduct the studies needed to show efficacy. Such interventions would be of little interest to the pharmaceutical industry because of the time and expense needed for clinical trials and expiration of patents. Similarly, lifestyle interventions that need to be sustained from early life would be difficult to validate in humans or to implement.

INITIAL CLINICAL TRIALS SCENARIOS

Translating interventions found to be effective in experimental animals into clinical appli-

cation can be time-consuming and expensive. Even in fields with an established translational tradition, such as oncology, it can take <10 years to complete translation. The process involves preclinical basic studies to determine efficacy, safety, and pharmacokinetics in mammals, usually in at least two species and following the good laboratory practice (GLP) conditions specified by regulatory agencies (Steinmetz and Spack 2009). Clinical outcome measures need to be selected that have been validated clinically, are reproducible, can be measured in a short time frame, are as noninvasive as possible, are accepted by regulatory agencies, and for which there are acceptable animal models. Iterative bench-to-bedside coupled to "reverse translational" bedside-to-bench developmental phases are often necessary, particularly for investigational new drugs (INDs). This requires a close partnership between biologists and clinicians with a strong basic biology background.

Although recent progress in developing successful interventions that target fundamental aging mechanisms in rodents has been impressive, the path forward to translation into humans has not yet been fully mapped. Innovative new solutions to the limitations in experimental paradigms, infrastructure, strategies, and personnel are required to translate these interventions into the clinic successfully.

Elderly subjects are generally excluded from clinical trials, particularly those involving INDs. However, early proof-of-concept clinical trials have begun using some of the interventions that have been found to increase life or health span in rodents with drugs already approved for other indications. Rather than determining their effect on life span, which is impractical in humans, these drugs are being used to intervene against particular age-related disorders. For example, a brief course of rapalogs before influenza immunization has been shown to increase immune responses in elderly subjects (Mannick et al. 2014). At least three clinical trials of rapamycin in Alzheimer's disease and another for frailty are underway or about to begin. Resveratrol congeners are being developed to treat type 2 diabetes (Baur et al. 2012).

At least six potential drug development paradigms can be envisaged, in which drugs that target basic aging processes could be tested in small-scale, relatively short proof-of-concept trials. These include treatment of (1) multiple comorbidities, (2) otherwise fatal conditions, (3) geriatric syndromes such as frailty, (4) decreased resilience, (5) localized diseases related to basic aging processes, and (6) accelerated aging-like states.

MULTIPLE COMORBIDITIES

Most clinical studies have involved younger subjects with a single disease and excluded subjects with comorbidities. However, multiple age-related chronic diseases frequently occur within the same patients. If the geroscience hypothesis is correct, agents that target fundamental aging mechanisms would be expected to alleviate different age-related chronic diseases within the same older subjects at the same time. Testing whether this occurs will require novel clinical study paradigms. One scenario for initial proof-of-principle trials would be to study effects of interventions in elderly subjects with combinations of two or more of diabetes, atherosclerosis, hypertension, memory impairment, chronic lung disease, renal dysfunction, or other age-related disorders. Outcomes could be surrogate endpoints already accepted by regulatory agencies, such as blood pressure, psychometric tests of cognitive function, fasting glucose, circulating lipids, cardiac function, etc. Endpoints could be combined into a composite score, although this carries the risk that an effective drug may appear less than effective if one of the score components is affected in a direction opposite to that expected. For example, rapamycin might result in improvements in a number of measures of function while also causing insulin resistance (Lamming et al. 2012).

OTHERWISE FATAL CONDITIONS

Agents that target basic aging processes could be tested in trials for severe or fatal conditions that share pathogenic mechanisms with aging pro-

cesses, and for which no effective treatments are available. Among these conditions are certain cancers, cancer predisposition syndromes, idiopathic pulmonary fibrosis, and primary biliary cirrhosis, each of which are associated with cellular senescence (Kirkland and Tchkonia 2014; Calhoun et al. 2015). Caloric restriction affects multiple pathways related to chronological aging. There is evidence that brief caloric restriction might allow use of higher chemotherapy or radiation doses or enhance effectiveness of treatments for cancer (Lee et al. 2012), suggesting that agents that mimic effects of caloric restriction may merit testing for this indication.

GERIATRIC SYNDROMES

The geriatric syndromes include frailty, sarcopenia, immobility, falls, gait disturbances, mild cognitive impairment, incontinence, and other conditions. As considered above, frailty is a syndrome involving weakness, loss of function, and decreased resilience (Fried et al. 2001; Walston et al. 2002, 2006, 2009; Bandeen-Roche et al. 2006, 2009; Rockwood et al. 2006; Leng et al. 2007; Kanapuru and Ershler 2009; Qu et al. 2009). Frailty can be identified by clinical scales that are somewhat but not perfectly sensitive and specific. These involve measures of weight loss, activity, weakness, fatigue, and numbers of chronic diseases and disabilities (Fried et al. 2001; Walston et al. 2002, 2006, 2009; Bandeen-Roche et al. 2006; Lucicesare et al. 2010; Rockwood and Mitnitski 2011). The prevalence of frailty increases with aging (Fried et al. 2001; Bandeen-Roche et al. 2006, 2009; Rockwood et al. 2006; Leng et al. 2007; Kanapuru and Ershler 2009; Qu et al. 2009; Lucicesare et al. 2010; Rockwood and Mitnitski 2011). Frailty is associated with chronic diseases, loss of independence, high mortality, sarcopenia, immobility, falling, cachexia, depression, confusion, and chronic inflammation.

Subjects with mild or moderate degrees of sarcopenia or frailty would probably be better candidates for initial trials of agents that target basic aging processes than those with severe, likely irreversible frailty. Timed walking distances, strength measurements, inflammation and

SASP markers, such as circulating IL-6, and pulmonary and renal function tests are among the outcome measures associated with frailty that are predictive of mortality and may soon be recognized by regulatory authorities (Villareal et al. 2011; Tchkonia et al. 2013; McLean et al. 2014).

RESILIENCE

Resilience, the ability to recover from perturbations, such as surgery, anesthesia, chemotherapy, radiation, a heart attack, fracture, or stroke, diminishes with aging and is associated with concurrent or subsequent clinically overt frailty. Studies of resilience could be scenarios for initial proof-of-principle studies of interventions that target aging processes in relatively short clinical trials. The stressor could be intense, such as time taken to recover from a hip fracture after a fall, or less intense, such as immune response to a routine vaccination. The stressor could be planned, such as chemotherapy, therapeutic radiation, elective surgery, or immunization, or unplanned, for example a heart attack, stroke, or pneumonia. For elective stressors, the candidate drug could be given before the stress occurs and for either elective or unplanned stressors, after the stress-inducing event during recovery.

Some of these types of resilience trials have already been reported. Influenza vaccine responses were demonstrated to be enhanced a couple of weeks after one to three doses of a rapalog (Mannick et al. 2014). Responses to chemotherapy are enhanced and side effects reduced if chemotherapy is preceded by brief caloric restriction (Lee et al. 2012). Effects of interventions that target fundamental aging processes could be tested in multiple clinical trials to determine effects on resilience, such as (1) nausea, appetite, strength, or endurance after chemotherapy; (2) time for recovery after myocardial infarction in older subjects (secondary endpoints could include measures of function, comorbidity, and blood or tissue biomarkers of fundamental aging processes); or (3) time needed for return of function (home vs. nursing home), wound healing, delirium, or discharge disposition after elective surgery.

LOCALIZED DISEASES RELATED TO FUNDAMENTAL AGING PROCESSES

Some localized disorders involve pathogenic processes similar to those associated with chronological aging. Local administration of agents that target basic aging processes by aerosols, injections, patches, or topical skin solutions could alleviate these conditions. Idiopathic pulmonary fibrosis associated with senescent cell accumulation in the lung (Minagawa et al. 2011; Takasaka et al. 2014) is a sometimes fatal condition without very effective treatment options. Studies with aerosolized senolytic or SASP-protective agents could be considered in patients with this condition if testing in animals, for example, mice with fibrosis induced by aerosolized bleomycin (Moore and Hogaboam 2008), suggest success. Another example is osteoarthritis, an age-related inflammatory condition that can affect multiple joints, which is associated with senescent cell accumulation in joints (Ryu et al. 2014). Osteoarthritis is currently treated with oral analgesics, anti-inflammatories, and intra-articular steroid injections. The oral agents have to be administered frequently, often at least daily. Repeated steroid injections can eventually worsen joint damage. Possibly, systemic or injected drugs that target fundamental aging processes, such as senescence, may have more sustained beneficial and fewer adverse effects than currently used treatments.

CONDITIONS RESEMBLING ACCELERATED AGING

Several disorders lead to an accelerated aging-like state, such as obesity, diabetes, long-term effects of chemotherapy or radiation, or progeroid syndromes (Kirkland 2013a; Tchkonia et al. 2013; Kirkland and Tchkonia 2014; Ness et al. 2014). It might be possible to alleviate these disorders with agents that target fundamental aging processes. For example, obesity and diabetes are associated with early onset of age-associated disorders, including frailty, sarcopenia, atherosclerosis, vascular dysfunction, early menopause, cognitive impairment, dementia,

Cite this article as *Cold Spring Harb Perspect Med* doi: 10.1101/cshperspect.a025908

and cancers (Tchkonia et al. 2010; De Felice and Ferreira 2014; Joost 2014). Survivors of childhood cancers who had been treated with chemotherapy or radiation sometimes develop frailty, sarcopenia, diabetes, cardiac disease, cognitive impairment, and unrelated second malignancies by mid-adulthood (Hudson et al. 2013; Ness et al. 2014). Progeroid syndromes have been associated with increased senescent cell burden (Benson et al. 2010). In short-term proof-of-principle studies, effects of candidate drugs on muscle strength, metabolic, cardiovascular, cognitive, or other functional measures could be tested in these subjects.

PRECLINICAL STUDIES

Experimental animal models are needed that reflect the human disorders, which are potential indications for drugs targeting aging processes. Studies in these models could provide data required so that proof-of-principle human trials can be initiated. For some indications, genetically modified mice or disease-inducing manipulations have been devised. For other indications, there are no models or only ones that are imperfect. Certain genetically modified mice have been developed that resemble human progerias and other syndromes that are caused by single-gene mutations (Baker et al. 2008; Chen et al. 2013; Tchkonia et al. 2013; Eren et al. 2014). In such mice, it is possible to evaluate drug candidates for effectiveness in alleviating aging-like or health-span phenotypes, such as insulin resistance, weakness, decreased endurance, reduced activity, or neurological dysfunction.

Many chronic diseases in humans are multifactorial, polygenic, and clinically manifest in later life. To model these diseases, animals with single-gene mutations have been bred that acquire superficially similar syndromes, but usually in early life. This approach has drawbacks for drug development. Single-gene mutations that lead to AD-like phenotypes in young animals do not fully mimic human AD in older patients (Shineman et al. 2011). Animals in which dysfunction-provoking mutations can be induced in later life could be better models

for drug development, because an aging tissue microenvironment would exist. It could also be useful to test more than one mutant line, each differing in their disease-inducing mutation, to offset at least partly the problems caused by the strategy of using single-gene mutant mice to model a polygenic disease. More mammalian experimental species beyond mice would help in testing generalizability and meeting regulatory requirements for new chemical entities. More research is required to improve and validate animal models of human age-related diseases.

Manipulations, such as high fat feeding, radiation, pharmacological interventions (e.g., chemotherapy, streptozotocin, PD-inducing agents, inhaled bleomycin or cigarette smoke), cancer xenografts, skin wounding, or surgically induced arthritis can be used in experimental animals to induce age-associated disorders or clinical stresses modeling those in humans. Effects of candidate drugs on a panel of such experimental animal models could provide data useful for deciding about potential clinical applications for each new agent. In some instances, such as AD, rather than using currently available animal models, tests of drugs might be faster and provide better information using human cell culture systems that model disease pathology more closely (Choi et al. 2014). For new chemical entities, new paradigms need to be developed for medicinal chemical optimization and testing toxicology and pharmacokinetics (absorption, distribution, metabolism, and excretion) that use cell culture and animal models in the context of an aging microenvironment.

BIOMARKERS

Three types of biomarkers are valuable during clinical trials: surrogate endpoint biomarkers, drug activity biomarkers, and biomarkers indicative of molecular and cellular mechanisms. Surrogate endpoint biomarkers are potential primary outcomes of clinical trials that can be substituted for clinical event outcomes. An example would be use of blood pressure as a predictor for risk of stroke, rather than numbers of strokes themselves, as the primary outcome of a clinical trial of a drug intended to reduce stroke risk. These types of surrogate endpoints can only be accepted by the medical community and regulators after years or decades of clinical studies. These studies need to show that the surrogate endpoint is sufficiently predictive of a clinical outcome and that it can be used in place of that clinical outcome. It would take many decades to validate surrogate biomarkers that predict life span or even health span in humans. Some work has been started to search for and validate surrogate biomarkers for frailty- or age-related disabilities, such as tests of muscle strength or circulating cytokines instead of clinical endpoints, such as time to nursing home admission or death. Although frailty, health span, and resilience biomarkers are some way from being accepted by regulators and first need to be refined in experimental animal and human studies, they are much closer to application than biomarkers predictive of longevity.

The second types of biomarkers are those that can be used to monitor drug delivery, activity, or efficacy, such as assays of blood concentrations of the drug to monitor compliance or tissue mTOR to follow pharmacodynamic activity in the case of rapamycin. These assays can be valuable for optimizing dosing of the drug.

The third types of biomarkers are those that elucidate mechanism of action of the drug in humans. In the case of studies of drugs intended to target fundamental aging processes, a range of parameters related to known aging mechanisms could be assayed to determine whether the drug actually does target aging mechanisms. Assays of parameters to confirm molecular mechanism of action can be informative, but precise information about mechanism of action, although academically interesting, is not always required for drug development. In the cases of many drugs currently used clinically, the molecular mechanism of action was not known before clinical use. Metformin is an example. Its mechanism of action has still not been completely delineated. Knowledge of mechanism of action might help to predict side effects, but in many cases this has not been true. Knowledge gained about molecular mechanisms during human studies may prompt reverse transla-

tional laboratory studies that inform future clinical trials, promote discovery of new drug targets and agents, and are helpful in further optimizing preclinical animal models.

CLINICAL TRIALS

Clinical trials are needed to test whether new interventions are safe, efficacious, and effective. The clinical trial component of drug development is divided into phases. Phase 0 studies are those that determine whether new chemical entities indeed act in humans in the same way as expected from preclinical animal studies, to acquire preliminary information about pharmacokinetics and pharmacodynamics, to identify the most promising lead candidates, or to delineate biodistribution characteristics. Phase 1 trials provide data about metabolism and pharmacologic actions of the drug in human subjects, side effects appearing with escalating doses, and initial evidence of effectiveness. Phase 1 trials can include healthy subjects or patients with the disease being studied. Phase 2 studies are controlled clinical trials to evaluate effectiveness of the candidate drug in subjects with the condition under study and to identify the more common short-term side effects. Some trials combine phases 1 and 2, investigating efficacy and toxicity simultaneously. Phase 3 trials are larger controlled or uncontrolled studies that are designed to generate additional information about risk–benefit relationships and to acquire information needed for drug labeling. Usually, phase 3 trials compare the new drug to drugs already being used to treat that condition. Phase 4 studies are conducted after clinical use has begun. These studies generate additional information about risks, benefits, comparative effectiveness, and optimal use. Phase 4 studies assess effectiveness of the approved drug in the general population and provide information about adverse effects that become apparent once the drug is in widespread use.

Information collected during initial proof-of-concept and later phase trials can be used to enable extended follow-up of subjects. This allows evaluation of long-term clinical outcomes and acquisition of information about whether the drug delays or prevents other chronic diseases, dysfunction, frailty, or loss of resilience.

TRANSLATION

Translation refers to research that takes discoveries from the laboratory to the clinic (bench to bedside), reverse translates findings from human studies back to the laboratory so mechanisms can be elucidated or additional applications studied (bedside to bench), or enables adoption into medical practice (bedside to practice). The first, T1, phase of translational research (bench to bedside) brings a basic discovery to initial application. T1 research sometimes involves pilot studies for future larger clinical trials. The T2 phase, bedside to practice, involves ascertaining the value of the drug in large-scale clinical trials. The T3 phase brings evidence-based guidelines developed during phase T2 into clinical practice. The T4 phase involves determining the real-world health effects of the intervention. It often takes more than a decade to progress through these translational steps before new drugs are widely accepted and prescribed.

INVESTIGATORS NEEDED FOR TRANSLATIONAL AGING RESEARCH

To shepherd drugs that target fundamental aging mechanisms through translation, there will be a need for investigators with an understanding of the basic biology of aging, clinical geriatrics, and conducting clinical trials (Kirkland and Peterson 2009; Kirkland 2013a). However, there is a shortage of such investigators. Few geriatricians have basic biology training or IND experience. Conversely, few basic biologists understand the nuances of clinical geriatrics, how to conduct clinical trials, or regulatory processes, unlike in other areas, such as infectious diseases or oncology. Investigators need to be trained in the basic biology of aging that have a thorough understanding of translational strategies and clinical geriatrics. This could take many years, even if we begin training these investigators now. Until then, teams of basic biologists, clinical geriatricians, and clinical trial investigators will need to work together to translate agents that target basic aging processes into the clinic. Clinical trial networks could help in advancing the field, perhaps emulating approaches taken by networks in the cancer field.

CONCLUSIONS

A new era in the basic biology of aging may be beginning, a time in which we can begin to translate findings from the basic biology of aging into a range of clinical applications. Although interventions that delay age-related dysfunction in experimental animals are at hand, development of new clinical trial paradigms with relevant, measureable outcomes accepted by regulators, funding, and personnel with new skills are needed soon. All this must be performed without diverting resources away from the current discovery and mechanistic studies in the aging biology field that are at the start of the pipeline that leads to these important advances.

Some age-related changes involve progenitor depletion, some chronic inflammation, and other extracellular matrix or structural changes (e.g., cataracts, osteoporosis). Different basic aging processes could operate to different extents in different organs and cell types. This suggests that a range of therapeutic strategies may need to be combined to achieve maximal effects, because no one is likely to be a panacea. At least two steps may be needed to alleviate age-related diseases or dysfunction, based on the long-standing principle of first removing bad tissue and then replacing with good tissue. The first step could include targeting senescent cells, the senescence-associated secretory phenotype, inflammation, or damaged macromolecules. The second step to repopulate damaged tissues might involve using agents that restore endogenous progenitor function or transplanting stem cells, differentiated cells, tissues, or organs. Advances are being made in each of these areas, and effective combined approaches may be within reach.

If any or all of the advances in the basic biology of the aging field can be translated into clinical application, it may become feasible to delay, prevent, alleviate, or reverse chronic age-related diseases and disabilities as a group, instead of targeting them one at a time, and even extend health or life span. If these speculations are realized, health care as we know it would be transformed.

ACKNOWLEDGMENTS

The author is grateful for advice and ideas shared by members of the Geroscience Network supported by National Institutes of Health (NIH) Grant R24AG044396 as well as support from NIH Grants AG13925, AG41122, and the Noaber, Ellison, and Glenn Foundations.

REFERENCES

Alliance for Aging Research. 2012. *Chronic disease and medical innovation in an aging nation.* Silver Book, Washington, DC.

Anderson RM, Weindruch R. 2012. The caloric restriction paradigm: Implications for healthy human aging. *Am J Hum Biol* 24: 101–106.

Anfossi G, Russo I, Bonomo K, Trovati M. 2010. The cardiovascular effects of metformin: Further reasons to consider an old drug as a cornerstone in the therapy of type 2 diabetes mellitus. *Curr Vasc Pharmacol* 8: 327–337.

Anisimov VN, Bartke A. 2013. The key role of growth hormone-insulin-IGF-1 signaling in aging and cancer. *Crit Rev Oncol Hematol* 87: 201–223.

Anisimov VN, Berstein LM, Egormin PA, Piskunova TS, Popovich IG, Zabezhinski MA, Tyndyk ML, Yurova MV, Kovalenko IG, Poroshina TE, et al. 2008. Metformin slows down aging and extends life span of female SHR mice. *Cell Cycle* 7: 2769–2773.

Anisimov VN, Egormin PA, Piskunova TS, Popovich IG, Tyndyk ML, Yurova MN, Zabezhinski MA, Anikin IV, Karkach AS, Romanyukha AA. 2010. Metformin extends life span of HER-2/neu transgenic mice and in combination with melatonin inhibits growth of transplantable tumors in vivo. *Cell Cycle* 9: 188–197.

Anisimov VN, Berstein LM, Popovich IG, Zabezhinski MA, Egormin PA, Piskunova TS, Semenchenko AV, Tyndyk ML, Yurova MN, Kovalenko IG, et al. 2011. If started early in life, metformin treatment increases life span and postpones tumors in female SHR mice. *Aging (Milano)* 3: 148–157.

Baker DJ, Perez-Terzic C, Jin F, Pitel KS, Niederlander NJ, Jeganathan K, Yamada S, Reyes S, Rowe L, Hiddinga HJ, et al. 2008. Opposing roles for p16^{Ink4a} and p19^{Arf} in senescence and ageing caused by BubR1 insufficiency. *Nat Cell Biol* 10: 825–836.

Baker DJ, Wijshake T, Tchkonia T, LeBrasseur NK, Childs BG, van de Sluis B, Kirkland JL, van Deursen JM. 2011. Clearance of p16^{Ink4a}-positive senescent cells delays ageing-associated disorders. *Nature* 479: 232–236.

Bandeen-Roche K, Xue QL, Ferrucci L, Walston J, Guralnik JM, Chaves P, Zeger SL, Fried LP. 2006. Phenotype of frailty: Characterization in the women's health and aging studies. *J Gerontol A Biol Sci Med Sci* 61: 262–266.

Bandeen-Roche K, Walston JD, Huang Y, Semba RD, Ferrucci L. 2009. Measuring systemic inflammatory regulation in older adults: Evidence and utility. *Rejuvenation Res* 12: 403–410.

Bannister CA, Holden SE, Jenkins-Jones S, Morgan CL, Halcox JP, Schernthaner G, Mukherjee J, Currie CJ. 2014. Can people with type 2 diabetes live longer than those without? A comparison of mortality in people initiated with metformin or sulphonylurea monotherapy and matched, non-diabetic controls. *Diabetes Obes Metab* 16: 1165–1173.

Bartke A. 2011. Single-gene mutations and healthy ageing in mammals. *Philos Trans R Soc Lond B Biol Sci* 366: 28–34.

Basso N, Paglia N, Stella I, de Cavanagh EM, Ferder L, del Rosario Lores Arnaiz M, Inserra F. 2005. Protective effect of the inhibition of the renin-angiotensin system on aging. *Regul Pept* 128: 247–252.

Basso N, Cini R, Pietrelli A, Ferder L, Terragno NA, Inserra F. 2007. Protective effect of long-term angiotensin II inhibition. *Am J Physiol Heart Circ Physiol* 293: H1351–H1358.

Baur JA, Ungvari Z, Minor RK, Le Couteur DG, de Cabo R. 2012. Are sirtuins viable targets for improving healthspan and lifespan? *Nat Rev Drug Discov* 11: 443–461.

Benson EK, Lee SW, Aaronson SA. 2010. Role of progerin-induced telomere dysfunction in HGPS premature cellular senescence. *J Cell Sci* 123: 2605–2612.

Brown DW, Giles WH, Croft JB. 2001. White blood cell count: An independent predictor of coronary heart disease mortality among a national cohort. *J Clin Epidemiol* 54: 316–322.

Bruunsgaard H, Pedersen BK. 2003. Age-related inflammatory cytokines and disease. *Immunol Allergy Clin North Am* 23: 15–39.

Bruunsgaard H, Andersen-Ranberg K, Hjelmborg JB, Pedersen BK, Jeune B. 2003. Elevated levels of tumor necrosis factor α and mortality in centenarians. *Am J Med* 115: 278–283.

Calhoun C, Shivshankar P, Saker M, Sloane LB, Livi CB, Sharp ZD, Orihuela CJ, Adnot S, White ES, Richardson A, et al. 2015. Senescent cells contribute to the physiological remodeling of aged lungs. *J Gerontol A Biol Sci Med Sci* doi: 10.1093/gerona/glu241.

Cesari M, Penninx BW, Newman AB, Kritchevsky SB, Nicklas BJ, Sutton-Tyrrell K, Tracy RP, Rubin SM, Harris TB, Pahor M. 2003. Inflammatory markers and cardiovascular disease (The Health, Aging and Body Composition [Health ABC] Study). *Am J Cardiol* 92: 522–528.

Chen Q, Liu K, Robinson AR, Clauson CL, Blair HC, Robbins PD, Niedernhofer LJ, Ouyang H. 2013. DNA damage drives accelerated bone aging via an NF-κB-dependent mechanism. *J Bone Miner Res* 28: 1214–1228.

Chiasson JL, Josse RG, Gomis R, Hanefeld M, Karasik A, Laakso M. 2002. Acarbose for prevention of type 2 diabetes mellitus: The STOP-NIDDM randomised trial. *Lancet* 359: 2072–2077.

Choi SH, Kim YH, Hebisch M, Sliwinski C, Lee S, D'Avanzo C, Chen H, Hooli B, Asselin C, Muffat J, et al. 2014. A three-dimensional human neural cell culture model of Alzheimer's disease. *Nature* **515:** 274–278.

Conboy IM, Conboy MJ, Wagers AJ, Girma ER, Weissman IL, Rando TA. 2005. Rejuvenation of aged progenitor cells by exposure to a young systemic environment. *Nature* **433:** 760–764.

Crandall J, Schade D, Ma Y, Fujimoto WY, Barrett-Connor E, Fowler S, Dagogo-Jack S, Andres R. 2006. The influence of age on the effects of lifestyle modification and metformin in prevention of diabetes. *J Gerontol A Biol Sci Med Sci* **61:** 1075–1081.

Davies NM, Kehoe PG, Ben-Shlomo Y, Martin RM. 2011. Associations of anti-hypertensive treatments with Alzheimer's disease, vascular dementia, and other dementias. *J Alzheimers Dis* **26:** 699–708.

De Felice FG, Ferreira ST. 2014. Inflammation, defective insulin signaling, and mitochondrial dysfunction as common molecular denominators connecting type 2 diabetes to Alzheimer disease. *Diabetes* **63:** 2262–2272.

Dykens JA, Moos WH, Howell N. 2005. Development of 17α-estradiol as a neuroprotective therapeutic agent: Rationale and results from a phase I clinical study. *Ann NY Acad Sci* **1052:** 116–135.

Eren M, Boe AE, Murphy SB, Place AT, Nagpal V, Morales-Nebreda L, Urich D, Quaggin SE, Budinger GR, Mutlu GM, et al. 2014. PAI-1-regulated extracellular proteolysis governs senescence and survival in *Klotho* mice. *Proc Natl Acad Sci* **111:** 7090–7095.

Ferder L, Inserra F, Romano L, Ercole L, Pszenny V. 1993. Effects of angiotensin-converting enzyme inhibition on mitochondrial number in the aging mouse. *Am J Physiol* **265:** C15–C18.

Ferrucci L, Harris TB, Guralnik JM, Tracy RP, Corti MC, Cohen HJ, Penninx B, Pahor M, Wallace R, Havlik RJ. 1999. Serum IL-6 level and the development of disability in older persons. *J Am Geriatr Soc* **47:** 639–646.

Fried LP, Tangen CM, Walston J, Newman AB, Hirsch C, Gottdiener J, Seeman T, Tracy R, Kop WJ, Burke G, et al. 2001. Frailty in older adults: Evidence for a phenotype. *J Gerontol A Biol Sci Med Sci* **56:** M146–M156

Fried LP, Xue QL, Cappola AR, Ferrucci L, Chaves P, Varadhan R, Guralnik JM, Leng SX, Semba RD, Walston JD, et al. 2009. Nonlinear multisystem physiological dysregulation associated with frailty in older women: Implications for etiology and treatment. *J Gerontol A Biol Sci Med Sci* **64:** 1049–1057.

Fry PS. 2000. Whose quality of life is it anyway? Why not ask seniors to tell us about it? *Int J Aging Hum Dev* **50:** 361–383.

Fry PS, Debats DL. 2006. Sources of life strengths as predictors of late-life mortality and survivorship. *Int J Aging Hum Dev* **62:** 303–334.

Goldfine AB, Fonseca V, Jablonski KA, Chen YD, Tipton L, Staten MA, Shoelson SE. 2013. Salicylate (salsalate) in patients with type 2 diabetes: A randomized trial. *Ann Intern Med* **159:** 1–12.

Goldman DP, Cutler D, Rowe JW, Michaud PC, Sullivan J, Peneva D, Olshansky SJ. 2013. Substantial health and economic returns from delayed aging may warrant a new focus for medical research. *Health Aff (Millwood)* **32:** 1698–1705.

Hanon O, Berrou JP, Negre-Pages L, Goch JH, Nadhazi Z, Petrella R, Sedefdjian A, Sevenier F, Shlyakhto EV, Pathak A. 2008. Effects of hypertension therapy based on eprosartan on systolic arterial blood pressure and cognitive function: Primary results of the Observational Study on Cognitive Function and Systolic Blood Pressure Reduction Open-Label Study. *J Hypertens* **26:** 1642–1650.

Harris TB, Ferrucci L, Tracy RP, Corti MC, Wacholder S, Ettinger WH Jr, Heimovitz H, Cohen HJ, Wallace R. 1999. Associations of elevated interleukin-6 and C-reactive protein levels with mortality in the elderly. *Am J Med* **106:** 506–512.

Harrison DE, Strong R, Sharp ZD, Nelson JF, Astle CM, Flurkey K, Nadon NL, Wilkinson JE, Frenkel K, Carter CS, et al. 2009. Rapamycin fed late in life extends lifespan in genetically heterogeneous mice. *Nature* **460:** 392–395.

Harrison DE, Strong R, Allison DB, Ames BN, Astle CM, Atamna H, Fernandez E, Flurkey K, Javors MA, Nadon NL, et al. 2014. Acarbose, 17-α-estradiol, and nordihydroguaiaretic acid extend mouse lifespan preferentially in males. *Aging Cell* **13:** 273–282.

Hong J, Zhang Y, Lai S, Lv A, Su Q, Dong Y, Zhou Z, Tang W, Zhao J, Cui L, et al. 2013. Effects of metformin versus glipizide on cardiovascular outcomes in patients with type 2 diabetes and coronary artery disease. *Diab Care* **36:** 1304–1311.

Howren MB, Lamkin DM, Suls J. 2009. Associations of depression with C-reactive protein, IL-1, and IL-6: A meta-analysis. *Psychosom Med* **71:** 171–186.

Hu FB, Meigs JB, Li TY, Rifai N, Manson JE. 2004. Inflammatory markers and risk of developing type 2 diabetes in women. *Diabetes* **53:** 693–700.

Huang CC, Chan WL, Chen YC, Chen TJ, Lin SJ, Chen JW, Leu HB. 2011. Angiotensin II receptor blockers and risk of cancer in patients with systemic hypertension. *Am J Cardiol* **107:** 1028–1033.

Hudson MM, Ness KK, Gurney JG, Mulrooney DA, Chemaitilly W, Krull KR, Green DM, Armstrong GT, Nottage KA, Jones KE, et al. 2013. Clinical ascertainment of health outcomes among adults treated for childhood cancer. *J Am Med Assoc* **309:** 2371–2381.

Johnson JA, Simpson SH, Toth EL, Majumdar SR. 2005. Reduced cardiovascular morbidity and mortality associated with metformin use in subjects with type 2 diabetes. *Diabet Med* **22:** 497–502.

Joost HG. 2014. Diabetes and cancer: Epidemiology and potential mechanisms. *Diab Vasc Dis Res* **11:** 390–394.

Kanapuru B, Ershler WB. 2009. Inflammation, coagulation, and the pathway to frailty. *Am J Med* **122:** 605–613.

Karnevi E, Said K, Andersson R, Rosendahl AH. 2013. Metformin-mediated growth inhibition involves suppression of the IGF-I receptor signalling pathway in human pancreatic cancer cells. *BMC Cancer* **13:** 235.

Katsimpardi L, Litterman NK, Schein PA, Miller CM, Loffredo FS, Wojtkiewicz GR, Chen JW, Lee RT, Wagers AJ, Rubin LL. 2014. Vascular and neurogenic rejuvenation of the aging mouse brain by young systemic factors. *Science* **344:** 630–634.

Kennedy BK, Pennypacker JK. 2014. Drugs that modulate aging: The promising yet difficult path ahead. *Transl Res* **163:** 456–465.

Kirkland JL. 2013a. Translating advances from the basic biology of aging into clinical application. *Exp Gerontol* **48:** 1–5.

Kirkland JL. 2013b. Inflammation and cellular senescence: Potential contribution to chronic diseases and disabilities with aging. *Public Policy Aging Rep* **23:** 12–15.

Kirkland JL, Peterson C. 2009. Healthspan, translation, and new outcomes for animal studies of aging. *J Gerontol A Biol Sci Med Sci* **64:** 209–212.

Kirkland JL, Tchkonia T. 2014. Clinical strategies and animal models for developing senolytic agents. *Exp Gerontol* **68:** 19–25.

Lamanna C, Monami M, Marchionni N, Mannucci E. 2011. Effect of metformin on cardiovascular events and mortality: A meta-analysis of randomized clinical trials. *Diabetes Obes Metab* **13:** 221–228.

Lamming DW, Ye L, Katajisto P, Goncalves MD, Saitoh M, Stevens DM, Davis JG, Salmon AB, Richardson A, Ahima RS, et al. 2012. Rapamycin-induced insulin resistance is mediated by mTORC2 loss and uncoupled from longevity. *Science* **335:** 1638–1643.

Lamming DW, Ye L, Sabatini DM, Baur JA. 2013. Rapalogs and mTOR inhibitors as anti-aging therapeutics. *J Clin Invest* **123:** 980–989.

Landman GW, Kleefstra N, van Hateren KJ, Groenier KH, Gans RO, Bilo HJ. 2010. Metformin associated with lower cancer mortality in type 2 diabetes: ZODIAC-16. *Diab Care* **33:** 322–326.

Lavasani M, Robinson AR, Lu A, Song M, Feduska JM, Ahani B, Tilstra JS, Feldman CH, Robbins PD, Niedernhofer LJ, et al. 2012. Muscle-derived stem/progenitor cell dysfunction limits healthspan and lifespan in a murine progeria model. *Nat Commun* **3:** 608.

Lee MS, Hsu CC, Wahlqvist ML, Tsai HN, Chang YH, Huang YC. 2011. Type 2 diabetes increases and metformin reduces total, colorectal, liver and pancreatic cancer incidences in Taiwanese: A representative population prospective cohort study of 800,000 individuals. *BMC Cancer* **11:** 20.

Lee C, Raffaghello L, Brandhorst S, Safdie FM, Bianchi G, Martin-Montalvo A, Pistoia V, Wei M, Hwang S, Merlino A, et al. 2012. Fasting cycles retard growth of tumors and sensitize a range of cancer cell types to chemotherapy. *Sci Transl Med* **4:** p124ra127.

Leng SX, Xue QL, Tian J, Walston JD, Fried LP. 2007. Inflammation and frailty in older women. *J Am Geriatr Soc* **55:** 864–871.

Li NC, Lee A, Whitmer RA, Kivipelto M, Lawler E, Kazis LE, Wolozin B. 2010. Use of angiotensin receptor blockers and risk of dementia in a predominantly male population: Prospective cohort analysis. *BMJ* **340:** b5465.

Libby G, Donnelly LA, Donnan PT, Alessi DR, Morris AD, Evans JM. 2009. New users of metformin are at low risk of incident cancer: A cohort study among people with type 2 diabetes. *Diab Care* **32:** 1620–1625.

Linz W, Jessen T, Becker RH, Scholkens BA, Wiemer G. 1997. Long-term ACE inhibition doubles lifespan of hypertensive rats. *Circulation* **96:** 3164–3172.

Linz W, Heitsch H, Scholkens BA, Wiemer G. 2000. Long-term angiotensin II type 1 receptor blockade with fonsartan doubles lifespan of hypertensive rats. *Hypertension* **35:** 908–913.

Lipton RB, Hirsch J, Katz MJ, Wang C, Sanders AE, Verghese J, Barzilai N, Derby CA. 2010. Exceptional parental longevity associated with lower risk of Alzheimer's disease and memory decline. *J Am Geriatr Soc* **58:** 1043–1049.

Liu B, Fan Z, Edgerton SM, Yang X, Lind SE, Thor AD. 2011. Potent anti-proliferative effects of metformin on trastuzumab-resistant breast cancer cells via inhibition of erbB2/IGF-1 receptor interactions. *Cell Cycle* **10:** 2959–2966.

Lucicesare A, Hubbard RE, Searle SD, Rockwood K. 2010. An index of self-rated health deficits in relation to frailty and adverse outcomes in older adults. *Aging Clin Exp Res* **22:** 255–260.

Majumder S, Caccamo A, Medina DX, Benavides AD, Javors MA, Kraig E, Strong R, Richardson A, Oddo S. 2012. Lifelong rapamycin administration ameliorates age-dependent cognitive deficits by reducing IL-1β and enhancing NMDA signaling. *Aging Cell* **11:** 326–335.

Mannick JB, Del Giudice G, Lattanzi M, Valiante NM, Praestgaard J, Huang B, Lonetto MA, Maecker HT, Kovarik J, Carson S, et al. 2014. mTOR inhibition improves immune function in the elderly. *Sci Transl Med* **6:** 268ra179.

Margolis KL, Manson JE, Greenland P, Rodabough RJ, Bray PF, Safford M, Grimm RH Jr, Howard BV, Assaf AR, Prentice R. 2005. Leukocyte count as a predictor of cardiovascular events and mortality in postmenopausal women: The Women's Health Initiative Observational Study. *Arch Intern Med* **165:** 500–508.

Martin-Montalvo A, Mercken EM, Mitchell SJ, Palacios HH, Mote PL, Scheibye-Knudsen M, Gomes AP, Ward TM, Minor RK, Blouin MJ, et al. 2013. Metformin improves healthspan and lifespan in mice. *Nat Commun* **4:** 2192.

McLean RR, Shardell MD, Alley DE, Cawthon PM, Fragala MS, Harris TB, Kenny AM, Peters KW, Ferrucci L, Guralnik JM, et al. 2014. Criteria for clinically relevant weakness and low lean mass and their longitudinal association with incident mobility impairment and mortality: The Foundation for the National Institutes of Health (FNIH) Sarcopenia Project. *J Gerontol A Biol Sci Med Sci* **69:** 576–583.

Mercken EM, Mitchell SJ, Martin-Montalvo A, Minor RK, Almeida M, Gomes AP, Scheibye-Knudsen M, Palacios HH, Licata JJ, Zhang Y, et al. 2014. SRT2104 extends survival of male mice on a standard diet and preserves bone and muscle mass. *Aging Cell* **13:** 787–796.

Minagawa S, Araya J, Numata T, Nojiri S, Hara H, Yumino Y, Kawaishi M, Odaka M, Morikawa T, Nishimura SL, et al. 2011. Accelerated epithelial cell senescence in IPF and the inhibitory role of SIRT6 in TGF-β-induced senescence of human bronchial epithelial cells. *Am J Physiol Lung Cell Mol Physiol* **300:** L391–L401.

Mitchell SJ, Martin-Montalvo A, Mercken EM, Palacios HH, Ward TM, Abulwerdi G, Minor RK, Vlasuk GP, Ellis JL, Sinclair DA, et al. 2014. The SIRT1 activator SRT1720 extends lifespan and improves health of mice fed a standard diet. *Cell Rep* **6:** 836–843.

Monami M, Colombi C, Balzi D, Dicembrini I, Giannini S, Melani C, Vitale V, Romano D, Barchielli A, Marchionni N, et al. 2011. Metformin and cancer occurrence in insulin-treated type 2 diabetic patients. *Diab Care* **34:** 129–131.

Moore BB, Hogaboam CM. 2008. Murine models of pulmonary fibrosis. *Am J Physiol Lung Cell Mol Physiol* **294:** L152–160.

Ness KK, Armstrong GT, Kundu M, Wilson CL, Tchkonia T, Kirkland JL. 2014. Frailty in childhood cancer survivors. *Cancer* **121:** 1540–1547.

O'Connor PM, Lapointe TK, Beck PL, Buret AG. 2010. Mechanisms by which inflammation may increase intestinal cancer risk in inflammatory bowel disease. *Inflamm Bowel Dis* **16:** 1411–1420.

Olshansky SJ, Carnes BA, Cassel C. 1990. In search of Methuselah: Estimating the upper limits to human longevity. *Science* **250:** 634–640.

Pai JK, Pischon T, Ma J, Manson JE, Hankinson SE, Joshipura K, Curhan GC, Rifai N, Cannuscio CC, Stampfer MJ, et al. 2004. Inflammatory markers and the risk of coronary heart disease in men and women. *N Engl J Med* **351:** 2599–2610.

Pasternak B, Svanstrom H, Callreus T, Melbye M, Hviid A. 2011. Use of angiotensin receptor blockers and the risk of cancer. *Circulation* **123:** 1729–1736.

Pradhan AD, Manson JE, Rifai N, Buring JE, Ridker PM. 2001. C-reactive protein, interleukin 6, and risk of developing type 2 diabetes mellitus. *J Amer Med Assoc* **286:** 327–334.

Qu T, Walston JD, Yang H, Fedarko NS, Xue QL, Beamer BA, Ferrucci L, Rose NR, Leng SX. 2009. Upregulated ex vivo expression of stress-responsive inflammatory pathway genes by LPS-challenged CD14$^+$ monocytes in frail older adults. *Mech Ageing Dev* **130:** 161–166.

Quinn BJ, Dallos M, Kitagawa H, Kunnumakkara AB, Memmott RM, Hollander MC, Gills JJ, Dennis PA. 2013. Inhibition of lung tumorigenesis by metformin is associated with decreased plasma IGF-I and diminished receptor tyrosine kinase signaling. *Cancer Prev Res (Phila)* **6:** 801–810.

Richardson A, Galvan V, Lin AL, Oddo S. 2014. How longevity research can lead to therapies for Alzheimer's disease: The rapamycin story. *Exp Gerontol* **68:** 51–58.

Rockwood K, Mitnitski A. 2011. Frailty defined by deficit accumulation and geriatric medicine defined by frailty. *Clin Geriatr Med* **27:** 17–26.

Rockwood K, Mitnitski A, Song X, Steen B, Skoog I. 2006. Long-term risks of death and institutionalization of elderly people in relation to deficit accumulation at age 70. *J Am Geriatr Soc* **54:** 975–979.

Roumie CL, Hung AM, Greevy RA, Grijalva CG, Liu X, Murff HJ, Elasy TA, Griffin MR. 2012. Comparative effectiveness of sulfonylurea and metformin monotherapy on cardiovascular events in type 2 diabetes mellitus: A cohort study. *Ann Intern Med* **157:** 601–610.

Ryu JH, Moua T, Daniels CE, Hartman TE, Yi ES, Utz JP, Limper AH. 2014. Idiopathic pulmonary fibrosis: Evolving concepts. *Mayo Clin Proc* **89:** 1130–1142.

Salani B, Maffioli S, Hamoudane M, Parodi A, Ravera S, Passalacqua M, Alama A, Nhiri M, Cordera R, Maggi D. 2012. Caveolin-1 is essential for metformin inhibitory effect on IGF1 action in non-small-cell lung cancer cells. *FASEB J* **26:** 788–798.

Santos EL, de Picoli Souza K, da Silva ED, Batista EC, Martins PJ, D'Almeida V, Pesquero JB. 2009. Long term treatment with ACE inhibitor enalapril decreases body weight gain and increases life span in rats. *Biochem Pharmacol* **78:** 951–958.

Schetter AJ, Heegaard NH, Harris CC. 2010. Inflammation and cancer: Interweaving microRNA, free radical, cytokine and p53 pathways. *Carcinogenesis* **31:** 37–49.

Seibel SA, Chou KH, Capp E, Spritzer PM, von Eye Corleta H. 2008. Effect of metformin on IGF-1 and IGFBP-1 levels in obese patients with polycystic ovary syndrome. *Eur J Obstet Gynecol Reprod Biol* **138:** 122–124.

Shineman DW, Basi GS, Bizon JL, Colton CA, Greenberg BD, Hollister BA, Lincecum J, Leblanc GG, Lee LB, Luo F, et al. 2011. Accelerating drug discovery for Alzheimer's disease: Best practices for preclinical animal studies. *Alzheimers Res Ther* **3:** 28.

Sinha M, Jang YC, Oh J, Khong D, Wu EY, Manohar R, Miller C, Regalado SG, Loffredo FS, Pancoast JR, et al. 2014. Restoring systemic GDF11 levels reverses age-related dysfunction in mouse skeletal muscle. *Science* **344:** 649–652.

Smith DL Jr, Elam CF Jr, Mattison JA, Lane MA, Roth GS, Ingram DK, Allison DB. 2010. Metformin supplementation and life span in Fischer-344 rats. *J Gerontol A Biol Sci Med Sci* **65:** 468–474.

Spranger J, Kroke A, Mohlig M, Hoffmann K, Bergmann MM, Ristow M, Boeing H, Pfeiffer AF. 2003. Inflammatory cytokines and the risk to develop type 2 diabetes: Results of the prospective population-based European Prospective Investigation into Cancer and Nutrition (EPIC)-Potsdam Study. *Diabetes* **52:** 812–817.

Srikrishna G, Freeze HH. 2009. Endogenous damage-associated molecular pattern molecules at the crossroads of inflammation and cancer. *Neoplasia* **11:** 615–628.

Steinmetz KL, Spack EG. 2009. The basics of preclinical drug development for neurodegenerative disease indications. *BMC Neurol* **9:** pS2.

Strong R, Miller RA, Astle CM, Floyd RA, Flurkey K, Hensley KL, Javors MA, Leeuwenburgh C, Nelson JF, Ongini E. 2008. Nordihydroguaiaretic acid and aspirin increase lifespan of genetically heterogeneous male mice. *Aging Cell* **7:** 641–650.

Takasaka N, Araya J, Hara H, Ito S, Kobayashi K, Kurita Y, Wakui H, Yoshii Y, Yumino Y, Fujii S, et al. 2014. Autophagy induction by SIRT6 through attenuation of insulin-like growth factor signaling is involved in the regulation of human bronchial epithelial cell senescence. *J Immunol* **192:** 958–968.

Tatar M. 2009. Can we develop genetically tractable models to assess healthspan (rather than life span) in animal models? *J Gerontol A Biol Sci Med Sci* **64:** 161–163.

Tchkonia T, Morbeck DE, von Zglinicki T, van Deursen J, Lustgarten J, Scrable H, Khosla S, Jensen MD, Kirkland JL. 2010. Fat tissue, aging, and cellular senescence. *Aging Cell* **9:** 667–684.

Tchkonia T, Zhu Y, van Deursen J, Campisi J, Kirkland JL. 2013. Cellular senescence and the senescent secretory phenotype: Therapeutic opportunities. *J Clin Invest* **123**: 966–972.

Tosca L, Rame C, Chabrolle C, Tesseraud S, Dupont J. 2010. Metformin decreases IGF1-induced cell proliferation and protein synthesis through AMP-activated protein kinase in cultured bovine granulosa cells. *Reproduction* **139**: 409–418.

Tseng CH. 2012. Diabetes, metformin use, and colon cancer: A population-based cohort study in Taiwan. *Eur J Endocrinol* **167**: 409–416.

Tuomisto K, Jousilahti P, Sundvall J, Pajunen P, Salomaa V. 2006. C-reactive protein, interleukin-6 and tumor necrosis factor alpha as predictors of incident coronary and cardiovascular events and total mortality. A population-based, prospective study. *Thromb Haemost* **95**: 511–518.

Villareal DT, Chode S, Parimi N, Sinacore DR, Hilton T, Armamento-Villareal R, Napoli N, Qualls C, Shah K. 2011. Weight loss, exercise, or both and physical function in obese older adults. *N Engl J Med* **364**: 1218–1229.

Walston J, McBurnie MA, Newman A, Tracy RP, Kop WJ, Hirsch CH, Gottdiener J, Fried LP. 2002. Frailty and activation of the inflammation and coagulation systems with and without clinical comorbidities: Results from the Cardiovascular Health Study. *Arch Intern Med* **162**: 2333–2341.

Walston J, Hadley E, Ferrucci L, Guralnick JM, Newman AB, Studenski SA, Ershler WB, Harris T, Fried LP. 2006. Research agenda for frailty in older adults: Toward a better understanding of physiology and etiology: Summary from the American Geriatrics Society/National Institute on Aging research conference on frailty in older adults. *J Am Geriatr Soc* **54**: 991–1001.

Walston JD, Matteini AM, Nievergelt C, Lange LA, Fallin DM, Barzilai N, Ziv E, Pawlikowska L, Kwok P, Cummings SR, et al. 2009. Inflammation and stress-related candidate genes, plasma interleukin-6 levels, and longevity in older adults. *Exp Gerontol* **44**: 350–355.

Whittington HJ, Hall AR, McLaughlin CP, Hausenloy DJ, Yellon DM, Mocanu MM. 2013. Chronic metformin associated cardioprotection against infarction: Not just a glucose lowering phenomenon. *Cardiovasc Drugs Ther* **27**: 5–16.

Zeymer U, Schwarzmaier-D'assie A, Petzinna D, Chiasson JL. 2004. Effect of acarbose treatment on the risk of silent myocardial infarctions in patients with impaired glucose tolerance: Results of the randomised STOP-NIDDM trial electrocardiography substudy. *Eur J Cardiovasc Prev Rehabil* **11**: 412–415.

Interventions for Human Frailty: Physical Activity as a Model

Linda P. Fried

Columbia University Medical Center, New York, New York 10032
Correspondence: lpfried@columbia.edu

In the last 100 years, populations in developed countries have experienced an unprecedented addition of 30 years to life expectancy. Developing countries are now experiencing this same phenomenon, but over a shorter time frame. With this success comes the challenge of maximizing health and vitality across these added years. The compression of morbidity to the latest point in the human life span could unleash a sustained third demographic dividend that benefits all of society. To accomplish this, society needs to invest in the prevention and treatment of frailty, as well as in the prevention of chronic diseases at every age and stage of life. A model intervention, physical activity, may offer a road map.

COMPLETING THE TRANSITION TO A WORLD OF LONGER LIVES: OPTIMIZING HEALTH ACROSS OUR LONGER LIFE COURSE

In the last 100 years, developed countries have experienced the unprecedented addition of 30—and more—years to life expectancy. Many developing countries are now experiencing this same extension in life expectancy, but generally over a much more compressed time frame of approximately 40 years. In all of these societies, this unprecedented addition of years of life has been because of public health interventions that have targeted infectious diseases and decreased early childhood and maternal mortality. Having now added, essentially, a new stage of life, the critical challenge is to understand how to maximize health and vitality across these longer years. Accomplishing this is essential to unleashing the opportunities and benefits of our longer lives—for older adults themselves and for all ages (Fried 2015). This goal has been expressed as the need to "compress morbidity" to the latest point in the human life span so that people live free of the diseases and conditions of old age for much of that old age (Fries 2001). If accomplished, a compression of morbidity could unleash a sustained third demographic dividend for all of society. However, to accomplish this requires society investing in creating what we could term a new life stage in which a healthy older population is able to make significant contributions to societal well-being and productivity (Fried in press a).

Accomplishing this compression of morbidity requires two significant components. First is investing in the prevention of the development of chronic diseases at every age and stage of life, which is described as a life course approach to prevention (Fried in press a). There is evidence that people who arrive at age 70 in good health are at high likelihood of remaining healthy into their later years (Fried in press a). On the other hand, those with cumulative exposures to adverse risk factors across childhood and adulthood manifest single and then multiple diseases as they get older, and are likely to become disabled secondary to these diseases as early as their 50s and 60s (Wolinsky et al. 1983; Levine and Crimmins 2014; Geronimus et al. 2015; Fried and Ferrucci, in press).

The second approach focuses on optimizing health and preventing or slowing the onset of frailty, dementia, multimorbidity, and disability in older age itself. The remainder of this article considers the issue of frailty and the opportunities for delaying, preventing, or treating frailty, considering insights offered from a model intervention: physical activity.

AGING AND DISEASE

Population-based research is now offering evidence, across multiple studies, that the process of aging can be distinguished from the processes and consequences of diseases. The first stage of such research, over the last 50 years, showed that diseases of specific organ systems, such as atherosclerotic disease of the heart, brain, and peripheral vasculature, could be understood

mechanistically as processes distinct from aging itself. These observations led to substantial research that showed that many chronic diseases are, in fact, preventable into the oldest ages. The implications, of course, are that these diseases were not inevitable with aging or, at least, not tied to a chronological age. Other innovative observations have suggested that a focus on investigating individual end-organ diseases misses a major conclusion of mechanistic research: that there are shared risk factors and etiologic pathways for multiple chronic diseases. At the behavioral level, these risk factors include physical activity, smoking, and diet, and, at the intrinsic level, there are mechanisms in common for multiple chronic diseases, such as inflammation (Tinetti and Fried 2004; Fried and Ferrucci, in press).

Parallel research has been increasingly seeking to characterize, in humans, the independent contributions of aging itself to health outcomes independent of superimposed chronic diseases. Interestingly, in multivariate analyses in a population-based cohort study, the Cardiovascular Health Study, specific chronic diseases are the independent predictors of 5-year mortality into the oldest ages; however, after 85 years, age becomes an additional independent predictor (Fig. 1). This suggests the influence of biologic

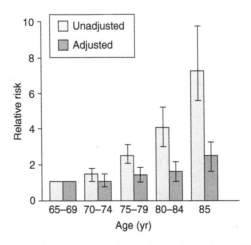

Figure 1. Age emerges as an independent predictor of mortality, independent of chronic diseases, beginning at age 85 years, based on analyses in community-dwelling men and women 65 years and older in four U.S. communities participating in the Cardiovascular Health Study. (From Fried et al. 1998; reprinted, with permission.)

aging as a force for mortality, emerging independently of disease in the oldest old. Additionally, new research suggests that biologic aging can be characterized using population-based approaches as the composite of altered physiologic regulation and function across multiple systems. With this approach, longitudinally, young as well as older adults show differentiation between biologic and chronologic age, potentially as young as the mid-30s (Cohen et al. 2014; Belsky et al. 2015; Cohen et al. 2015).

THE EMERGENCE OF FRAILTY IN OLDER AGE

With increasing age, after about age 70, a new phenotype emerges, which is distinct from any single chronic disease and is an independent predictor of mortality in the short (3 yr) and intermediate (7 yr) term: that of frailty. This phenotype (Table 1) meets the criteria for a medical syndrome (Fried et al. 2001; Bandeen-Roche et al. 2006; Morley et al. 2013). Criteria consist of a critical mass of three to five clinical components, including muscle weakness, slowed gait, low physical activity, sense of low energy or exhaustion, and unintentional weight loss. For 90% of those who become frail, frailty is chronic and progressive in its development (Xue 2011). Those with a subclinical presentation of only one or two components are at 2.5-fold increased risk of progressing to having three to five components present in 3 yr with full manifestation of the frailty phenotype (Fried et al. 2001). The prevalence of frailty increases with age after age 65, and is twice as prevalent in women as men and African-Americans as whites (Fried et al. 2001; Hirsch et al. 2006). By age 85, ~25% of older adults in the community manifest this phenotype; that is, they show three, four, or five components of frailty (Fried et al. 2001).

The characteristics of the phenotype were originally hypothesized to be a vicious cycle resulting from dysregulated energetics (Fig. 2). Such a cycle can be potentially initiated at any point, but the evidence is that in most older adults who become frail, it begins with declines in strength, walking speed, and/or physical activity and then progresses to also include self-reported low energy or exhaustion, and significant unintended weight loss (Xue 2011).

Notably, this phenotypic presentation has the characteristics of a medical syndrome: a constellation of symptoms and signs, when present in a critical mass, which predicts both characteristic outcomes and identifies a distinct underlying pathophysiology. The critical mass of criteria that are present have greater specificity for both outcomes and pathophysiology than any one or two factors (Fried et al. 2001; Bandeen-Roche et al. 2006). This constellation of factors has been shown to have the dynamics of a clinical syndrome in which there are no distinct subsets of criteria that cluster together, and any three (of five) factors that are present synergistically predict outcomes of mortality, disability onset and progression, hospitalization, falls, and nursing home admission, significantly more than any one or two factors (Bandeen-Roche et al. 2006). It has recently been recommended, in an international consensus conference, that frailty be considered a

Table 1. Criteria that define frailty[a]

Characteristic	Criteria for frailty[b]
Weight loss	Lost >10 pounds unintentionally last year
Exhaustion	Felt last week that "everything I did was an effort" or "I could not get going"
Slowness	Time to walk 15 feet (cutoff depends on sex and height)
Low activity level	Expends <270 kcal/week (calculated from activity scale incorporating episodes of walking, household chores, yard work, etc.)
Weakness	Grip strength measured using hand dynamometer (cutoff depends on sex and body mass index [BMI])

From Fried in press b; reprinted, with permission.
[a]Syndrome present when more than three characteristics are identified.
[b]For specific measures and details for determining frailty criteria, see Fried et al. 2001.

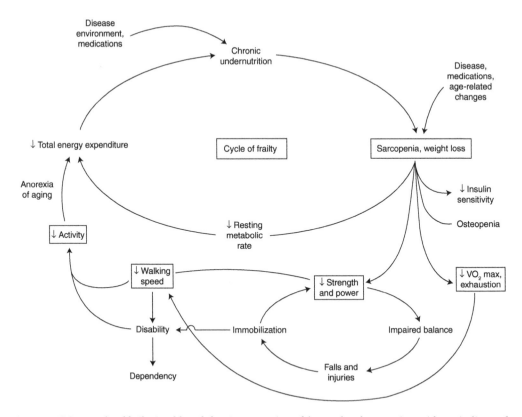

Figure 2. Vicious cycle of frailty in older adults. An expression of dysregulated energetics, evidence indicates that numerous exposures and chronic diseases, as well as aging-related processes, can initiate this cycle at any point. However, the early manifestations in the main are declines in muscle strength, walking speed, and/or physical activity, which predict development of exhaustion and—at end stages—significant unintentional weight loss. (From Fried et al. 2001; reprinted with permission.)

new medical syndrome (Morley et al. 2013) and that clinicians screen for frailty in people 70 and older as a basis for intervening to prevent or alleviate further decline.

MECHANISMS UNDERLYING THE CLINICAL SYNDROME OF FRAILTY

Consistent with the definition of a syndrome, the clinical presentation, or phenotype, identifies a specific etiologic pathophysiology. The extant evidence is briefly summarized here, considering both physiologic and biologic genesis.

First, there are a number of individual physiologic systems that have been identified as dysregulated in association with frailty. As background, there are many data that indicate declines with age in function of multiple func-

tional and regulatory systems in the body, such as circulating levels of specific hormones (e.g., estrogens, testosterone, and insulin-like growth factor 1 [IGF-1]) (Kuchel et al. 2001), impaired insulin sensitivity (Davidson 1979; DeFronzo 1981; Shimokata et al. 1991; Scheen 2005; Metter et al. 2008), along with elevations of markers of inflammation (e.g., interleukin 6 [IL-6] and C-reactive protein [CRP]) with aging (Ershler and Keller 2000; Reuben et al. 2002; Cohen et al. 2003; Joseph et al. 2005). Dynamic interactions between declines in gonadal hormones and increases in inflammatory mediators affect bone mass, as well as mobility (Cappola et al. 2003), whereas the impaired insulin sensitivity is predictive of both frailty and diabetes (Kalyani et al. 2012b). Notably, observations on insulin resistance indicate that, although this becomes apparent in the vast majority of women >85, frailty is associated with highest levels of dysregulation. For example, as exemplified for insulin resistance-homeostatic model assessment (IR-HOMA) (Barzilay et al. 2007) or glycated hemoglobin (HgbA1C) (Kalyani et al. 2012b), an HgbA1C of 6.5% and 8.0% or greater, respectively, are associated with frailty prevalence (Blaum et al. 2009) and threefold higher incidence (Kalyani et al. 2012b). In the latter study, the association of HgbA1C with frailty incidence is nonlinear. This suggests that the change in physiologic function with aging is associated with frailty past a threshold level of severity.

With this pattern in mind, I briefly summarize here the physiologic and biologic systems shown, to date, to be associated—when dysregulated—with the syndrome of frailty. For example, frail men and women 65 and older have lower muscle density and muscle mass and higher fat mass than do nonfrail persons (Cesari et al. 2006). Phenotypic frailty is also associated with altered resting metabolic rate (Weiss et al. 2012). Elevation of the following physiologic systems have been associated with the phenotype of frailty: elevated glucose and insulin in response to glucose tolerance test and elevated HgbA1C (Kalyani et al. 2012b), IR-HOMA, and inflammation (Walston et al. 2002; Barzilay et al. 2007); higher evening cortisol levels (Varadhan et al. 2008; Johar et al. 2014); elevated white blood cell counts (Baylis et al. 2013); and increased heart rate variability in association with frailty (Chaves et al. 2005). Along with the direct association between level of inflammation and frailty, there is an increase in frailty with the number of inflammatory diseases present (Chang et al. 2012).

Systems that are lower in association with frailty are: IGF-1, testosterone, didehydroepiandrosterone sulfate (DHEA-S), markers of clotting (Walston et al. 2002; Kalyani et al. 2012a), hemoglobin (Chaves et al. 2005), and serum carotenoid and micronutrient concentrations, with greater likelihood of multiple micronutrient deficiencies (Michelon et al. 2006). Low daily energy intake (<21 kcal/kg) is associated with 24% increased risk of frailty (Bartali et al. 2006). Low protein and vitamins D, C, and folate intakes are also associated with frailty, along with low intake of more than three nutrients.

Further, there is mounting evidence that biologic systems that are altered with aging also predict frailty. For example, building on the hypothesis that dysregulated energetics may underlie the development of frailty (Fried 2001), increasing attention is being focused on mitochondrial dysfunction. There is now strong evidence for such dysfunction with aging itself. The decline in mitochondrial function with aging limits production of adenosine triphosphate (ATP) (Chistiakov et al. 2014), which most prominently affects the function of tissues with high energy demand, for example, the nervous system, heart, and skeletal muscle. This has been shown to predict sarcopenia and loss of physical fitness and an increase in central obesity, diabetes, fatigue, and inflammation (Coen et al. 2013), all of which contribute to frailty. Now there is more specific implication of a mitochondrial genetic variant in frailty (Moore et al. 2010), and evidence that mitochondrial DNA (mtDNA) levels and number might form part of the biological component of the phenotype (Ashar et al. 2015). In separate studies, mtDNA copy number has been associated with ATP production rate (Short et al. 2005), and worsening in frailty status is associated with a significant decrease in global DNA methylation levels (Bellizzi et al. 2012). These findings support the theory that aging-associated decrease in mitochondrial function, with alterations in production of energy, might underlie both the phenotype and dysregulations of frailty.

Second, the whole is greater than the sum of the parts. It is clear that these individual systems are not independent of each other. For example, sarcopenia results, in part, both from hormonal deficiency and cytokine excess (Newman et al. 2001; Morley et al. 2005). Postmenopausal declines in serum estrogen and androgen levels contribute to increased bone loss (Joseph et al. 2005). Higher serum testosterone in older women is associated with insulin resistance (IR) and metabolic syndrome (Patel et al. 2009). Both IR and diabetes have been associated with excessive loss of lean body mass and muscle strength (Park et al. 2007, 2009). Further, IR is associated with skeletal muscle mitochondrial dysfunction (Phielix et al. 2008). Many physiological declines are mediated by proinflammatory cytokines, decline in hormones (especially, testosterone [T] and IGF-1), and IR; this has been well summarized by Morley et al. (2005). Further, inflammation is associated with lower leg muscle mass and strength, IR, frailty, and disability (Guralnik et al. 1994; Ferrucci et al. 2002; Visser et al. 2002; Hubbard et al. 2010).

With regard to frailty, prevalence and incidence of frailty is a product of the multiplicity of dysregulated systems, more than any one system. Overall, it has been shown that there is a nonlinear association of the number of systems dysregulated with frailty (Fried et al. 2009), and even within systems, the number of abnormal hormones (Cappola et al. 2003) or micronutrients (Semba et al. 2006) associated with frailty. Further, higher cortisol, DHEA-S ratio, as well as lower levels of DHEA-S and higher white blood cell counts predicted frailty over 10-yr follow-up, indicating interactions between the immune and endocrine axis (Baylis et al. 2013). One summary has stated that the "presence of decreased gonadal hormones and IGF-1, combined with unusually high peripheral levels of cytokines, inflammatory mediators, and coagulation markers all enhance risk of sarcopenia and frailty" (Joseph et al. 2005).

The nonlinear association of the number of systems dysregulated with the probability of frailty is consistent with frailty emerging from severe dysregulation of the complex dynamical system that maintains organismal homeostasis and resilience (Fried et al. 2009). As shown in Figure 3, there is a nonlinear increase in the prevalence of the phenotype of frailty as the number of abnormal physiologic systems increases in women 70–79 in the Women's Health and Aging Studies. The systems considered here include: hematological (hemoglobin >12 mg/dL); inflammatory (IL-6 top tertile); hormonal (deficiencies in IGF-1, DHEA-S, and HgbA1C); adiposity (lowest quartile, triceps skinfold thickness); neuromuscular (slow fine motor speed); and two or more micronutrients deficient. Further, the population subdivided into two classes: 70% with one or no systems abnormal, and 30% with multiple systems (2.9, on average) abnormal; thus, three or more systems abnormal predict being frail. Notably, there were no dominating patterns of systems involved in association with frailty. However, the second class showed an odds ratio of 2.6 ($p < 0.05$) in association with being frail, adjusting for comorbid diseases, age, race, and education (Fried et al. 2009). Class 2 was associated with being frail after individually controlling for each of the eight systems, and none of the eight systems were individually significant except fine motor speed. Notably, the number

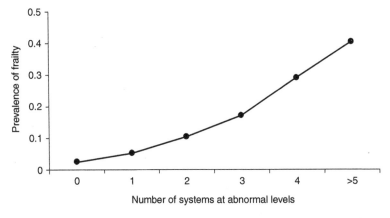

Figure 3. Association of number of physiological systems at abnormal levels with prevalence of being frail, women aged 70–79 ($p < 0.01$ for quantitative trend). (From Fried et al. 2009; reprinted, with permission.)

of comorbid diseases was also associated with frailty, but independent of multisystem abnormalities; the strength of association approximated that of class 2. These findings support the biology underlying the phenotype of frailty as being that of a complex system, with the dominating predictor being the number of systems dysregulated, not any particular system.

The picture gets even more complex when one considers that the physiological systems implicated in frailty each affect two or more of the five components of the frailty phenotype and further mutually affect or regulate each other. The substantial network of physiologic connections implied here is consistent with the interconnected systems of feedforward and -back loops that are elements of a complex dynamical system. It has been suggested that the aggregate effect of senescent processes, such as those in the systems studied here, have a nonlinear contribution to a complex system (Yates 2002), even if their decline over the adult life course displays a quasilinear property individually. The findings summarized above are consistent with this theory, and with the concept of a complex adaptive system in which dysregulation of a threshold number of systems may culminate in critical loss of homeostatic adaptive capacity. This emergent property of frailty as the manifestation of a dysregulated complex adaptive system supports the concept of frailty as a biologically coherent behavior (Kitano 2002), and provides insight into reasons for the vulnerability of older adults who are in this emergent state of frailty to adverse health outcomes (e.g., death, falls, and disability). It is further supported by emerging work proposing underlying processes of alterations of networks of biomarkers with aging, with the decrements in whole networks predicting frailty (Cohen et al. 2015).

In summary, the phenotype of frailty appears to emerge in relation to latent, apparently etiologic, changes of increasing dysregulation of multiple physiologic systems. Frailty emerges when a threshold level of dysregulation develops. The severity of their individual and aggregate dysregulation predicts the severity of the phenotype, and both dysregulation and the emergence of the phenotype are age associated

and predict mortality, but do not predict specific chronic diseases (Cohen et al. 2003, 2013, 2015). These changes involve, first and centrally, systems that mutually regulate homeostasis and response to stress and injury. Second, the findings are consistent with the dysregulation of a physiologic complex adaptive system, which is essential for resiliency (Fried et al. 2009).

Third, there is evidence that this dysregulated complex system is particularly manifest in the response to a challenge or stressor, and may well underlie the high risk of adverse outcomes in frailty. In a series of studies of response to an oral glucose tolerance test administered to 85- to 95-yr-old women in the Women's Health and Aging Study II, >70% had compromised response to the glucose tolerance test; only 27% had a normal fasting glucose and normal oral glucose tolerance test (OGTT). However, those who were frail had a differentially worse response. Specifically, women who were frail had 67 mg/dL greater increase in glucose in response to OGTT, and markedly delayed return of glucose and insulin back to baseline, compared with those who were nonfrail or prefrail (Kalyani et al. 2012b).

Further, frail women showed abnormal hormonal stimulus-response dynamics in energy metabolism response after oral glucose load (Kalyani et al. 2012a), with a pattern of elevated glucose-raising hormones and decreased glucose-lowering hormones in the frail compared with the nonfrail. Overall, the findings suggested that frailty status may represent a threshold of dysregulation in energy metabolism hormones, particularly for ghrelin. These findings support the theory that the entire physiological network of hormones that mutually regulate glucose homeostasis are dysregulated in frailty. This further supports the line of reasoning that intrinsic homeostatic mechanisms to maintain energy balance may be blunted with aging (Wilson and Morley 2003).

Overall, these findings are consistent with frailty as a marker of a severe end stage of biologic aging, with underlying dysregulation of the complex dynamical systems that maintain a resilient organism, and in which dysregulation of the system will become evident in the presence of stressors, with consequences, such as adverse health outcomes, which occur when the system is challenged. This theory is consistent with the basic tenets of biological complex systems (Csete and Doyle 2002; Kitano 2002). This complex system could well be considered a hallmark of aging, adding to the nine laid out by López-Otín et al. (2013).

IMPLICATIONS OF FRAILTY AS A DYSREGULATED COMPLEX SYSTEM FOR THERAPEUTIC OR PREVENTIVE APPROACH

Ultimately, the evidence of frailty as a complex system suggests why interventions targeting only one of many dysregulated systems, such as hormonal supplementation, have not been found to prevent or ameliorate frail states.

For example, many trials of monotherapies, such as replacement of estrogen or testosterone (Snyder et al. 1999; Taaffe et al. 2005; Ronkainen et al 2009; Kenny et al. 2010), for mitigation of frailty have not been successful. There are many other dimensions that need to be understood beyond this; for example, why is it that estrogen-replacement therapy does not protect against skeletal muscle loss with aging in women, whereas testosterone does predict muscle mass (Kenny et al. 2003)? With these observations as background, for the rest of this article we focus on the evidence of frailty as a complex system problem as the basis for a differential approach to prevention or treatment.

A particular focus to intervention within this framework could be the role of energy dysregulation as a critical entry point to development of frailty. Certainly, across the evidence summarized above, there are key elements of energy dysregulation now identified as components or drivers of frailty, from the altered energy intake in diet, to the phenotype itself, including unintentional weight loss, physical exhaustion and sarcopenia, and muscle weakness (Fig. 2) (Fried et al. 2001), to altered glucose metabolism and muscle efficiency, to mitochondrial function. Considering the evidence, this would suggest that therapeutic opportunities need to focus on improving energy production and use, and shifting the entire complex system of function.

PHYSICAL ACTIVITY AS A MODEL FOR INTERVENTIONS THAT MIGHT PREVENT, DELAY, OR TREAT FRAILTY

Physical activity, as an intervention, offers an intriguing model for the type of intervention that could prevent or mitigate frailty. This article concludes by offering the reasoning for this statement.

First, if one considers the association of physical activity with frailty, declines in physical activity both predict, early on, the development of the rest of the phenotype (Xue et al. 2012) and likely exacerbate both frailty-related outcomes and the underlying dysregulation. Consider in terms of outcomes that, although frailty has been shown to predict mortality, falls, and disability more powerfully than physical activity alone, low levels of physical activity, into the oldest ages, independently predict these outcomes, including mortality (Kushi et al 1997; Fried et al. 1998; Andersen et al. 2000; Katzmarzyk et al. 2003; Hu et al. 2004). Physical activity is consistently a predictor of lower mortality, as well as of greater muscle strength, into the oldest ages. In observational studies of older adults, for example, Xue et al. (2012) has shown in a 12-yr follow-up of women 70–79 at baseline in the Women's Health and Aging Study II, that those who rapidly declined in physical activity over 12 yr or who were always sedentary (adjusting for chronic diseases, disability, obesity, and other confounders) had hazard ratios for death of 2.34 and 3.34, compared with those who were always active. Further, the findings suggested that physical activity did not have to be vigorous to have benefit (Xue et al. 2012).

Second, physical activity is known to preserve or improve the function of many of the physiologic systems at abnormal levels in frail older adults: for example, sarcopenia and protein synthesis (Zampieri et al. 2015), muscle force and function, and fiber morphology. Gómez-Cabello et al. (2014) showed that, in people 65 and older, physical activity in women was associated with both upper and lower body strength, whereas in men it was associated with strength of knee and grip. Further, exercise maintains or improves glucose metabolism, in-

Cite this article as *Cold Spring Harb Perspect Med* doi: 10.1101/cshperspect.a025916

flammation, and anemia, as well as exercise tolerance underlying the feeling of "exhaustion" or low energy in the phenotype.

Third, declines in energy availability to the individual because of impairment in ATP production with aging may depend on mitochondrial dysfunction (Trounce et al. 1989; Shigenaga et al. 1994), which itself is up-regulated by physical activity. Exercise has been shown to maintain mitochondrial function in aging and prevent release of reactive oxygen species (ROS). Regular exercise over adulthood not only preserves muscle force, function, and morphology, but preserves ultrastructure of intracellular organelles involved in calcium handling and ATP production, and lowers expression of genes related to autophagy and ROS detoxification, important for clearance of damaged organelles and proteins. Conversely, inactive aging results in drastic reduction in number of calcium release units and volume and number of mitochondria in muscle, compared with young individuals (Zampieri et al. 2015), as well as structure and preservation of organelle positioning—all dimensions that could directly influence ATP production and, thus, be important for overall muscle performance (Zampieri et al. 2015).

Notably, it appears possible to maintain a tuned energetics system with maintenance of physical activity over the life course and into the oldest ages. As reported by Zampieri et al. (2015), skeletal muscle of well-trained seniors with a 30-year history of regular exercise is more like that of young adults than that of age-matched sedentary individuals, and signaling pathways controlling muscle mass and metabolism are differently modulated in senior sportsmen to support maintenance of skeletal muscle structure (preservation of slow-type fibers), function, bioenergetics characteristics (improved oxidative muscle metabolism), and phenotype. The general organization of the metabolic apparatus is far better preserved in individuals who exercised regularly. For example, the frequency of mitochondria is higher in athletic than sedentary seniors, with mitochondria being even more positively affected than the excitation–contraction coupling apparatus.

Zampieri et al. (2015) have shown that the frequency of calcium release unit (CRU)-mitochondrial pairs is threefold higher in senior sportsmen; this functional coupling up-regulates ATP production when muscle is active (Brookes et al. 2004; Shkryl and Shirokova 2006; Mosole et al. 2014). Notably, there is also significant up-regulation of miR-206 expression in senior sportsmen, compared with young adults and healthy sedentary seniors; this microRNA has been shown to play a specific role in the early events of regeneration (Cacchiarelli et al. 2010). Conversely, autophagy-related genes were significantly up-regulated in sedentary seniors compared with age-matched senior sportsmen and young adults. Thus, this suggests that signaling pathways that control muscle mass and metabolism are differently modulated in senior sportsmen to preserve bioenergetics characteristics and morphology, and that regular physical activity can now be seen as "a good strategy to attenuate age-related general decay of muscle structure and function" (Zampieri et al. 2015). Supporting this, as one outcome measure, the Health ABC study found in older participants followed for 5 yr that sedentary individuals had significantly increased risk of decline in gait speed to <0.60 m/sec and/or inability to rise from a chair without use of arms (Peterson et al. 2009) compared with those who actively exercised. This suggests that self-selected exercise activities may be independently associated with delaying onset and progression of frailty (Peterson et al. 2009).

Zampieri's conclusion was that "a specific, well-directed program of training could improve body balance, muscle structure, and contractile properties in elderly persons" (Zampieri et al. 2015). In fact, exercise trials of different intensities in older adults of different degrees of frailty have shown the benefit. Notably, one of the first studies in this arena was that by Fiatarone et al. (1990), which showed an almost twofold improvement of nonagenarian muscle strength and balance by a weight-lifting protocol. More recently, the LIFE Study randomized controlled trial showed that, among adults 70–89 at risk for disability (defined as sedentary, SPPB 9 or lower; could walk 400 m in <15 min; no major cognitive impairment), a structured, moderate intensity physical activity program that included aerobic (150 min per week of walking, with a goal of 30 min daily at intensity of 13), plus resistance and flexibility and balance training activities (compared with health education program) led to a significant difference in frailty prevalence at 12 mo (10%) compared with successful aging group (19%), and the mean number of frailty criteria was especially reduced for younger subjects, Blacks, participants with frailty, and those with multimorbidity (Pahor et al. 2014; Cesari et al. 2015).

IMPLICATIONS

Thus, trials of physical activity suggest that this offers a low-risk and effective preventive and therapeutic intervention to prevent or mitigate frailty. These findings are important in themselves. However, they offer intriguing implications in terms of criteria for other therapies to be considered in the future. These implications draw directly from further evidence, that physical activity "tunes"—or improves the function—of many systems in the body that are implicated in frailty etiology, as discussed above. These include muscle structure and function, hormonal, inflammatory, immune, hematologic, and other coregulatory functions, as well as supporting improved brain functions. Evidence ranges from trials of aerobic exercise training (for 6–9 mo), which showed improved heart rate variability in older adults (Levy et al. 1998; Schuit et al. 1999; Stein et al. 1999; Lipsitz 2002), to the metformin trial, which showed that the lifestyle intervention (with at least 150 min of physical activity per week) was more effective in restoring normal postload glucose values than was metformin or placebo, with

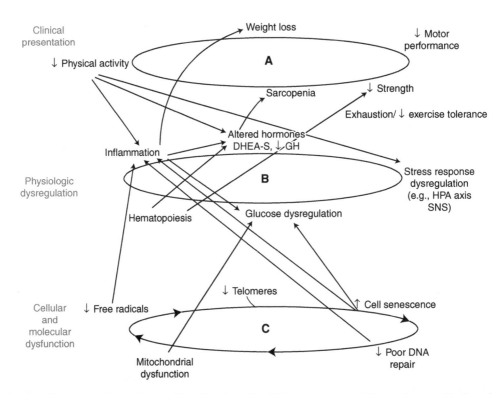

Figure 4. Physical activity positively affects function of multiple components of the syndrome of frailty at levels of (A) frailty phenotype, (B) physiologic dysregulation, and (C) cellular function. DHEA-S, didehydroepiandrosterone sulfate; GH, growth hormone; HPA, hypothalamic–pituitary–adrenal; SNS, sympathetic nervous system.

 Cite this article as *Cold Spring Harb Perspect Med* doi: 10.1101/cshperspect.a025916

the advantage of the lifestyle intervention being greater in older individuals (Nathan 2002); to the work of Lindholm et al. (2014), indicating that exercise of one leg in human volunteers leads to leg-specific epigenetic benefits in muscle: in addition to increased mass, epigenetic benefits, including new methylation patterns in >5000 sites on genome of muscle cells, particularly affected many muscle enhancers in genome that amplify expression of proteins by genes. Most of the genes affected play a role in energy metabolism, insulin response, and inflammation in muscles.

Further, exercise offers an intriguing model of a single intervention with multisystem benefits (Fig. 4). This confers great efficiency, but it also offers a model for types of interventions that may be effective in frailty as the outcome of dysregulation of the well-tuned complex dynamical system of a resilient organism. This is because physical activity simultaneously upregulates many systems that mutually regulate each other in combination. Thus, the whole organism could be retuned to a higher functional level. This offers a model intervention that matches well a complex system problem. If monotherapies are not sufficient, then finding an intervention that "tunes" a critical mass of systems would be critical. Kitano (2007) has asserted that frailty is the consequence of a trade-off between robustness and fragility (Kitano 2007), with robustness contributing to homeostasis and maintenance of functionality. Bortz (2008) has further stated that "Frailty is an assertion of the loss of dynamics, body-wise and system-wide" and that "It is in the tuning of the entire orchestra of genes to the cueing 'A' of the oboe where health resides." Physical activity offers both the first evidence as to effective approaches to preventing or treating frailty and a biologic model for future therapies.

REFERENCES

Andersen LB, Schnohr P, Schroll M, Hein HO. 2000. All-cause mortality associated with physical activity during leisure time, work, sports, and cycling to work. *Arch Intern Med* **160:** 1621–1628.

Ashar F, Moes A, Moore A, Grove M, Chaves PM, Coresh J, Newman A, Matteini A, Bandeen-Roche K, Boerwinkle E, et al. 2015. Association of mitochondrial DNA levels with frailty and all-cause mortality. *J Mol Med* **93:** 177–186.

Bandeen-Roche K, Xue QL, Ferrucci L, Walston J, Guralnik JM, Chaves P, Zeger SL, Fried LP. 2006. Phenotype of frailty: Characterization in the women's health and aging studies. *J Gerontol A Biol Sci Med Sci* **61:** 262–266.

Bartali B, Frongillo EA, Bandinelli S, Lauretani F, Semba RD, Fried LP, Ferrucci L. 2006. Low nutrient intake is an essential component of frailty in older persons. *J Gerontol A Biol Sci Med Sci* **61:** 589–593.

Barzilay JI, Blaum C, Moore T, Xue QL, Hirsch CH, Walston JD, Fried LP. 2007. Insulin resistance and inflammation as precursors of frailty: The Cardiovascular Health Study. *Arch Intern Med* **167:** 635–641.

Baylis D, Bartlett DB, Syddall HE, Ntani G, Gale CR, Cooper C, Lord JM, Sayer AA. 2013. Immune-endocrine biomarkers as predictors of frailty and mortality: A 10-year longitudinal study in community-dwelling older people. *Age (Dordr)* **35:** 963–971.

Bellizzi D, D'Aquila P, Montesanto A, Corsonello A, Mari V, Mazzei B, Lattanzio F, Passarino G. 2012. Global DNA methylation in old subjects is correlated with frailty. *Age (Dordr)* **34:** 169–179.

Belsky DW, Caspi A, Houts R, Cohen HJ, Corcoran DL, Danese A, Harrington H, Israel S, Levine ME, Schaefer JD, et al. 2015. Quantification of biological aging in young adults. *Proc Natl Acad Sci* **112:** E4104–E4110.

Blaum CS, Xue QL, Tian J, Semba RD, Fried LP, Walston J. 2009. Is hyperglycemia associated with frailty status in older women? *J Am Geriatr Soc* **57:** 840–847.

Bortz WM. 1993. The physics of frailty. *J Am Geriatr Soc* **41:** 1004–1008.

Bortz WM. 2008. Frailty. *Mech Ageing Dev* **129:** 680.

Brookes PS, Yoon Y, Robotham JL, Anders MW, Sheu SS. 2004. Calcium, ATP, and ROS: A mitochondrial love–hate triangle. *Am J Physiol Cell Physiol* **287:** C817–C833.

Cacchiarelli D, Martone J, Girardi E, Cesana M, Incitti T, Morlando M, Nicoletti C, Santini T, Sthandier O, Barberi L, et al. 2010. MicroRNAs involved in molecular circuitries relevant for the Duchenne muscular dystrophy pathogenesis are controlled by the dystrophin/nNOS pathway. *Cell Metab* **12:** 341–351.

Cappola AR, Xue QL, Ferrucci L, Guralnik JM, Volpato S, Fried LP. 2003. Insulin-like growth factor I and Interleukin-6 contribute synergistically to disability and mortality in older women. *J Clin Endocrinol Metab* **88:** 2019–2025.

Cesari M, Leeuwenburgh C, Lauretani F, Onder G, Bandinelli S, Maraldi C, Guralnik JM, Pahor M, Ferrucci L. 2006. Frailty syndrome and skeletal muscle: Results from the Invecchiare in Chianti study. *Am J Clin Nutr* **83:** 1142–1148.

Cesari M, Vellas B, Hsu FC, Newman AB, Doss H, King AC, Manini TM, Church T, Gill TM, Miller ME, et al. 2015. A physical activity intervention to treat the frailty syndrome in older persons—Results from the LIFE-P study. *J Gerontol A Biol Sci Med Sci* **70:** 216–222.

Chang SS, Weiss CO, Xue QL, Fried LP. 2012. Association between inflammatory-related disease burden and frailty: Results from the Women's Health and Aging Studies (WHAS) I and II. *Arch Gerontol Geriatr* **54:** 9–15.

Chaves PHM, Semba RD, Leng SX, Woodman RC, Ferrucci L, Guralnik JM, Fried LP. 2005. Impact of anemia and cardiovascular disease on frailty status of community-dwelling older women: The Women's Health and Aging Studies I and II. *J Gerontol A Biol Sci Med Sci* **60:** 729–735.

Chistiakov DA, Sobenin IA, Revin VV, Orekhov AN, Bobryshev YV. 2014. Mitochondrial aging and age-related dysfunction of mitochondria. *Biomed Res Int* **2014:** 238463.

Coen PM, Jubrias SA, Distefano G, Amati F, Mackey DC, Glynn NW, Manini TM, Wohlgemuth SE, Leeuwenburgh C, Cummings SR. 2013. Skeletal muscle mitochondrial energetics are associated with maximal aerobic capacity and walking speed in older adults. *J Gerontol A Biol Sci Med Sci* **68:** 447–455.

Cohen HJ, Harris T, Pieper CF. 2003. Coagulation and activation of inflammatory pathways in the development of functional decline and mortality in the elderly. *Am J Med* **114:** 180–187.

Cohen AA, Milot E, Yong J, Seplaki CL, Fulop T, Bandeen-Roche K, Fried LP. 2013. A novel statistical approach shows evidence for multi-system physiological dysregulation during aging. *Mech Ageing Dev* **134:** 110–117.

Cohen AA, Milot E, Li Q, Legault V, Fried LP, Ferrucci L. 2014. Cross-population validation of statistical distance as a measure of physiological dysregulation during aging. *Exp Gerontol* **57:** 203–210.

Cohen AA, Milot E, Li Q, Bergeron P, Poirier R, Dusseault-Bélanger F, Fülöp T, Leroux M, Legault V, Metter EJ. 2015. Detection of a novel, integrative aging process suggests complex physiological integration. *PloS ONE* **10:** e0116489.

Csete ME, Doyle JC. 2002. Reverse engineering of biological complexity. *Science* **295:** 1664–1669.

Davidson MB. 1979. The effect of aging on carbohydrate metabolism: A review of the English literature and a practical approach to the diagnosis of diabetes mellitus in the elderly. *Metabolism* **28:** 688–705.

DeFronzo RA. 1981. Glucose intolerance and aging. *Diabetes Care* **4:** 493–501.

Ershler WB, Keller ET. 2000. Age-associated increased interleukin-6 gene expression, late-life diseases, and frailty. *Annu Rev Med* **51:** 245–270.

Ferrucci L, Penninx BW, Volpato S, Harris TB, Bandeen-Roche K, Balfour J, Leveille SG, Fried LP, Md JM. 2002. Change in muscle strength explains accelerated decline of physical function in older women with high interleukin-6 serum levels. *J Am Geriatr Soc* **50:** 1947–1954.

Fiatarone MA, Marks EC, Ryan ND, Meredith CN, Lipsitz LA, Evans WJ. 1990. High-intensity strength training in nonagenarians: Effects on skeletal muscle. *JAMA* **263:** 3029–3034.

Fried LP. *Global report on aging and health.* World Health Organization, Geneva (in press a).

Fried LP. Frailty. In *Geriatrics review syllabus: A core curriculum in geriatric medicine,* 9th ed. (ed. Medina-Walpole A, et al.). American Geriatrics Society, New York (in press b).

Fried LP, Ferrucci L. Etiological role of aging in chronic diseases: From epidemiological evidence to the new gero-

science, Chap. 2. In *Advances in geroscience* (ed. Sierra F, Kohanski RA). Springer, New York (in press).

Fried LP, Kronmal RA, Newman AB, Bild DE, Mittelmark MB, Polak JF, Robbins JA, Gardin JM. 1998. Risk factors for 5-year mortality in older adults: The Cardiovascular Health Study. *JAMA* **279:** 585–592.

Fried LP, Tangen CM, Walston J, Newman AB, Hirsch C, Gottdiener J, Seeman T, Tracy R, Kop WJ, Burke G, et al. 2001. Frailty in older adults: Evidence for a phenotype. *J Gerontol A Biol Sci Med Sci* **56:** M146–M156.

Fried LP, Xue QL, Cappola AR, Ferrucci L, Chaves P, Varadhan R, Guralnik JM, Leng SX, Semba RD, Walston JD. 2009. Nonlinear multisystem physiological dysregulation associated with frailty in older women: Implications for etiology and treatment. *J Gerontol A Biol Sci Med Sci* **64:** 1049–1057.

Fries J. 2001. Aging, cumulative disability, and the compression of morbidity. *Compr Ther* **27:** 322–329.

Geronimus AT, Pearson JA, Linnenbringer E, Schulz AJ, Reyes AG, Epel ES, Lin J, Blackburn EH. 2015. Race-ethnicity, poverty, urban stressors, and telomere length in a Detroit community-based sample. *J Health Soc Behav* **56:** 199–224.

Gómez-Cabello A, Carnicero JA, Alonso-Bouzón C, Tresguerres JÁ, Alfaro-Acha A, Ara I, Rodriguez-Mañas L, García-García FJ. 2014. Age and gender, two key factors in the associations between physical activity and strength during the ageing process. *Maturitas* **78:** 106–112.

Green DR, Galluzzi L, Kroemer G. 2011. Mitochondria and the autophagy–inflammation–cell death axis in organismal aging. *Science* **333:** 1109–1112.

Guralnik JM, Simonsick EM, Ferrucci L, Glynn RJ, Berkman LF, Blazer DG, Scherr PA, Wallace RB. 1994. A short physical performance battery assessing lower extremity function: Association with self-reported disability and prediction of mortality and nursing home admission. *J Gerontol* **49:** M85–M94.

Hirsch C, Anderson ML, Newman A, Kop W, Jackson S, Gottdiener J, Tracy R, Fried LP. 2006. The association of race with frailty: The cardiovascular health study. *Ann Epidemiol* **16:** 545–553.

Hu FB, Willett WC, Li T, Stampfer MJ, Colditz GA, Manson JE. 2004. Adiposity as compared with physical activity in predicting mortality among women. *N Engl J Med* **351:** 2694–2703.

Hubbard RE, Lang IA, Llewellyn DJ, Rockwood K. 2010. Frailty, body mass index, and abdominal obesity in older people. *J Gerontol A Biol Sci Med Sci* **65:** 377–381.

Johar H, Emeny RT, Bidlingmaier M, Reincke M, Thorand B, Peters A, Heier M, Ladwig KH. 2014. Blunted diurnal cortisol pattern is associated with frailty: A cross-sectional study of 745 participants aged 65 to 90 years. *J Clin Endocrinol Metab* **99:** E464–E468.

Joseph C, Kenny AM, Taxel P, Lorenzo JA, Duque G, Kuchel GA. 2005. Role of endocrine-immune dysregulation in osteoporosis, sarcopenia, frailty and fracture risk. *Mol Aspects Med* **26:** 181–201.

Kalyani R, Varadhan R, Weiss CO, Fried LP, Cappola AR. 2012a. Frailty status and altered dynamics of circulating energy metabolism hormones after oral glucose in older women. *J Nutr Health Aging* **16:** 679–686.

Cite this article as *Cold Spring Harb Perspect Med* doi: 10.1101/cshperspect.a025916

Kalyani RR, Tian J, Xue QL, Walston J, Cappola AR, Fried LP, Brancati FL, Blaum CS. 2012b. Hyperglycemia and incidence of frailty and lower extremity mobility limitations in older women. *J Am Geriatr Soc* **60:** 1701–1707.

Kalyani RR, Varadhan R, Weiss CO, Fried LP, Cappola AR. 2012c. Frailty status and altered glucose-insulin dynamics. *J Gerontol A Biol Sci Med Sci* **67:** 1300–1306.

Katzmarzyk PT, Janssen I, Ardern CI. 2003. Physical inactivity, excess adiposity and premature mortality. *Obes Rev* **4:** 257–290.

Kenny AM, Dawson L, Kleppinger A, Iannuzzi-Sucich M, Judge JO. 2003. Prevalence of sarcopenia and predictors of skeletal muscle mass in nonobese women who are long-term users of estrogen-replacement therapy. *J Gerontol A Biol Sci Med Sci* **58:** M436–M440.

Kenny AM, Kleppinger A, Annis K, Rathier M, Browner B, Judge JO, McGee D. 2010. Effects of transdermal testosterone on bone and muscle in older men with low bioavailable testosterone levels, low bone mass, and physical frailty. *J Am Geriatr Soc* **58:** 1134–1143.

Kern H, Pelosi L, Coletto L, Musarò A, Sandri M, Vogelauer M, Trimmel L, Cvecka J, Hamar D, Kovarik J, et al. 2011. Atrophy/hypertrophy cell signaling in muscles of young athletes trained with vibrational-proprioceptive stimulation. *Neurol Res* **33:** 998–1009.

Kitano H. 2002. Computational systems biology. *Nature* **420:** 206–210.

Kitano H. 2007. Towards a theory of biological robustness. *Mol Syst Biol* **3:** 137.

Kobrosly RW, van Wijngaarden E, Seplaki CL, Cory-Slechta DA, Moynihan J. 2014. Depressive symptoms are associated with allostatic load among community-dwelling older adults. *Physiol Behav* **123:** 223–230.

Kuchel GA, Tannenbaum C, Greenspan SL, Resnick NM. 2001. Can variability in the hormonal status of elderly women assist in the decision to administer estrogens? *J Womens Health Gend Based Med* **10:** 109–116.

Kushi LH, Fee RM, Folsom AR, Mink PJ, Anderson KE, Sellers TA. 1997. Physical activity and mortality in postmenopausal women. *JAMA* **277:** 1287–1292.

Levine ME, Crimmins EM. 2014. Evidence of accelerated aging among African Americans and its implications for mortality. *Soc Sci Med* **118:** 27–32.

Levy WC, Cerqueira MD, Harp GD, Johannessen KA, Abrass IB, Schwartz RS, Stratton JR. 1998. Effect of endurance exercise training on heart rate variability at rest in healthy young and older men. *Am J Cardiol* **82:** 1236–1241.

Lindholm ME, Marabita F, Gomez-Cabrero D, Rundqvist H, Ekström TJ, Tegnér J, Sundberg CJ. 2014. An integrative analysis reveals coordinated reprogramming of the epigenome and the transcriptome in human skeletal muscle after training. *Epigenetics* **9:** 1557–1569.

Lipsitz LA. 2002. Dynamics of stability: The physiologic basis of functional health and frailty. *J Gerontol A Biol Sci Med Sci* **57:** B115–B125.

López-Otín C, Blasco MA, Partridge L, Serrano M, Kroemer G. 2013. The hallmarks of aging. *Cell* **153:** 1194–1217.

Metter EJ, Windham BG, Maggio M, Simonsick EM, Ling SM, Egan JM, Ferrucci L. 2008. Glucose and insulin measurements from the oral glucose tolerance test and mortality prediction. *Diabetes Care* **31:** 1026–1030.

Michelon E, Blaum C, Semba RD, Xue QL, Ricks MO, Fried LP. 2006. Vitamin and carotenoid status in older women: Associations with the frailty syndrome. *J Gerontol A Biol Sci Med Sci* **61:** 600–607.

Moore AZ, Biggs ML, Matteini A, O'Connor A, McGuire S, Beamer BA, Fallin MD, Fried LP, Walston J, Chakravarti A, et al. 2010. Polymorphisms in the mitochondrial DNA control region and frailty in older adults. *PLoS ONE* **5:** e11069.

Morley JE, Kim MJ, Haren MT, Kevorkian R, Banks WA. 2005. Frailty and the aging male. *Aging Male* **8:** 135–140.

Morley JE, Vellas B, Abellan van Kan G, Anker SD, Bauer JM, Bernabei R, Cesari M, Chumlea WC, Doehner W, Evans J, et al. 2013. Frailty consensus: A call to action. *J Am Med Dir Assoc* **14:** 392–397.

Mosole S, Carraro U, Kern H, Loefler S, Fruhmann H, Vogelauer M, Burggraf S, Mayr W, Krenn M, Paternostro-Sluga T. 2014. Long-term high-level exercise promotes muscle reinnervation with age. *J Neuropathol Exp Neurol* **73:** 284–294.

Musarò A, McCullagh K, Paul A, Houghton L, Dobrowolny G, Molinaro M, Barton ER, Sweeney HL, Rosenthal N. 2001. Localized Igf-1 transgene expression sustains hypertrophy and regeneration in senescent skeletal muscle. *Nat Genet* **27:** 195–200.

Musarò A, Giacinti C, Dobrowolny G, Pelosi L, Rosenthal N. 2004. The role of IGF-I on muscle wasting: A therapeutic approach. *Basic Appl Myol* **14:** 29–33.

Nathan DM. 2002. Initial management of glycemia in type 2 diabetes mellitus. *N Engl J Med* **347:** 1342–1349.

Newman AB, Gottdiener JS, McBurnie MA, Hirsch CH, Kop WJ, Tracy R, Walston JD, Fried LP. 2001. Associations of subclinical cardiovascular disease with frailty. *J Gerontol A Biol Sci Med Sci* **56:** M158–M166.

Nicklett EJ, Semba RD, Xue QL, Tian J, Sun K, Cappola AR, Simonsick EM, Ferrucci L, Fried LP. 2012. Fruit and vegetable intake, physical activity, and mortality in older community-dwelling women. *J Am Geriatr Soc* **60:** 862–868.

Pahor M, Guralnik JM, Ambrosius WT, Blair S, Bonds DE, Church TS, Espeland MA, Fielding RA, Gill TM, Groessl EJ, et al. 2014. Effect of structured physical activity on prevention of major mobility disability in older adults: The life study randomized clinical trial. *JAMA* **311:** 2387–2396.

Park SW, Goodpaster BH, Strotmeyer ES, Kuller LH, Broudeau R, Kammerer C, de Rekeneire N, Harris TB, Schwartz AV, Tylavsky FA, et al. 2007. Accelerated loss of skeletal muscle strength in older adults with type 2 diabetes: The health, aging, and body composition study. *Diabetes Care* **30:** 1507–1512.

Park SW, Goodpaster BH, Lee JS, Kuller LH, Boudreau R, de Rekeneire N, Harris TB, Kritchevsky S, Tylavsky FA, Nevitt M, et al. 2009. Excessive loss of skeletal muscle mass in older adults with type 2 diabetes. *Diabetes Care* **32:** 1993–1997.

Patel SM, Ratcliffe SJ, Reilly MP, Weinstein R, Bhasin S, Blackman MR, Cauley JA, Sutton-Tyrrell K, Robbins J, Fried LP, et al. 2009. Higher serum testosterone concentration in older women is associated with insulin resistance, metabolic syndrome, and cardiovascular disease. *J Clin Endocrinol Metab* **94:** 4776–4784.

Pelosi L, Giacinti C, Nardis C, Borsellino G, Rizzuto E, Nicoletti C, Wannenes F, Battistini L, Rosenthal N, Molinaro M, et al. 2007. Local expression of IGF-1 accelerates muscle regeneration by rapidly modulating inflammatory cytokines and chemokines. *FASEB J* 21: 1393–1402.

Peterson MJ, Giuliani C, Morey MC, Pieper CF, Evenson KR, Mercer V, Cohen HJ, Visser M, Brach JS, Kritchevsky SB, et al. 2009. Physical activity as a preventative factor for frailty: The health, aging, and body composition study. *J Gerontol A Biol Sci Med Sci* 64A: 61–68.

Phielix E, Schrauwen-Hinderling VB, Mensink M, Lenaers E, Meex R, Hoeks J, Kooi ME, Moonen-Kornips E, Sels JP, Hesselink MK, et al. 2008. Lower intrinsic ADP-stimulated mitochondrial respiration underlies in vivo mitochondrial dysfunction in muscle of male type 2 diabetic patients. *Diabetes* 57: 2943–2949.

Reuben DB, Cheh AI, Harris TB, Ferrucci L, Rowe JW, Tracy RP, Seeman TE. 2002. Peripheral blood markers of inflammation predict mortality and functional decline in high-functioning community-dwelling older persons. *J Am Geriatr Soc* 50: 638–644.

Ronkainen PH, Kovanen V, Alen M, Pöllänen E, Palonen EM, Ankarberg-Lindgren C, Hämäläinen E, Turpeinen U, Kujala UM, Puolakka J. 2009. Postmenopausal hormone replacement therapy modifies skeletal muscle composition and function: A study with monozygotic twin pairs. *J Appl Physiol* 107: 25–33.

Scheen AJ. 2005. Diabetes mellitus in the elderly: Insulin resistance and/or impaired insulin secretion? *Diabetes Metab* 31: 5S27–25S34.

Schuit AJ, van Amelsvoort LG, Verheij TC, Rijneke RD, Maan AC, Swenne CA, Schouten EG. 1999. Exercise training and heart rate variability in older people. *Med Sci Sports Exerc* 31: 816–821.

Semba RD, Bartali B, Zhou J, Blaum C, Ko CW, Fried LP. 2006. Low serum micronutrient concentrations predict frailty among older women living in the community. *J Gerontol A Biol Sci Med Sci* 61: 594–599.

Shigenaga MK, Hagen TM, Ames BN. 1994. Oxidative damage and mitochondrial decay in aging. *Proc Natl Acad Sci* 91: 10771–10778.

Shimokata H, Muller DC, Fleg JL, Sorkin J, Ziemba AW, Andres R. 1991. Age as independent determinant of glucose tolerance. *Diabetes* 40: 44–51.

Shkryl VM, Shirokova N. 2006. Transfer and tunneling of Ca^{2+} from sarcoplasmic reticulum to mitochondria in skeletal muscle. *J Biol Chem* 281: 1547–1554.

Short KR, Bigelow ML, Kahl J, Singh R, Coenen-Schimke J, Raghavakaimal S, Nair KS. 2005. Decline in skeletal muscle mitochondrial function with aging in humans. *Proc Natl Acad Sci* 102: 5618–5623.

Snyder PJ, Peachey H, Hannoush P, Berlin JA, Loh L, Lenrow DA, Holmes JH, Dlewati A, Santanna J, Rosen CJ, et al. 1999. Effect of testosterone treatment on body composition and muscle strength in men over 65 years of age. *J Clin Endocrinol Metab* 84: 2647–2653.

Stein PK, Ehsani AA, Domitrovich PP, Kleiger RE, Rottman JN. 1999. Effect of exercise training on heart rate variability in healthy older adults. *Am Heart J* 138: 567–576.

Szanton S, Allen J, Seplaki C, Bandeen-Roche K, Fried L. 2009. Allostatic load and frailty in the women's health and aging studies. *Biol Res Nurs* 10: 248–256.

Taaffe DR, Sipila S, Cheng S, Puolakka J, Toivanen J, Suominen H. 2005. The effect of hormone replacement therapy and/or exercise on skeletal muscle attenuation in postmenopausal women: A yearlong intervention. *Clin Physiol Funct Imaging* 25: 297–304.

Tinetti ME, Fried T. 2004. The end of the disease era. *Am J Med* 116: 179–185.

Trounce I, Byrne E, Marzuki S. 1989. Decline in skeletal muscle mitochondrial respiratory chain function: Possible factor in ageing. *Lancet* 1: 637–639.

Umanskaya A, Santulli G, Xie W, Andersson DC, Reiken SR, Marks AR. 2014. Genetically enhancing mitochondrial antioxidant activity improves muscle function in aging. *Proc Natl Acad Sci* 111: 15250–15255.

Varadhan R, Walston J, Cappola AR, Carlson MC, Wand GS, Fried LP. 2008. Higher levels and blunted diurnal variation of cortisol in frail older women. *J Gerontol A Biol Sci Med Sci* 63: 190–195.

Visser M, Kritchevsky SB, Goodpaster BH, Newman AB, Nevitt M, Stamm E, Harris TB. 2002. Leg muscle mass and composition in relation to lower extremity performance in men and women aged 70 to 79: The health, aging and body composition study. *J Am Geriatr Soc* 50: 897–904.

Walston J, McBurnie M, Newman A, Tracy RP, Kop WJ, Hirsch CH, Gottdiener J, Fried LP, Cardiovascular Health Study. 2002. Frailty and activation of the inflammation and coagulation systems with and without clinical comorbidities: Results from the cardiovascular health study. *Arch Int Med* 162: 2333–2341.

Weiss CO, Cappola AR, Varadhan R, Fried LP. 2012. Resting metabolic rate in old-old women with and without frailty: Variability and estimation of energy requirements. *J Am Geriatr Soc* 60: 1695–1700.

Wilson MMG, Morley JE. 2003. Invited review: Aging and energy balance. *J Appl Physiol* 95: 1728–1736.

Wolinsky FD, Coe RM, Miller DK, Prendergast JM, Creel MJ, Chavez MN. 1983. Health services utilization among the noninstitutionalized elderly. *J Health Soc Behav* 24: 325–337.

Xue QL. 2011. The frailty syndrome: Definition and natural history. *Clin Geriatr Med* 27: 1–15.

Xue QL, Bandeen-Roche K, Mielenz TJ, Seplaki CL, Szanton SL, Thorpe RJ, Kalyani RR, Chaves PH, Dam TT, Ornstein K, et al. 2012. Patterns of 12-year change in physical activity levels in community-dwelling older women: Can modest levels of physical activity help older women live longer? *Am J Epidemiol* 176: 534–543.

Yates FE. 2002. Complexity of a human being: Changes with age. *Neurobiol Aging* 23: 17–19.

Zampieri S, Pietrangelo L, Loefler S, Fruhmann H, Vogelauer M, Burggraf S, Pond A, Grim-Stieger M, Cvecka J, Sedliak M, et al. 2015. Lifelong physical exercise delays age-associated skeletal muscle decline. *J Gerontol A Biol Sci Med Sci* 70: 163–173.

The Aging Heart

Ying Ann Chiao and Peter S. Rabinovitch

Department of Pathology, University of Washington, Seattle, Washington 98195

Correspondence: petersr@u.washington.edu

Aging results in progressive deteriorations in the structure and function of the heart and is a dominant risk factor for cardiovascular diseases, the leading cause of death in Western populations. Although the phenotypes of cardiac aging have been well characterized, the molecular mechanisms of cardiac aging are just beginning to be revealed. With the continuously growing elderly population, there is a great need for interventions in cardiac aging. This article will provide an overview of the phenotypic changes of cardiac aging, the molecular mechanisms underlying these changes, and some of the recent advances in the development of interventions to delay or reverse cardiac aging.

Cardiovascular diseases are the leading cause of death in most developed nations. Although it has received the least public attention, aging is by far the dominant risk factor for development cardiovascular diseases, as the prevalence of cardiovascular diseases increases dramatically with increasing age. The prevalence rate of cardiovascular diseases is >70% for Americans 60 to 79 years of age and >80% for Americans >80 years of age (Go et al. 2014). Even without associated systemic risk factors, intrinsic cardiac aging leads to structural and functional deteriorations of the heart in elderly individuals. Therefore, interventions to combat cardiac aging will not only improve healthspan of the elderly but can also extend lifespan by delaying cardiovascular disease-related deaths. Although there is presently no treatment for cardiac aging, recent advances in the understanding of the mechanisms of cardiac aging have provided new insights, and we are now poised on the threshold of development of new interventions to attenuate or reverse cardiac aging.

CARDIAC AGING IN HUMAN AND ANIMAL MODELS

The Framingham Heart Study and the Baltimore Longitudinal Study on Aging (BLSA) have shown that, in healthy individuals without concomitant cardiovascular diseases, aging results in an increase in the prevalence of left ventricular (LV) hypertrophy, a decline in diastolic function, and relatively preserved systolic function at rest but a decline in exercise capacity, as well as an increase in the prevalence of atrial fibrillation (Lakatta and Levy 2003b). These changes can be independent of conventional risk factors for heart disease (smoking, hypertension, blood lipid levels, diabetes, etc.) and, thus, may be considered to be part of intrinsic cardiac aging. At rest, systolic function measured by the ejection fraction (EF) remains steady in older populations. However, on exercise, maximum heart rate and EF are lower in older populations, indicating reduced cardiac reserve (Lakatta 2002). An age-dependent increase in myocardial performance index (MPI) has also been shown (Spencer et al. 2004). An increase in MPI indicates that a greater fraction of systole is spent to cope with the pressure changes during isovolumetric phases, and has been shown to reflect both LV systolic and/or diastolic dysfunction (Tei et al. 1997). Because of impaired early diastolic filling and an increased contribution of atrial contraction to LV filling, the peak early filling velocity and the ratio of the early and late (E/A ratio) filling velocity decrease with age; the early component is larger than the late atrial component of filling in young persons, but, when this reverses, it is an indicator of diastolic dysfunction (Downes et al. 1989; Swinne et al. 1992; Kitzman 2002; Choi et al. 2009). Diastolic dysfunction is increasingly seen in the elderly in the absence of systolic heart failure, a condition that has been given the designation of heart failure with preserved ejection fraction (HFpEF). It is especially prevalent in aged women (Brouwers et al. 2012) and is an increasing cause of hospital admissions (Oktay et al. 2013).

Rodents, particularly the mouse model, are widely used in cardiac aging studies. Murine cardiac aging phenotypes closely recapitulate the phenotypes of human cardiac aging (Lakatta and Levy 2003a). Echocardiography performed on a mouse longevity cohort showed that left ventricular mass index (LVMI) and left atrial dimension significantly increased with age. Diastolic function measured by tissue Doppler declines with age, whereas systolic function showed a modest reduction from young adult to the oldest group. The MPI also worsens with age, which is consistent with the age-related declines in systolic and diastolic function (Barger et al. 2008). In addition to the similar cardiac aging phenotypes, the relatively short lifespan and the availability of genetically modified mice are the advantages of using mouse model in the study of the molecular mechanisms of cardiac aging. Despite sharing similar cardiac aging phenotypes as human, laboratory mice do not develop elevated blood pressure or adverse blood glucose and lipid profiles (Zheng et al. 2003; Dai et al. 2009). This allows the intrinsic cardiac changes of aging to be investigated without the added complications of cardiovascular risk factors, including hypertension and diabetes.

MOLECULAR MECHANISMS OF CARDIAC AGING

Recent studies have shown the involvements of multiple molecular mechanisms in the pathogenesis of cardiac aging. These mechanisms are summarized in Figure 1 and discussed below.

Altered Nutrient and Growth Signaling

Cardiac hypertrophy is a hallmark of cardiac aging. Deregulation of nutrient and growth signaling pathways, including mechanistic target of rapamycin (mTOR) and insulin-like growth factor-1 (IGF-1) signaling, have been implicated in cardiac hypertrophy and aging. mTOR integrates nutrient and hormonal signals to regulate growth and is a major modulator of aging and age-related disease (Kennedy et al. 2007). Previous studies in *Drosophila* and mouse models have shown that increased mTOR signaling impairs and reduced mTOR signaling improves resistance to cardiac aging. Bodmer's laboratory initially showed that inhibition of the mTOR pathway could attenuate the age-related decline in cardiac function in *Drosophila* (Luong et al. 2006). They later showed that eukaryotic translation-initiation factor 4E (eIF4E)-binding protein (4EBP) overexpression attenuates the age-related decline to a similar extent as overexpression of the TOR antagonist tuberous sclerosis complex (TSC), and overexpression of eIF4E leads to an accelerated decline of myocardial function (Wessells et al. 2009). These findings implicate a major role of mTOR/eIF4E signaling in cardiac aging in *Drosophila*. In addition, Meikle et al. (2005) showed that mice with cardiac-specific deletion of TSC1, a model of increased mTOR signaling, develop dilated cardiomyopathy and have a median lifespan of 6 mo. Although there is no evidence yet on the

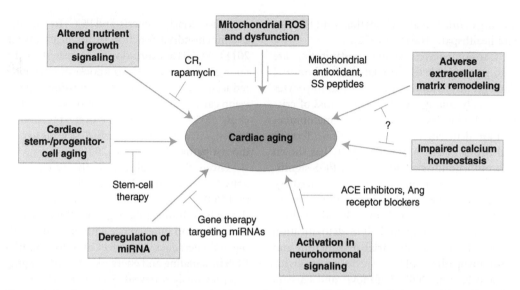

Figure 1. A schematic summary of the molecular mechanisms of cardiac aging and potential cardiac aging interventions. ROS, Reactive oxygen species; CR, calorie restriction; SS, Szeto–Schiller; miRNA, microRNA; ACE, angiotensin-converting enzyme; Ang, angiotensin.

beneficial effects of genetic manipulation to decrease mTOR activity in the aging mammalian heart, inhibition of mTOR signaling by caloric restriction (CR) or rapamycin (see below) has been shown to protect against cardiac aging.

Insulin/IGF-1 signaling is one of the best-characterized pathways of lifespan regulation in animal models. Deficiency in insulin/IGF-1 signaling improves cardiac performance at advanced age in *Drosophila* and attenuates age-related cardiomyocyte dysfunction in mice (Wessells et al. 2004; Li et al. 2008). However, in humans, an age-dependent decline in serum IGF-1 levels (Corpas et al. 1993) correlates with an increased risk of heart failure among elderly patients without prior history of heart disease (Vasan et al. 2003), and interventions that increase IGF-1 signaling, such as growth hormone therapy, may be beneficial in heart failure (Broglio et al. 1999; Khan et al. 2002). The beneficial effects of IGF-1 on cardiovascular disease may be conferred by mitochondrial protection mechanisms. One study showed that in vitro treatment of endothelial cells and cardiomyocytes with IGF-1 decreased mitochondrial superoxide production (Csiszar et al. 2008).

Furthermore, low-plasma levels of growth hormone (GH) and IGF-1 in Ames dwarf mice are associated with increased mitochondrial oxidative stress in the vasculature and the heart, which is responsible for the impaired contractile function (Ren and Brown-Borg 2002). Recent studies show that IGF-1 treatment of aged rats protects against mitochondrial oxidative stress (Puche et al. 2008), and other studies suggest that interventions that increase circulating IGF-1 levels exert cardiovascular protective effects in aging (Rivera et al. 2005; Groban et al. 2006; Lopez-Lopez et al. 2007). The roles of mitochondrial oxidative stress in cardiac aging are discussed further below.

Mitochondrial Oxidative Damage and Mitochondrial Dysfunction

The mitochondrial free-radical theory of aging proposes that excessive mitochondrial reactive oxygen species (ROS) damages mitochondrial DNA and redox-sensitive mitochondrial proteins, causing mitochondrial dysfunction and further increase in ROS production (the "vicious cycle" of ROS-induced ROS release), and that this oxidative damage leads to cellular and organ functional declines that limit lifespan and healthspan (Harman 1972).

Cardiomyocytes, being postmitotic, are highly susceptible to age-related mitochondrial damage. Mitochondria in aged cardiomyocytes are usually enlarged with swelling, loss of cristae, and even destruction of inner membranes and are deficient in ATP production (Terman and Brunk 2004). A previous study has shown that mitochondrial production of ROS significantly increases in the heart with advanced age (Judge et al. 2005). Also, increasing evidence suggests that abnormal mitochondrial ROS production and impaired ROS detoxification contribute to mitochondrial dysfunction and cardiomyopathy in old age (reviewed in Terzioglu and Larsson 2007; Trifunovic and Larsson 2008; Mammucari and Rizzuto 2010).

Direct evidence supporting the role of mitochondrial oxidative damage in cardiac aging was provided by mice overexpressing catalase targeted to the mitochondria (mCAT) (Schriner et al. 2005; Dai et al. 2009). In addition to an extension of median and maximum lifespan, mCAT mice displayed greatly attenuated cardiac aging phenotypes, including reduced cardiac hypertrophy and improved diastolic function and myocardial performance (Dai et al. 2009). The attenuated cardiac aging phenotypes in mCAT mice were accompanied by significantly reduced mitochondrial protein oxidative damage and mitochondrial DNA mutation and deletion frequencies, suggesting prevention of mitochondrial oxidative damage as a strategy for protection from cardiac aging. Additional evidence is provided by mice with homozygous mutation of mitochondrial polymerase γ ($Polg^{m/m}$), which have substantial increases in mtDNA mutations and deletions with age (Trifunovic et al. 2004; Kujoth et al. 2005). $Polg^{m/m}$ mice have a shortened lifespan and develop cardiomyopathy in middle age (13–14 mo) (Trifunovic et al. 2004; Dai et al. 2010). Interestingly, mCAT partially rescues the mitochondrial damage and cardiomyopathy in $Polg^{m/m}$ mice, further supporting the role of mitochondrial ROS in cardiac aging (Dai et al. 2010).

Peroxisome proliferator-activated receptor γ coactivator 1α (PCG-1α) is the key regulator of mitochondrial biogenesis, and PCG-1α enhances mitochondrial function in the heart (Wenz 2011). PCG-1α expression is repressed in the failing heart, and PCG-1α knockout mice have reduced mitochondrial gene expression and develop cardiac dysfunction at 7 mo of age (Arany et al. 2005). Cardiac-specific overexpression of PCG-1α in adult mice nevertheless leads to cardiomyopathy (Russell et al. 2004).

Given the complexity of the systems involved, mitochondrial dysfunction and aberrant ROS production likely contribute to aging through both direct damage to cellular macromolecules and interference with normal signaling and energetics. The effect of mitochondrial ROS in signaling and energetics in cardiac aging was previously reviewed (Dai et al. 2014a).

Adverse Extracellular Matrix (ECM) Remodeling

ECM is a complex collection of proteins located outside the cells and provides structural and biological supports to the surrounding cells. Cardiac fibroblasts are the primary sources of cardiac ECM proteins, including collagen types I, II, III, IV, V, and VI, elastin, fibronectin, laminin, and fibrinogen (DeQuach et al. 2010). Cardiac ECM aligns cardiomyocytes and provides structural support to the heart; however, excessive ECM deposition increases the stiffness of the myocardium and mediates diastolic dysfunction (Ouzounian et al. 2008). ECM composition is dynamically remodeled by the balance of the synthesis and degradation of ECM proteins by matrix metalloproteinases (MMPs) and other proteases. Cardiac aging is associated with myocardial fibrosis, and deregulation of ECM synthesis and degradation has both been observed in aging hearts.

Transforming growth factor-β (TGF-β) is a profibrotic factor that has been shown to induce the expression of ECM proteins and inhibit matrix degradation by MMPs (Bujak and Frangogiannis 2007). Reduced TGF-β1 expression results in reduced myocardial fibrosis and improved LV compliance in 24-mo-old TGF-β1 heterozygous mice (Brooks and Conrad 2000). Connective tissue growth factor (CTGF), another profibrotic factor, is a downstream mediator from TGF-β and its expression increases with age (Wang et al. 2010). Mice overexpressing CTGF in a cardiomyocyte-specific manner show accelerated cardiac aging and begin to develop age-related cardiac dysfunction at 7 mo of age (Panek et al. 2009). The role of ECM synthesis in cardiac aging is also implicated by the accelerated myocardial fibrosis that accompanies higher TGF-β and CTGF levels in senescence-accelerated mice that display diastolic dysfunction at 6 mo of age (Reed et al. 2011). In another study, Bradshaw et al. (2010) showed that expression of a matricellular protein, secreted protein acidic and rich in cysteine (SPARC), increased with age, and that deletion of SPARC resulted in reduced fibrillar collagen content in the LV and decreased LV diastolic stiffness. Together, this evidence suggests that increased ECM synthesis is an important mediator of diastolic dysfunction with age and that reduced ECM synthesis can improve cardiac aging.

MMPs are a family of 25 zinc-dependent enzymes that regulate ECM degradation; tissue inhibitors of matrix metalloproteinase (TIMPs) are a family composed of TIMP-1, -2, -3, and -4, which regulate MMP proteolytic activity in the tissue (Tayebjee et al. 2005). The expression levels of MMPs and TIMPs are differentially regulated by age (Lindsey et al. 2005; Bonnema et al. 2007), but the roles of most MMPs and TIMPs in cardiac aging have not been established. Spinale and colleagues showed that cardiac-specific MT1-MMP overexpression accelerated cardiac aging responses in mice and that MT1-MMP-overexpressing mice have increased myocardial fibrosis and LV dysfunction at middle age (Spinale et al. 2009). More recently, Chiao et al. (2011) showed that MMP-9 levels increase in the LV and plasma of aged C57Bl6 mice, and that aged MMP-9 knockout mice display attenuated cardiac aging phenotypes, including reduced collagen deposition and preserved diastolic function (Chiao et al. 2012). The attenuated cardiac aging phenotypes in MMP-9 knockout mice are accompanied by reduced expression of profibrotic proteins, periostin, and CTGF and a compensatory increase in

MMP-8 levels in the LV (Chiao et al. 2012). Together, these findings suggest a role of ECM degradation that is under complex regulation by MMPs in cardiac aging.

Impaired Calcium Homeostasis

One mechanism underlying age-dependent diastolic dysfunction is impaired active relaxation of cardiomyocytes (Zile and Brutsaert 2002; Kass et al. 2004). During relaxation, calcium ions dissociate from the actin–myosin complex and are taken up into the sarcoplasmic reticulum (SR) or extruded outside the cardiomyocyte. Impaired Ca^{2+} cycling, increased myofilament stiffness, reduced Ca^{2+} sensitivity of myofilament proteins, and alterations in actin or myosin properties can lead to impaired cardiomyocyte relaxation (Zile and Brutsaert 2002; Kass et al. 2004; Borlaug and Kass 2006). Aged mouse hearts have reduced sarco/endoplasmic reticulum Ca^{2+}-ATPase (SERCA2) expression (Dai et al. 2009) and activity (Janczewski and Lakatta 2010), with compensatory increase in the levels of the Na^+/Ca^{2+} exchanger (Koban et al. 1998). Studies suggest that the aged heart uses the compensatory increase in the L-type Ca^{2+} currents (Josephson et al. 2002) and prolongation of action potential duration to preserve SR loading and to maintain intracellular Ca^{2+}-transients and contractions in old cardiomyocytes (Janczewski et al. 2002). Posttranslational modifications of SERCA2, including age-related oxidation and nitration, have also been shown (Knyushko et al. 2005; Sharov et al. 2006), but their roles in cardiac aging are as yet unclear.

Chronic Activation in Neurohormonal Signaling

The renin angiotensin aldosterone system (RAAS) is the key endocrine system regulating hypertension and stress-induced cardiac hypertrophy. Angiotensin II (Ang II) infusion induces cardiomyocyte hypertrophy, increases cardiac fibrosis, and impairs cardiomyocyte relaxation (Domenighetti et al. 2005); these responses closely recapitulate the age-related changes in the heart (Dai et al. 2009). Studies have showed that Ang II concentration significantly increased in the aged rodent heart (Groban et al. 2006; Dai et al. 2009) potentially caused by increased tissue levels of angiotensin-converting enzyme (ACE) (Lakatta 2003). Moreover, long-term inhibition of Ang signaling by ACE inhibitors, angiotensin receptor blockers, or genetic disruption of Ang II receptor type I extend lifespan and delay age-dependent cardiac pathology in rodents (Basso et al. 2007; Benigni et al. 2009).

Activation of the β-adrenergic signaling increases heart rate, contractility, blood pressure, wall stress, and metabolic demand of the heart, and chronic stimulation of β-adrenergic signaling is deleterious to the heart. Deletion of adenylate cyclase type 5 (AC5), a key enzyme downstream from β-adrenergic signaling, extends murine lifespan and is protective against age-dependent cardiac hypertrophy, systolic dysfunction, apoptosis, and fibrosis (Yan et al. 2007).

Other Potential Mechanisms

Increasing evidence suggests that microRNAs (miRNAs) are important regulators of aging and cardiovascular diseases (Smith-Vikos and Slack 2012; Quiat and Olson 2013), and several recent studies implicated the roles of miRNAs in cardiac aging. van Almen et al. (2011) showed that the expression of the miR-17-92 cluster (consisting of miR-18a, miR19a, and miR-19b) decreases, whereas the expression of their targets, CTGF and ECM protein thrombospondin-1 (TSP-1) increases, in heart-failure-prone C57Bl6 × 129Sv mice. In aged cardiomyocyte cultures, these investigators showed that miR-18a and miR-19b regulated expression of CTGF, TSP-1, and collagen, suggesting that these miRNAs mediate age-related ECM remodeling in the hearts. In C57Bl6 mice, Jazbutyte et al. (2013) detected an age-related increase in miR-22 in hearts and showed that miR-22 regulates cardiac fibroblast senescence. In a recent study, Boon et al. (2013) showed that expression of miR-34a increased in aged mouse hearts, and in vivo silencing of miR-34a for 1 wk rescued the increase in cardiomyocyte cell death in aged mice. They also showed that aged miR-34a

knockout mice had improved contractile function and reduced cardiac hypertrophy compared to wild-type littermates. This evidence suggests that increased miR-34a expression in the aged heart contributes to cardiac aging.

Previous studies have shown that cardiac stem cells and progenitor cells may regenerate the adult heart to some extent (Beltrami et al. 2003; Hsieh et al. 2007). Cardiac stem cells in the aged heart may have impaired regenerative capacity, either by senescence intrinsic to the stem cells or by an extrinsic hostile microenvironment associated with advanced age (reviewed in Anversa et al. 2005; Ballard and Edelberg 2007). Torella and colleagues (2004) observed an increased proportion of senescent cardiac stem cells (which express senescent marker p16[ink4a] and have reduced telomere length) in old wild-type mice and showed that IGF-1 overexpression can prevent senescence of cardiac stem cells. A recent study also showed that attenuation of the IGF-1/IGF-1-receptor and hepatocyte growth factor/mesenchymal-epithelial transition factor (c-Met) systems mediates aging of cardiac progenitor cells (Gonzalez et al. 2008). In another study, Bergmann et al. (2009) measured ^{14}C from nuclear bomb tests in genomic DNA of human myocardial cells and used this method to show that the turnover or renewal rate of cardiomyocytes is reduced from 1% in young adult hearts to 0.45% in the hearts of the elderly. These results suggest that the regenerative capacity of cardiac stem cells declines with aging and that such declines may mediate the impaired myocardial repair in aged hearts.

RECENT ADVANCES IN INTERVENTIONS FOR CARDIAC AGING

The improved understanding of the pathogenesis of cardiac aging may greatly advance the development of interventions that target specific mechanisms to delay or treat cardiac aging (Fig. 1).

Calorie Restriction (CR)

CR is the most well-studied longevity intervention and has been shown to increase lifespan in a wide array of model organisms, from yeast and nematodes to mice, rats, and (perhaps) rhesus monkeys (Colman et al. 2009; Cruzen and Colman 2009; Fowler et al. 2010; Kastman et al. 2010; McKiernan et al. 2011). CR is protective against a variety of age-related pathologies, including cardiovascular disease, in rodents and nonhuman primates (Cruzen and Colman 2009; Niemann et al. 2010; Shinmura et al. 2011a). An early study by Taffett et al. (1997) showed that CR had a large positive effect on age-related impairment of diastolic function in mice. Kemi et al. (2000) then found that moderate dietary restriction (35% reduction in calorie intake) can attenuate age-related cardiomyopathy in male Sprague–Dawley rats. A later study showed that human volunteers undertaking CR for a mean of 6.5 yr had reduced blood pressure and systemic oxidative stress, and improved diastolic function (Meyer et al. 2006). Similar improvements in diastolic function have been reproduced in individuals maintained on 1-yr CR (Riordan et al. 2008). In addition to the protective effects of long-term CR, our laboratory recently showed that CR for 10 wk was able to reverse the preexisting cardiac hypertrophy and diastolic dysfunction in old mice, and that this was accompanied by proteomic and metabolomic remodeling to a more youthful state (Dai et al. 2014b).

Multiple mechanisms have been implicated in the beneficial effects of CR including inhibition of mTOR signaling, normalization of mitochondrial biogenesis (Lopez-Lluch et al. 2006), attenuation of mitochondrial ROS production and the subsequent ROS-induced signaling (Nisoli et al. 2005; Ungvari et al. 2008; Csiszar et al. 2009; Shinmura et al. 2011b), and increased SIRT1 signaling (Lopez-Lluch et al. 2006, 2008). These have proven to be fertile areas for the study of pharmacologic interventions to enhance healthspan.

Rapamycin

Although a large body of evidence supports the protective effects of CR in age-related pathologies including cardiac aging, the use of CR in humans would be challenging. Developing CR mimetics that mimic the beneficial effects of CR by targeting cellular metabolic and stress response pathways without actual restriction on calorie intake has been of great interest to the aging research community. mTOR plays a principle role in nutrient signaling and rapamycin is a well-established inhibitor of mTOR. The National Institute on Aging (NIA) Intervention Testing Program has recently shown that long-term rapamycin treatment initiated at 9 or 18 mo of age extended lifespan in mice with a mixed genetic background; these results were reproducible in three independent laboratories (Harrison et al. 2009; Miller et al. 2011). Subsequent studies have confirmed these results and have extended the studies to other measures of healthspan (Wilkinson et al. 2012).

With increasing evidence supporting the role of mTOR in aging and healthspan, the effects of rapamycin on cardiac aging are undoubtedly of interest to the aging research community. Rapamycin treatment for 1 yr initiated at mid-life attenuated the increased LV dimensions in aged hearts, but failed to show any effect on systolic function in male mice (Neff et al. 2013). Flynn et al. (2013) showed that rapamycin treatment for 12 wk initiated at late life can attenuate age-related cardiac hypertrophy and marginally improve systolic function in female mice, accompanied by a reduction in age-related inflammation. Recently, our laboratory showed that short-term rapamycin (10 wk) recapitulated the effect of CR and substantially improved both diastolic function and LV hypertrophy in old mice (Dai et al. 2014b). The reversal of cardiac aging phenotypes appeared to be mechanistically linked to proteomic and metabolic remodeling to increase mitochondrial protein content and reverse the age-related metabolic shift from fatty acid oxidation (FAO) to glycolysis and gluconeogenesis.

Further investigations on the mechanisms and kinetics of rapamycin benefits will be required to evaluate the potential of rapamycin as a pharmacological intervention to prevent, or even reverse, cardiac aging and its concomitant negative physiological consequences.

Mitochondrial Intervention

Mitochondrial dysfunction and mitochondrial ROS are critical mechanisms in the age-dependent decline in cardiac function; therefore, interventions combating mitochondrial ROS and improving mitochondrial function are attractive targets for interventions in cardiac aging. The success of mCAT protection in cardiac aging, but not of peroxisomal catalase or the nontargeted antioxidant N-acetylcysteine, underscored the importance of mitochondrial specificity in antioxidant intervention (Dai et al. 2011). One approach for targeting antioxidants to mitochondria is to use the negative potential gradient across the inner mitochondrial membrane (IMM). The negative potential gradient across IMM allows lipophilic cations to penetrate the IMM and accumulate in the mitochondrial matrix. Triphenylalkylphosphonium ions (TPP^+) have been conjugated to coenzyme Q (MitoQ) and plastoquinone (SkQ1) (Skulachev et al. 2009; Smith et al. 2012) to deliver these redox-active compounds into the mitochondrial matrix. Although their effects on cardiac aging have not been established, MitoQ and SkQ1 have been shown to have beneficial effects in models of ischemia reperfusion and cardiac hypertrophy (Adlam et al. 2005; Bakeeva et al. 2008; Graham et al. 2009; Dikalova et al. 2010). A major limitation of these TPP^+-conjugated antioxidants is their dependence on mitochondrial potential, which is often compromised in pathological conditions. MitoQ and SkQ have also been shown to inhibit respiration and disrupt mitochondrial potential at concentrations above 25 μM, which limits their uptake (Kelso et al. 2001; Antonenko et al. 2008). Moreover, MitoQ is reduced to a semiquinone radical at complex I and can increase superoxide production (O'Malley et al. 2006; Murphy and Smith 2007; Scatena et al. 2007); this pro-oxidant activity of MitoQ must be carefully evaluated when considering this intervention.

Another approach for targeting an intervention to mitochondria can be performed by utilizing an affinity to a mitochondrial component. The Szeto–Schiller (SS) compounds are tetrapeptides with an alternating aromatic-cationic amino acids motif. Studies have shown that SS peptides preferentially concentrate in the IMM over 1000-fold compared with the cytosolic concentration (Zhao et al. 2004; Doughan and Dikalov 2007; Bakeeva et al. 2008). In contrast to MitoQ and SkQ1, the mitochondrial uptake of SS peptides is not dependent on mitochondrial potential, and they can concentrate even in depolarized mitochondria (Zhao et al. 2004; Doughan and Dikalov 2007). The most-studied SS peptide, SS-31 (D-Arg-2′, 6′-dimethyltyrosine-Lys-Phe-NH_2), was originally thought to exert its beneficial effect by the free radical scavenging activity of dimethyltyrosine (Graham et al. 2009). However, recent studies have revealed that SS-31 selectively binds to cardiolipin on the inner mitochondrial (Birk et al. 2013a,b; Szeto 2014). The binding of SS-31 to cardiolipin alters the interaction of cardiolipin with cytochrome c, and favors its electron carrier function while inhibiting its peroxidase activity (Birk et al. 2013a; Szeto 2014). SS-31 treatment increases ATP production, inhibits ROS generation, and prevents cardiolipin peroxidation and loss of cristae (Birk et al. 2013a). These findings suggest that the mitochondrial protective properties of SS-31 may be attributed to ROS-independent mechanisms, such as improved energetics, with reduction of ROS production as a secondary benefit. Our laboratory showed that the mitochondrial protective peptide SS-31 prevents pressure overload-induced cardiac hypertrophy as well as failure in a highly parallel manner to mCAT overexpression (Dai et al. 2011, 2012, 2013). Although the effects of SS-31 on cardiac aging have not been reported, recent studies from our laboratory have shown that 8-wk treatments of SS-31 can reverse age-related diastolic dysfunction in old mice (Chiao and PS Rabinovitch, unpubl.), supporting the therapeutic potential of SS-31 in cardiac aging.

Inhibition of Renin Angiotensin Aldosterone Signaling

As noted above, Ang II concentrations increase in aged hearts, and Ang II infusion induces structural, functional, and molecular changes similar to cardiac aging (Groban et al. 2006; Dai et al. 2009), highlighting the therapeutic potential of inhibition of Ang II signaling in cardiac aging. Basso et al. (2007) showed that long-term blockade of Ang II signaling by the angiotensin-converting enzyme inhibitor enalapril or by the angiotensin receptor type I inhibitor losartan can extend the lifespan of male Wistar rats and substantially attenuate age-related cardiovascular pathologies. In an earlier study, Inserra and colleagues (1995) showed that life-long treatment of enalapril can attenuate cardiac hypertrophy and interstitial fibrosis in hearts of 24-mo-old mice without significant changes in blood pressure. A later study by Stein et al. (2010) showed that long-term (10-mo) treatment with losartan, beginning at 12 mo of age, can also reduce myocardial fibrosis and fibrosis-related arrhythmias in aged mice. Groban et al. (2012) recently compared the effects of low-dose (non-blood-pressure lowering) enalapril and losartan for 6 mo initiated at 24 mo of age in male Fischer 344 × Brown Norway rats. They showed that, although low-dose enalapril and losartan both reduced cardiac oxidative stress, only enalapril was able to mitigate diastolic dysfunction, and they suggested that this may be mediated by a lowered ratio of phospholamban to SERCA2.

GDF-11

Recently, Loffredo et al. (2013) showed, by heterochronic parabiosis, that the circulation of young mice can regress cardiac hypertrophy in aged mouse hearts. By proteomic analysis of plasma samples from young and old mice, they identified that growth differentiation factor 11 (GDF11) is a circulating factor that declines with age and may be responsible for the reversal of age-related hypertrophy in heterochronic parabiosis. Importantly, restoring circulating GDF11 levels of old mice to young levels, by daily intraperitoneal injection of recombinant GDF11 (rGDF11) for 30 days can also reverse age-related hypertrophy. Treatment with rGDF11 reduced hypertrophic markers (ANP and BNP) expression and increased

SERCA-2 expression, recapitulating the molecular changes mediated by parabiosis (Loffredo et al. 2013). The precise mechanism of GDF11 action, its effect on age-related diastolic dysfunction, and the role of GDF11 in human cardiac aging remain to be investigated; however, the results from the mouse model suggest an exciting therapeutic potential of GDF11 in cardiac aging.

Therapies Targeting miRNAs

As discussed above, recent studies suggest that miRNAs are important regulators of cardiac aging. With age, the expression of the miR-17-92 cluster (miR-18a, miR19a, and miR-19b) decreases, whereas the expressions miR-22 and miR-34a increase in hearts (Boon et al. 2013). Boon et al. (2013) showed that in vivo silencing of miR-34a by injection of antisense oligonucleotides (antagomirs) or locked nucleic acid (LNA)-based anti-miRs can reduce expression of miR-34a and partially rescue cardiac phenotypes in mice. This finding supports the potential of gene therapy to reverse the age-related changes in miRNAs to treat cardiac aging. However, as one miRNA is likely to have multiple targets, gene therapy targeting miRNA may trigger undesirable side effects. An alternative approach is to identify the miRNA targets that mediate cardiac aging responses and manipulate the specific target genes as treatment strategy.

Cardiac Stem-Cell or Progenitor-Cell Therapy

The recent discovery that the heart is able to regenerate although cardiac stem cells and cardiac progenitor cells has attracted enormous attention to the potential of stem-cell therapy for cardiovascular diseases and aging. Two approaches for stem-cell therapy are (1) direct delivery of cardiac stem cells/cardiac progenitor cells (with or without treatment to enhance cardiac differentiation or regenerative capacity) to the heart, and (2) delivery of agents that enhance the function of endogenous cardiac stem cells or progenitor cells (Ballard and Edelberg 2007). Potential therapeutic agents for enhancing endogenous stem-cell or progenitor-cell function include stromal-cell-derived factor (SDF)-1, platelet-derived growth factor (PDGF), vascular endothelial growth factor (VEGF), and IFG-1 (Ballard and Edelberg 2007). For direct stem-cell or progenitor-cell delivery, the therapeutic effects are limited by the proliferation, engraftment, survival, and persistence of the transplanted cells. In a recent study, Mohsin et al. (2012) used lentivirus transduction to overexpress Pim-1 kinase in cardiac progenitor cells isolated from a 68-yr-old heart failure patient, and showed enhanced survival, proliferation, differentiation, and persistence of the cardiac progenitor cells after transplanted into an immunocompromised mouse model of myocardial infarction. Strikingly, transplant of Pim-1-expressing progenitor cells significantly improved myocardial healing and function of the infarcted heart in 8 wk. The ability to rejuvenate human cardiac progenitor cells ex vivo by Pim-1 modification is highly encouraging to the development of stem-cell therapy for cardiac aging (Mohsin et al. 2013).

FUTURE PERSPECTIVES ON CARDIAC AGING INTERVENTIONS

The improved understanding on the molecular mechanisms of cardiac aging has led to promising advancements in the development of cardiac aging interventions. Recent studies have shown the potential of different therapeutic approaches to delay or treat cardiac aging, ranging from CR to pharmacologic interventions (rapamycin, enalapril, and SS-31), recombinant protein therapy (IGF-1 and GDF-11), gene therapy (miRNAs), and cardiac stem-cell therapy. However, future studies will be required to evaluate the translational potentials of these interventions.

As a general rule for cardiac aging interventions, short-term treatment(s) beginning at late life will have higher translational potential compared with long-term or life-long treatments. This is particularly relevant to systemic treatments or treatments that target multiple pathways, as a briefer treatment will lower the chances of irreversible side effects. Therefore,

it is important to study the kinetics and pharmacodynamics of the treatment to determine the minimal effective dose and duration, as well as the persistence of the treatment to determine the optimum therapeutic regimen.

Another issue for consideration is that there may be gender-specific differences in therapeutic responses. Many of the potential interventions noted above have been tested in only one gender of animals, and, therefore, potential gender-specific differences in beneficial effects remain unknown. For instance, rapamycin provided a greater lifespan extension in female mice at the initial dose (14 ppm) studied by the NIA Intervention Testing Program (Harrison et al. 2009), but a later study showed an improved effect in males at a higher dose, although the lifespan extension benefit was greater in females than males at each dose. This gender difference was associated with a sexual dimorphism of rapamycin levels in blood (Miller et al. 2014).

CONCLUDING REMARKS

As the elderly population in developed countries is expected to double in the next 25 years, there will be an urgent need for interventions to attenuate or reverse cardiac impairment and the concomitant negative physiological consequences in the elderly. Recent studies show promising results of multiple novel interventions to delay or reverse cardiac aging. More in-depth understanding of the molecular mechanisms of intrinsic cardiac aging and the mechanistic effects of these interventions will be required to guide the development and future translation of these novel therapies to clinical application. Mechanistic insights may also identify other more specific therapeutic targets and provide guidance toward interventions for other age-related pathologies.

ACKNOWLEDGMENTS

We acknowledge support from the Ellison Medical Foundation and the American Federation for Aging Research, as well as National Institutes of Health (NIH) Grants AG001751, AG038550, and HL101186. Y.A.C is a Glenn/AFAR postdoctoral fellow for the Translational Research on Aging.

REFERENCES

Adlam VJ, Harrison JC, Porteous CM, James AM, Smith RA, Murphy MP, Sammut IA. 2005. Targeting an antioxidant to mitochondria decreases cardiac ischemia-reperfusion injury. *FASEB J* **19:** 1088–1095.

Antonenko YN, Avetisyan AV, Bakeeva LE, Chernyak BV, Chertkov VA, Domnina LV, Ivanova OY, Izyumov DS, Khailova LS, Klishin SS, et al. 2008. Mitochondria-targeted plastoquinone derivatives as tools to interrupt execution of the aging program. 1: Cationic plastoquinone derivatives: Synthesis and in vitro studies. *Biochemistry (Mosc)* **73:** 1273–1287.

Anversa P, Rota M, Urbanek K, Hosoda T, Sonnenblick EH, Leri A, Kajstura J, Bolli R. 2005. Myocardial aging—A stem cell problem. *Basic Res Cardiol* **100:** 482–493.

Arany Z, He H, Lin J, Hoyer K, Handschin C, Toka O, Ahmad F, Matsui T, Chin S, Wu PH, et al. 2005. Transcriptional coactivator PGC-1α controls the energy state and contractile function of cardiac muscle. *Cell Metab* **1:** 259–271.

Bakeeva LE, Barskov IV, Egorov MV, Isaev NK, Kapelko VI, Kazachenko AV, Kirpatovsky VI, Kozlovsky SV, Lakomkin VL, Levina SB, et al. 2008. Mitochondria-targeted plastoquinone derivatives as tools to interrupt execution of the aging program. 2: Treatment of some ROS- and age-related diseases (heart arrhythmia, heart infarctions, kidney ischemia, and stroke). *Biochemistry (Mosc)* **73:** 1288–1299.

Ballard VL, Edelberg JM. 2007. Stem cells and the regeneration of the aging cardiovascular system. *Circ Res* **100:** 1116–1127.

Barger JL, Kayo T, Vann JM, Arias EB, Wang J, Hacker TA, Wang Y, Raederstorff D, Morrow JD, Leeuwenburgh C, et al. 2008. A low dose of dietary resveratrol partially mimics caloric restriction and retards aging parameters in mice. *PLoS ONE* **3:** e2264.

Basso N, Paglia N, Stella I, de Cavanagh EM, Ferder L, del Rosario Lores Arnaiz M, Inserra F. 2005. Protective effect of the inhibition of the renin-angiotensin system on aging. *Regul Pept* **128:** 247–252.

Basso N, Cini R, Pietrelli A, Ferder L, Terragno NA, Inserra F. 2007. Protective effect of long-term angiotensin II inhibition. *Am J Physiol Heart Circ Physiol* **293:** H1351–H1358.

Beltrami AP, Barlucchi L, Torella D, Baker M, Limana F, Chimenti S, Kasahara H, Rota M, Musso E, Urbanek K, et al. 2003. Adult cardiac stem cells are multipotent and support myocardial regeneration. *Cell* **114:** 763–776.

Benigni A, Corna D, Zoja C, Sonzogni A, Latini R, Salio M, Conti S, Rottoli D, Longaretti L, Cassis P, et al. 2009. Disruption of the Ang II type 1 receptor promotes longevity in mice. *J Clin Invest* **119:** 524–530.

Bergmann O, Bhardwaj RD, Bernard S, Zdunek S, Barnabe-Heider F, Walsh S, Zupicich J, Alkass K, Buchholz BA, Druid H, et al. 2009. Evidence for cardiomyocyte renewal in humans. *Science* **324:** 98–102.

Birk AV, Chao WM, Bracken WC, Warren JD, Szeto HH. 2013a. Targeting mitochondrial cardiolipin and the cytochrome *c*/cardiolipin complex to promote electron transport and optimize mitochondrial ATP synthesis. *Br J Pharmacol* **171:** 2017–2028.

Birk AV, Liu S, Soong Y, Mills W, Singh P, Warren JD, Seshan SV, Pardee JD, Szeto HH. 2013b. The mitochondrial-targeted compound SS-31 re-energizes ischemic mitochondria by interacting with cardiolipin. *J Am Soc Nephrol* **24:** 1250–1261.

Bonnema DD, Webb CS, Pennington WR, Stroud RE, Leonardi AE, Clark LL, McClure CD, Finklea L, Spinale FG, Zile MR. 2007. Effects of age on plasma matrix metalloproteinases (MMPs) and tissue inhibitor of metalloproteinases (TIMPs). *J Card Fail* **13:** 530–540.

Boon RA, Iekushi K, Lechner S, Seeger T, Fischer A, Heydt S, Kaluza D, Treguer K, Carmona G, Bonauer A, et al. 2013. MicroRNA-34a regulates cardiac ageing and function. *Nature* **495:** 107–110.

Borlaug BA, Kass DA. 2006. Mechanisms of diastolic dysfunction in heart failure. *Trends Cardiovasc Med* **16:** 273–279.

Bradshaw AD, Baicu CF, Rentz TJ, Van Laer AO, Bonnema DD, Zile MR. 2010. Age-dependent alterations in fibrillar collagen content and myocardial diastolic function: Role of SPARC in post-synthetic procollagen processing. *Am J Physiol Heart Circ Physiol* **298:** H614–H622.

Broglio F, Fubini A, Morello M, Arvat E, Aimaretti G, Gianotti L, Boghen MF, Deghenghi R, Mangiardi L, Ghigo E. 1999. Activity of GH/IGF-I axis in patients with dilated cardiomyopathy. *Clin Endocrinol (Oxf)* **50:** 417–430.

Brooks WW, Conrad CH. 2000. Myocardial fibrosis in transforming growth factor β1 heterozygous mice. *J Mol Cell Cardiol* **32:** 187–195.

Brouwers FP, Hillege HL, van Gilst WH, van Veldhuisen DJ. 2012. Comparing new onset heart failure with reduced ejection fraction and new onset heart failure with preserved ejection fraction: An epidemiologic perspective. *Curr Heart Fail Rep* **9:** 363–368.

Bujak M, Frangogiannis NG. 2007. The role of TGF-β signaling in myocardial infarction and cardiac remodeling. *Cardiovasc Res* **74:** 184–195.

Chiao YA, Dai Q, Zhang J, Lin J, Lopez EF, Ahuja SS, Chou YM, Lindsey ML, Jin YF. 2011. Multi-analyte profiling reveals matrix metalloproteinase-9 and monocyte chemotactic protein-1 as plasma biomarkers of cardiac aging. *Circ Cardiovasc Genet* **4:** 455–462.

Chiao YA, Ramirez TA, Zamilpa R, Okoronkwo SM, Dai Q, Zhang J, Jin YF, Lindsey ML. 2012. Matrix metalloproteinase-9 deletion attenuates myocardial fibrosis and diastolic dysfunction in ageing mice. *Cardiovasc Res* **96:** 444–455.

Choi SY, Chang HJ, Choi SI, Kim KI, Cho YS, Youn TJ, Chung WY, Chae IH, Choi DJ, Kim HS, et al. 2009. Long-term exercise training attenuates age-related diastolic dysfunction: Association of myocardial collagen cross-linking. *J Korean Med Sci* **24:** 32–39.

Colman RJ, Anderson RM, Johnson SC, Kastman EK, Kosmatka KJ, Beasley TM, Allison DB, Cruzen C, Simmons HA, Kemnitz JW, et al. 2009. Caloric restriction delays disease onset and mortality in rhesus monkeys. *Science* **325:** 201–204.

Corpas E, Harman SM, Blackman MR. 1993. Human growth hormone and human aging. *Endocr Rev* **14**: 20–39.

Cruzen C, Colman RJ. 2009. Effects of caloric restriction on cardiovascular aging in non-human primates and humans. *Clin Geriatr Med* **25**: 733–743, ix–x.

Csiszar A, Labinskyy N, Perez V, Recchia FA, Podlutsky A, Mukhopadhyay P, Losonczy G, Pacher P, Austad SN, Bartke A, et al. 2008. Endothelial function and vascular oxidative stress in long-lived GH/IGF-deficient Ames dwarf mice. *Am J Physiol Heart Circ Physiol* **295**: H1882–H1894.

Csiszar A, Labinskyy N, Jimenez R, Pinto JT, Ballabh P, Losonczy G, Pearson KJ, de Cabo R, Ungvari Z. 2009. Anti-oxidative and anti-inflammatory vasoprotective effects of caloric restriction in aging: Role of circulating factors and SIRT1. *Mech Ageing Dev* **130**: 518–527.

Dai DF, Santana LF, Vermulst M, Tomazela DM, Emond MJ, MacCoss MJ, Gollahon K, Martin GM, Loeb LA, Ladiges WC, et al. 2009. Overexpression of catalase targeted to mitochondria attenuates murine cardiac aging. *Circulation* **119**: 2789–2797.

Dai DF, Chen T, Wanagat J, Laflamme M, Marcinek DJ, Emond MJ, Ngo CP, Prolla TA, Rabinovitch PS. 2010. Age-dependent cardiomyopathy in mitochondrial mutator mice is attenuated by overexpression of catalase targeted to mitochondria. *Aging Cell* **9**: 536–544.

Dai DF, Chen T, Szeto H, Nieves-Cintron M, Kutyavin V, Santana LF, Rabinovitch PS. 2011. Mitochondrial targeted antioxidant peptide ameliorates hypertensive cardiomyopathy. *J Am Coll Cardiol* **58**: 73–82.

Dai DF, Hsieh EJ, Liu Y, Chen T, Beyer RP, Chin MT, MacCoss MJ, Rabinovitch PS. 2012. Mitochondrial proteome remodelling in pressure overload-induced heart failure: The role of mitochondrial oxidative stress. *Cardiovasc Res* **93**: 79–88.

Dai DF, Hsieh EJ, Chen T, Menendez LG, Basisty NB, Tsai L, Beyer RP, Crispin DA, Shulman NJ, Szeto HH, et al. 2013. Global proteomics and pathway analysis of pressure-overload-induced heart failure and its attenuation by mitochondrial-targeted peptides. *Circ Heart Fail* **6**: 1067–1076.

Dai DF, Chiao YA, Marcinek DJ, Szeto HH, Rabinovitch PS. 2014a. Mitochondrial oxidative stress in aging and healthspan. *Longev Healthspan* **3**: 6.

Dai DF, Karunadharma PP, Chiao YA, Basisty N, Crispin D, Hsieh EJ, Chen T, Gu H, Djukovic D, Raftery D, et al. 2014b. Altered proteome turnover and remodeling by short-term caloric restriction or rapamycin rejuvenate the aging heart. *Aging Cell* **13**: 529–539.

DeQuach JA, Mezzano V, Miglani A, Lange S, Keller GM, Sheikh F, Christman KL. 2010. Simple and high yielding method for preparing tissue specific extracellular matrix coatings for cell culture. *PLoS ONE* **5**: e13039.

Dikalova AE, Bikineyeva AT, Budzyn K, Nazarewicz RR, McCann L, Lewis W, Harrison DG, Dikalov SI. 2010. Therapeutic targeting of mitochondrial superoxide in hypertension. *Circ Res* **107**: 106–116

Domenighetti AA, Wang Q, Egger M, Richards SM, Pedrazzini T, Delbridge LM. 2005. Angiotensin II–mediated phenotypic cardiomyocyte remodeling leads to age-dependent cardiac dysfunction and failure. *Hypertension* **46**: 426–432.

Doughan AK, Dikalov SI. 2007. Mitochondrial redox cycling of mitoquinone leads to superoxide production and cellular apoptosis. *Antioxid Redox Signal* **9**: 1825–1836.

Downes TR, Nomeir AM, Smith KM, Stewart KP, Little WC. 1989. Mechanism of altered pattern of left ventricular filling with aging in subjects without cardiac disease. *Am J Cardiol* **64**: 523–527.

Flynn JM, O'Leary MN, Zambataro CA, Academia EC, Presley MP, Garrett BJ, Zykovich A, Mooney SD, Strong R, Rosen CJ, et al. 2013. Late-life rapamycin treatment reverses age-related heart dysfunction. *Aging Cell* **12**: 851–862.

Fowler CG, Chiasson KB, Leslie TH, Thomas D, Beasley TM, Kemnitz JW, Weindruch R. 2010. Auditory function in rhesus monkeys: Effects of aging and caloric restriction in the Wisconsin monkeys five years later. *Hear Res* **261**: 75–81.

Go AS, Mozaffarian D, Roger VL, Benjamin EJ, Berry JD, Blaha MJ, Dai S, Ford ES, Fox CS, Franco S, et al. 2014. Heart disease and stroke statistics—2014 update: A report from the American Heart Association. *Circulation* **129**: e28–e292.

Gonzalez A, Rota M, Nurzynska D, Misao Y, Tillmanns J, Ojaimi C, Padin-Iruegas ME, Muller P, Esposito G, Bearzi C, et al. 2008. Activation of cardiac progenitor cells reverses the failing heart senescent phenotype and prolongs lifespan. *Circ Res* **102**: 597–606.

Graham D, Huynh NN, Hamilton CA, Beattie E, Smith RA, Cocheme HM, Murphy MP, Dominiczak AF. 2009. Mitochondria-targeted antioxidant MitoQ$_{10}$ improves endothelial function and attenuates cardiac hypertrophy. *Hypertension* **54**: 322–328.

Groban L, Pailes NA, Bennett CD, Carter CS, Chappell MC, Kitzman DW, Sonntag WE. 2006. Growth hormone replacement attenuates diastolic dysfunction and cardiac angiotensin II expression in senescent rats. *J Gerontol A Biol Sci Med Sci* **61**: 28–35.

Groban L, Lindsey S, Wang H, Lin MS, Kassik KA, Machado FS, Carter CS. 2012. Differential effects of late-life initiation of low-dose enalapril and losartan on diastolic function in senescent Fischer 344 × Brown Norway male rats. *Age (Dordr)* **34**: 831–843.

Harman D. 1972. The biologic clock: The mitochondria? *J Am Geriatr Soc* **20**: 145–147.

Harrison DE, Strong R, Sharp ZD, Nelson JF, Astle CM, Flurkey K, Nadon NL, Wilkinson JE, Frenkel K, Carter CS, et al. 2009. Rapamycin fed late in life extends lifespan in genetically heterogeneous mice. *Nature* **460**: 392–395.

Hsieh PC, Segers VF, Davis ME, MacGillivray C, Gannon J, Molkentin JD, Robbins J, Lee RT. 2007. Evidence from a genetic fate-mapping study that stem cells refresh adult mammalian cardiomyocytes after injury. *Nat Med* **13**: 970–974.

Inserra F, Romano L, Ercole L, de Cavanagh EM, Ferder L. 1995. Cardiovascular changes by long-term inhibition of the renin-angiotensin system in aging. *Hypertension* **25**: 437–442.

Janczewski AM, Lakatta EG. 2010. Modulation of sarcoplasmic reticulum Ca^{2+} cycling in systolic and diastolic heart failure associated with aging. *Heart Fail Rev* 15: 431–445.

Janczewski AM, Spurgeon HA, Lakatta EG. 2002. Action potential prolongation in cardiac myocytes of old rats is an adaptation to sustain youthful intracellular Ca^{2+} regulation. *J Mol Cell Cardiol* 34: 641–648.

Jazbutyte V, Fiedler J, Kneitz S, Galuppo P, Just A, Holzmann A, Bauersachs J, Thum T. 2013. MicroRNA-22 increases senescence and activates cardiac fibroblasts in the aging heart. *Age (Dordr)* 35: 747–762.

Josephson IR, Guia A, Stern MD, Lakatta EG. 2002. Alterations in properties of L-type Ca channels in aging rat heart. *J Mol Cell Cardiol* 34: 297–308.

Judge S, Jang YM, Smith A, Hagen T, Leeuwenburgh C. 2005. Age-associated increases in oxidative stress and antioxidant enzyme activities in cardiac interfibrillar mitochondria: Implications for the mitochondrial theory of aging. *FASEB J* 19: 419–421.

Kass DA, Bronzwaer JG, Paulus WJ. 2004. What mechanisms underlie diastolic dysfunction in heart failure? *Circ Res* 94: 1533–1542.

Kastman EK, Willette AA, Coe CL, Bendlin BB, Kosmatka KJ, McLaren DG, Xu G, Canu E, Field AS, Alexander AL, et al. 2010. A calorie-restricted diet decreases brain iron accumulation and preserves motor performance in old rhesus monkeys. *J Neurosci* 30: 7940–7947.

Kelso GF, Porteous CM, Coulter CV, Hughes G, Porteous WK, Ledgerwood EC, Smith RA, Murphy MP. 2001. Selective targeting of a redox-active ubiquinone to mitochondria within cells: Antioxidant and antiapoptotic properties. *J Biol Chem* 276: 4588–4596.

Kemi M, Keenan KP, McCoy C, Hoe CM, Soper KA, Ballam GC, van Zwieten MJ. 2000. The relative protective effects of moderate dietary restriction versus dietary modification on spontaneous cardiomyopathy in male Sprague–Dawley rats. *Toxicol Pathol* 28: 285–296.

Kennedy BK, Steffen KK, Kaeberlein M. 2007. Ruminations on dietary restriction and aging. *Cell Mol Life Sci* 64: 1323–1328.

Khan AS, Sane DC, Wannenburg T, Sonntag WE. 2002. Growth hormone, insulin-like growth factor-1 and the aging cardiovascular system. *Cardiovasc Res* 54: 25–35.

Kitzman DW. 2002. Diastolic heart failure in the elderly. *Heart Fail Rev* 7: 17–27.

Knyushko TV, Sharov VS, Williams TD, Schoneich C, Bigelow DJ. 2005. 3-Nitrotyrosine modification of SERCA2a in the aging heart: A distinct signature of the cellular redox environment. *Biochemistry* 44: 13071–13081.

Koban MU, Moorman AF, Holtz J, Yacoub MH, Boheler KR. 1998. Expressional analysis of the cardiac Na-Ca exchanger in rat development and senescence. *Cardiovasc Res* 37: 405–423.

Kujoth GC, Hiona A, Pugh TD, Someya S, Panzer K, Wohlgemuth SE, Hofer T, Seo AY, Sullivan R, Jobling WA, et al. 2005. Mitochondrial DNA mutations, oxidative stress, and apoptosis in mammalian aging. *Science* 309: 481–484.

Lakatta EG. 2002. Age-associated cardiovascular changes in health: Impact on cardiovascular disease in older persons. *Heart Fail Rev* 7: 29–49.

Lakatta EG. 2003. Arterial and cardiac aging: major shareholders in cardiovascular disease enterprises. Part III: Cellular and molecular clues to heart and arterial aging. *Circulation* 107: 490–497.

Lakatta EG, Levy D. 2003a. Arterial and cardiac aging: major shareholders in cardiovascular disease enterprises. Part I: Aging arteries—A "set up" for vascular disease. *Circulation* 107: 139–146.

Lakatta EG, Levy D. 2003b. Arterial and cardiac aging: Major shareholders in cardiovascular disease enterprises. Part II: The aging heart in health: Links to heart disease. *Circulation* 107: 346–354.

Li Q, Ceylan-Isik AF, Li J, Ren J. 2008. Deficiency of insulin-like growth factor 1 reduces sensitivity to aging-associated cardiomyocyte dysfunction. *Rejuvenation Res* 11: 725–733.

Lindsey ML, Goshorn DK, Squires CE, Escobar GP, Hendrick JW, Mingoia JT, Sweterlitsch SE, Spinale FG. 2005. Age-dependent changes in myocardial matrix metalloproteinase/tissue inhibitor of metalloproteinase profiles and fibroblast function. *Cardiovasc Res* 66: 410–419.

Loffredo FS, Steinhauser ML, Jay SM, Gannon J, Pancoast JR, Yalamanchi P, Sinha M, Dall'Osso C, Khong D, Shadrach JL, et al. 2013. Growth differentiation factor 11 is a circulating factor that reverses age-related cardiac hypertrophy. *Cell* 153: 828–839.

Lopez-Lluch G, Hunt N, Jones B, Zhu M, Jamieson H, Hilmer S, Cascajo MV, Allard J, Ingram DK, Navas P, et al. 2006. Calorie restriction induces mitochondrial biogenesis and bioenergetic efficiency. *Proc Natl Acad Sci* 103: 1768–1773.

Lopez-Lluch G, Irusta PM, Navas P, de Cabo R. 2008. Mitochondrial biogenesis and healthy aging. *Exp Gerontol* 43: 813–819.

Lopez-Lopez C, Dietrich MO, Metzger F, Loetscher H, Torres-Aleman I. 2007. Disturbed cross talk between insulin-like growth factor I and AMP-activated protein kinase as a possible cause of vascular dysfunction in the amyloid precursor protein/presenilin 2 mouse model of Alzheimer's disease. *J Neurosci* 27: 824–831.

Luong N, Davies CR, Wessells RJ, Graham SM, King MT, Veech R, Bodmer R, Oldham SM. 2006. Activated FOXO-mediated insulin resistance is blocked by reduction of TOR activity. *Cell Metab* 4: 133–142.

Mammucari C, Rizzuto R. 2010. Signaling pathways in mitochondrial dysfunction and aging. *Mech Ageing Dev* 131: 536–543.

McKiernan SH, Colman RJ, Lopez M, Beasley TM, Aiken JM, Anderson RM, Weindruch R. 2011. Caloric restriction delays aging-induced cellular phenotypes in rhesus monkey skeletal muscle. *Exp Gerontol* 46: 23–29.

Meikle L, McMullen JR, Sherwood MC, Lader AS, Walker V, Chan JA, Kwiatkowski DJ. 2005. A mouse model of cardiac rhabdomyoma generated by loss of Tsc1 in ventricular myocytes. *Hum Mol Genet* 14: 429–435.

Meyer TE, Kovacs SJ, Ehsani AA, Klein S, Holloszy JO, Fontana L. 2006. Long-term caloric restriction ameliorates the decline in diastolic function in humans. *J Am Coll Cardiol* 47: 398–402.

Miller RA, Harrison DE, Astle CM, Baur JA, Boyd AR, de Cabo R, Fernandez E, Flurkey K, Javors MA, Nelson JF, et al. 2011. Rapamycin, but not resveratrol or simvastatin, extends life span of genetically heterogeneous mice. *J Gerontol A Biol Sci Med Sci* **66**: 191–201.

Miller RA, Harrison DE, Astle CM, Fernandez E, Flurkey K, Han M, Javors MA, Li X, Nadon NL, Nelson JF, et al. 2014. Rapamycin-mediated lifespan increase in mice is dose and sex dependent and metabolically distinct from dietary restriction. *Aging Cell* **13**: 468–477.

Mohsin S, Khan M, Toko H, Bailey B, Cottage CT, Wallach K, Nag D, Lee A, Siddiqi S, Lan F, et al. 2012. Human cardiac progenitor cells engineered with Pim-I kinase enhance myocardial repair. *J Am Coll Cardiol* **60**: 1278–1287.

Mohsin S, Khan M, Nguyen J, Alkatib M, Siddiqi S, Hariharan N, Wallach K, Monsanto M, Gude N, Dembitsky W, et al. 2013. Rejuvenation of human cardiac progenitor cells with Pim-1 kinase. *Circ Res* **113**: 1169–1179.

Murphy MP, Smith RA. 2007. Targeting antioxidants to mitochondria by conjugation to lipophilic cations. *Annu Rev Pharmacol Toxicol* **47**: 629–656.

Neff F, Flores-Dominguez D, Ryan DP, Horsch M, Schroder S, Adler T, Afonso LC, Aguilar-Pimentel JA, Becker L, Garrett L, et al. 2013. Rapamycin extends murine lifespan but has limited effects on aging. *J Clin Invest* **123**: 3272–3291.

Niemann B, Chen Y, Issa H, Silber RE, Rohrbach S. 2010. Caloric restriction delays cardiac ageing in rats: Role of mitochondria. *Cardiovasc Res* **88**: 267–276.

Nisoli E, Tonello C, Cardile A, Cozzi V, Bracale R, Tedesco L, Falcone S, Valerio A, Cantoni O, Clementi E, et al. 2005. Calorie restriction promotes mitochondrial biogenesis by inducing the expression of eNOS. *Science* **310**: 314–317.

Oktay AA, Rich JD, Shah SJ. 2013. The emerging epidemic of heart failure with preserved ejection fraction. *Curr Heart Fail Rep* **10**: 401–410.

O'Malley Y, Fink BD, Ross NC, Prisinzano TE, Sivitz WI. 2006. Reactive oxygen and targeted antioxidant administration in endothelial cell mitochondria. *J Biol Chem* **281**: 39766–39775.

Ouzounian M, Lee DS, Liu PP. 2008. Diastolic heart failure: Mechanisms and controversies. *Nat Clin Pract Cardiovasc Med* **5**: 375–386.

Panek AN, Posch MG, Alenina N, Ghadge SK, Erdmann B, Popova E, Perrot A, Geier C, Dietz R, Morano I, et al. 2009. Connective tissue growth factor overexpression in cardiomyocytes promotes cardiac hypertrophy and protection against pressure overload. *PLoS ONE* **4**: e6743.

Puche JE, Garcia-Fernandez M, Muntane J, Rioja J, Gonzalez-Baron S, Castilla Cortazar I. 2008. Low doses of insulin-like growth factor-I induce mitochondrial protection in aging rats. *Endocrinology* **149**: 2620–2627.

Quiat D, Olson EN. 2013. MicroRNAs in cardiovascular disease: From pathogenesis to prevention and treatment. *J Clin Invest* **123**: 11–18.

Reed AL, Tanaka A, Sorescu D, Liu H, Jeong EM, Sturdy M, Walp ER, Dudley SC Jr, Sutliff RL. 2011. Diastolic dysfunction is associated with cardiac fibrosis in the senescence-accelerated mouse. *Am J Physiol Heart Circ Physiol* **301**: H824–H831.

Ren J, Brown-Borg HM. 2002. Impaired cardiac excitation–contraction coupling in ventricular myocytes from Ames

dwarf mice with IGF-I deficiency. *Growth Horm IGF Res* **12**: 99–105.

Riordan MM, Weiss EP, Meyer TE, Ehsani AA, Racette SB, Villareal DT, Fontana L, Holloszy JO, Kovacs SJ. 2008. The effects of caloric restriction- and exercise-induced weight loss on left ventricular diastolic function. *Am J Physiol Heart Circ Physiol* **294**: H1174–H1182.

Rivera EJ, Goldin A, Fulmer N, Tavares R, Wands JR, de la Monte SM. 2005. Insulin and insulin-like growth factor expression and function deteriorate with progression of Alzheimer's disease: Link to brain reductions in acetylcholine. *J Alzheimers Dis* **8**: 247–268.

Russell LK, Mansfield CM, Lehman JJ, Kovacs A, Courtois M, Saffitz JE, Medeiros DM, Valencik ML, McDonald JA, Kelly DP. 2004. Cardiac-specific induction of the transcriptional coactivator peroxisome proliferator-activated receptor γ coactivator-1α promotes mitochondrial biogenesis and reversible cardiomyopathy in a developmental stage-dependent manner. *Circ Res* **94**: 525–533.

Scatena R, Bottoni P, Botta G, Martorana GE, Giardina B. 2007. The role of mitochondria in pharmacotoxicology: A reevaluation of an old, newly emerging topic. *Am J Physiol Cell Physiol* **293**: C12–C21.

Schriner SE, Linford NJ, Martin GM, Treuting P, Ogburn CE, Emond M, Coskun PE, Ladiges W, Wolf N, Van Remmen H, et al. 2005. Extension of murine life span by overexpression of catalase targeted to mitochondria. *Science* **308**: 1909–1911.

Sharov VS, Dremina ES, Galeva NA, Williams TD, Schoneich C. 2006. Quantitative mapping of oxidation-sensitive cysteine residues in SERCA in vivo and in vitro by HPLC-electrospray-tandem MS: Selective protein oxidation during biological aging. *Biochem J* **394**: 605–615.

Shinmura K, Tamaki K, Sano M, Murata M, Yamakawa H, Ishida H, Fukuda K. 2011a. Impact of long-term caloric restriction on cardiac senescence: Caloric restriction ameliorates cardiac diastolic dysfunction associated with aging. *J Mol Cell Cardiol* **50**: 117–127.

Shinmura K, Tamaki K, Sano M, Nakashima-Kamimura N, Wolf AM, Amo T, Ohta S, Katsumata Y, Fukuda K, Ishiwata K, et al. 2011b. Caloric restriction primes mitochondria for ischemic stress by deacetylating specific mitochondrial proteins of the electron transport chain. *Circ Res* **109**: 396–406.

Skulachev VP, Anisimov VN, Antonenko YN, Bakeeva LE, Chernyak BV, Erichev VP, Filenko OF, Kalinina NI, Kapelko VI, Kolosova NG, et al. 2009. An attempt to prevent senescence: A mitochondrial approach. *Biochim Biophys Acta* **1787**: 437–461.

Smith RA, Hartley RC, Cocheme HM, Murphy MP. 2012. Mitochondrial pharmacology. *Trends Pharmacol Sci* **33**: 341–352.

Smith-Vikos T, Slack FJ. 2012. MicroRNAs and their roles in aging. *J Cell Sci* **125**: 7–17.

Spencer KT, Kirkpatrick JN, Mor-Avi V, Decara JM, Lang RM. 2004. Age dependency of the Tei index of myocardial performance. *J Am Soc Echocardiogr* **17**: 350–352.

Spinale FG, Escobar GP, Mukherjee R, Zavadzkas JA, Saunders SM, Jeffords LB, Leone AM, Beck C, Bouges S, Stroud RE. 2009. Cardiac-restricted overexpression of membrane type-1 matrix metalloproteinase in mice: Ef-

Cite this article as *Cold Spring Harb Perspect Med* doi: 10.1101/cshperspect.a025148

fects on myocardial remodeling with aging. *Circ Heart Fail* **2**: 351–360.

Stein M, Boulaksil M, Jansen JA, Herold E, Noorman M, Joles JA, van Veen TA, Houtman MJ, Engelen MA, Hauer RN, et al. 2010. Reduction of fibrosis-related arrhythmias by chronic renin-angiotensin-aldosterone system inhibitors in an aged mouse model. *Am J Physiol Heart Circ Physiol* **299**: H310–H321.

Swinne CJ, Shapiro EP, Lima SD, Fleg JL. 1992. Age-associated changes in left ventricular diastolic performance during isometric exercise in normal subjects. *Am J Cardiol* **69**: 823–826.

Szeto HH. 2014. First-in-class cardiolipin therapeutic to restore mitochondrial bioenergetics. *Br J Pharmacol* **171**: 2029–2050.

Taffet GE, Pham TT, Hartley CJ. 1997. The age-associated alterations in late diastolic function in mice are improved by caloric restriction. *J Gerontol A Biol Sci Med Sci* **52**: B285–B290.

Tayebjee MH, Lip GYH, Blann AD, MacFadyen RJ. 2005. Effects of age, gender, ethnicity, diurnal variation and exercise on circulating levels of matrix metalloproteinases (MMP)-2 and -9, and their inhibitors, tissue inhibitors of matrix metalloproteinases (TIMP)-1 and -2. *Thromb Res* **115**: 205–210.

Tei C, Nishimura RA, Seward JB, Tajik AJ. 1997. Noninvasive Doppler-derived myocardial performance index: correlation with simultaneous measurements of cardiac catheterization measurements. *J Am Soc Echocardiogr* **10**: 169–178.

Terman A, Brunk UT. 2004. Myocyte aging and mitochondrial turnover. *Exp Gerontol* **39**: 701–705.

Terzioglu M, Larsson NG. 2007. Mitochondrial dysfunction in mammalian ageing. *Novartis Found Symp* **287**: 197–208; discussion 208–2113.

Torella D, Rota M, Nurzynska D, Musso E, Monsen A, Shiraishi I, Zias E, Walsh K, Rosenzweig A, Sussman MA, et al. 2004. Cardiac stem cell and myocyte aging, heart failure, and insulin-like growth factor-1 overexpression. *Circ Res* **94**: 514–524.

Trifunovic A, Larsson NG. 2008. Mitochondrial dysfunction as a cause of ageing. *J Intern Med* **263**: 167–178.

Trifunovic A, Wredenberg A, Falkenberg M, Spelbrink JN, Rovio AT, Bruder CE, Bohlooly YM, Gidlof S, Oldfors A, Wibom R, et al. 2004. Premature ageing in mice expressing defective mitochondrial DNA polymerase. *Nature* **429**: 417–423.

Ungvari Z, Parrado-Fernandez C, Csiszar A, de Cabo R. 2008. Mechanisms underlying caloric restriction and life-span regulation: Implications for vascular aging. *Circ Res* **102**: 519–528.

van Almen GC, Verhesen W, van Leeuwen RE, van de Vrie M, Eurlings C, Schellings MW, Swinnen M, Cleutjens JP, van Zandvoort MA, Heymans S, et al. 2011. MicroRNA-18 and microRNA-19 regulate CTGF and TSP-1 expression in age-related heart failure. *Aging cell* **10**: 769–779.

Vasan RS, Sullivan LM, D'Agostino RB, Roubenoff R, Harris T, Sawyer DB, Levy D, Wilson PW. 2003. Serum insulin-like growth factor I and risk for heart failure in elderly individuals without a previous myocardial infarction: The Framingham Heart Study. *Ann Intern Med* **139**: 642–648.

Wang M, Zhang J, Walker SJ, Dworakowski R, Lakatta EG, Shah AM. 2010. Involvement of NADPH oxidase in age-associated cardiac remodeling. *J Mol Cell Cardiol* **48**: 765–772.

Wenz T. 2011. Mitochondria and PGC-1α in aging and age-associated diseases. *J Aging Res* **2011**: 810619.

Wessells RJ, Fitzgerald E, Cypser JR, Tatar M, Bodmer R. 2004. Insulin regulation of heart function in aging fruit flies. *Nat Genet* **36**: 1275–1281.

Wessells R, Fitzgerald E, Piazza N, Ocorr K, Morley S, Davies C, Lim HY, Elmen L, Hayes M, Oldham S, et al. 2009. d4eBP acts downstream of both dTOR and dFoxo to modulate cardiac functional aging in *Drosophila*. *Aging Cell* **8**: 542–552.

Wilkinson JE, Burmeister L, Brooks SV, Chan CC, Friedline S, Harrison DE, Hejtmancik JF, Nadon N, Strong R, Wood LK, et al. 2012. Rapamycin slows aging in mice. *Aging Cell* **11**: 675–682.

Yan L, Vatner DE, O'Connor JP, Ivessa A, Ge H, Chen W, Hirotani S, Ishikawa Y, Sadoshima J, Vatner SF. 2007. Type 5 adenylyl cyclase disruption increases longevity and protects against stress. *Cell* **130**: 247–258.

Zhao K, Zhao GM, Wu D, Soong Y, Birk AV, Schiller PW, Szeto HH. 2004. Cell-permeable peptide antioxidants targeted to inner mitochondrial membrane inhibit mitochondrial swelling, oxidative cell death, and reperfusion injury. *J Biol Chem* **279**: 34682–34690.

Zheng F, Plati AR, Potier M, Schulman Y, Berho M, Banerjee A, Leclercq B, Zisman A, Striker LJ, Striker GE. 2003. Resistance to glomerulosclerosis in B6 mice disappears after menopause. *Am J Pathol* **162**: 1339–1348.

Zile MR, Brutsaert DL. 2002. New concepts in diastolic dysfunction and diastolic heart failure. Part II: Causal mechanisms and treatment. *Circulation* **105**: 1503–1508.

Inhibition of the Mechanistic Target of Rapamycin (mTOR)–Rapamycin and Beyond

Dudley W. Lamming

Division of Endocrinology, Department of Medicine, University of Wisconsin-Madison and William S. Middleton Memorial Veterans Hospital, Madison, Wisconsin 53705

Correspondence: dlamming@medicine.wisc.edu

Rapamycin is a Food and Drug Administration (FDA)-approved immunosuppressant and anticancer agent discovered in the soil of Easter Island in the early 1970s. Rapamycin is a potent and selective inhibitor of the mechanistic target of rapamycin (mTOR) protein kinase, which acts as a central integrator of nutrient signaling pathways. During the last decade, genetic and pharmaceutical inhibition of mTOR pathway signaling has been found to promote longevity in yeast, worms, flies, and mice. In this article, we will discuss the molecular biology underlying the effects of rapamycin and its physiological effects, evidence for rapamycin as an antiaging compound, mechanisms by which rapamycin may extend life span, and the potential limitations of rapamycin as an antiaging molecule. Finally, we will discuss possible strategies that may allow us to inhibit mTOR signaling safely while minimizing side effects, and reap the health, social, and economic benefits from slowing the aging process.

The mechanistic target of rapamycin (mTOR) is a phosphatidylinositol-3-kinase (PI3K)-like serine/threonine protein kinase that is conserved in eukaryotes including yeast, worms, flies, and mammals. mTOR was discovered as a result of the search for the target of rapamycin, a polyketide produced by *Streptomyces hygroscopicus*, which originally attracted attention because of its ability to inhibit the growth of *Candida albicans* and other fungi (Vezina et al. 1975). It was soon determined that rapamycin also acts against mammalian cells, with effects on both cell size and proliferation, leading to its development as an immunosuppressant (Seto 2012). Its immunosuppressive effects led to very cautious exploration of the potential of rapamycin as an anticancer agent, but several rapamycin derivatives, including everolimus and temsirolimus, as well as rapamycin itself (sirolimus) are FDA-approved both as immunosuppressants and anticancer agents. Rapamycin has attracted significant interest with the finding in 2009 that rapamycin treatment can robustly extend the life span of mice (Harrison et al. 2009). In this article, we discuss the molecular biology of mTOR, research into the mechanism by which mTOR inhibition promotes life span, the side effects of rapamycin, and possible ways in which rapamycin or alternative strategies to inhibit mTOR signaling may enable us to extend human life span and health span.

Cite this article as *Cold Spring Harb Perspect Med* doi: 10.1101/cshperspect.a025924

MOLECULAR BIOLOGY OF RAPAMYCIN

The mTOR protein kinase is found in two evolutionarily conserved protein complexes with distinct functions, substrates, and sensitivity to rapamycin (Fig. 1). mTOR complex 1 (mTORC1) consists of the mTOR protein kinase, RAPTOR, and mLST8, along with the regulatory proteins PRAS40 and DEPTOR. mTORC1 plays a key role in the regulation of translation and cell growth through substrates that include S6 kinase 1 (S6K1) and the eukaryotic initiation factor eIF4E-binding protein 1 (4E-BP1) (reviewed in Caron et al. 2015). Other mTORC1 substrates include unc-51-like autophagy-activating kinase 1 (ULK-1), a key regulator of autophagy, transcription factor EB (TFEB), a regulator of lysosome biogenesis, and Grb-10, an insulin-receptor binding protein (Hsu et al. 2011; Kim et al. 2011; Settembre et al. 2012). The activity of mTORC1 toward many substrates is acutely sensitive to rapamycin, but mTORC1 also possess rapamycin-resistant activity toward certain substrates (Thoreen et al. 2012; Kang et al. 2013).

The activity of mTORC1 is dependent on its localization to the lysosome by the Rag/ragulator complex, where it can interact with its activator Rheb, but these proteins are not, strictly speaking, components of mTORC1 itself (Sancak et al. 2010; Bar-Peled and Sabatini 2014). The regulation of mTORC1 activity is extremely complex, but in brief, the Rag/ragulator complex recruits mTORC1 to the lysosome when amino acids and glucose are plentiful, whereas the tuberous sclerosis complex 1/2 (TSC1/2) complex, which negatively regulates mTORC1 signaling, departs from the lysosome in response to insulin (Bar-Peled and Sabatini 2014; Menon et al. 2014). The regulation of the Rag/ragulator complex has been a subject of intensive investigation, resulting in the identification of the additional upstream regulators of mTORC1 signaling, including the GATOR1/2 complex and Sestrin1–Sestrin3 (Bar-Peled et al. 2013; Chantranupong et al. 2014).

mTOR complex 2 (mTORC2) consists of mTOR, rapamycin-insensitive companion of mammalian target of rapamycin (RICTOR),

mLST8, PROTOR1/2, and mSin1, as well as the regulatory protein DEPTOR. In contrast to mTORC1, mTORC2 is relatively resistant to the effects of rapamycin both in vitro and in vivo, but can be disrupted by prolonged treatment (Sarbassov et al. 2006; Lamming et al. 2012). mTOR complex 2 (mTORC2) regulates a diverse set of substrates downstream from the insulin/insulin-like growth factor 1 (IGF-1) receptor, the best characterized of which include AKT on residues T450, S473, and S477/T479, serum/glucocorticoid-regulated kinase (SGK) S422, and protein kinase C α (PKC-α) (Guertin et al. 2006; Garcia-Martinez and Alessi 2008; Ikenoue et al. 2008; Liu et al. 2014a). More recently, mTORC2 has been shown to regulate control of other PKC family members, including PKC-δ and PKC-ζ, and mTORC2 also regulates the stability of insulin receptor substrate 1 (IRS1) via phosphorylation of the ubiquitin ligase subunit Fbw8 (Gan et al. 2012; Kim et al. 2012; Li and Gao 2014). It is apparent from this diverse set of substrates that mTORC2 is a key effector of the insulin signaling pathway.

Although the pathway that mediates activation of mTORC2 by the insulin/IGF-1 receptor is not fully understood, at least some mTORC2 localizes to the mitochondrial-associated endoplasmic reticulum membrane, and the activity of mTORC2 may be dependent on its association with ribosomal subunits (Zinzalla et al. 2011; Betz et al. 2013). A variety of other proteins, including TSC1/2, P-rex1, Rac1, sestrin3, and XPLN, have also been implicated in the regulation of mTORC2 (Hernandez-Negrete et al. 2007; Huang et al. 2008; Saci et al. 2011; Khanna et al. 2013; Tao et al. 2014). However, a cohesive model integrating all of these additional proteins is still lacking.

RAPAMYCIN TREATMENT EXTENDS LIFE SPAN

The role of the mTOR-signaling pathway in longevity was first discovered in 2003 in *Caenorhabditis elegans*; mutation of worm mTOR or RNAi against mTOR more than doubled the life span (Vellai et al. 2003). A similar effect was found in *Drosophila*, in which expression of

Cite this article as *Cold Spring Harb Perspect Med* doi: 10.1101/cshperspect.a025924

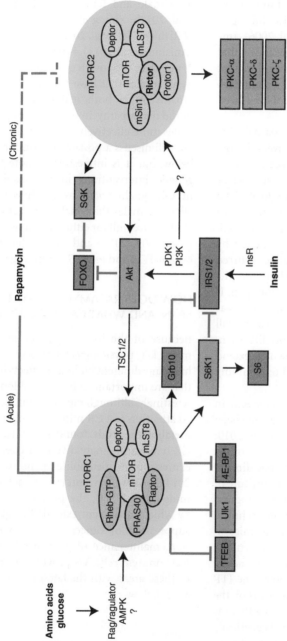

Figure 1. The mechanistic target of rapamycin (mTOR) signaling pathway. Rapamycin is an acute inhibitor of mTOR complex 1 (mTORC1), which phosphorylates substrates including S6 kinase 1 (S6K1), eIF4E-binding protein 1 (4E-BP1), transcription factor EB (TFEB), unc–51-like autophagy-activating kinase 1 (Ulk1), and growth factor receptor-bound protein 10 (GRB-10). Rapamycin dosed chronically also inhibits mTOR complex 2 (mTORC2), which regulates the phosphorylation of Akt, serum/glucocorticoid regulated kinase (SGK), and members of the protein kinase C (PKC) family. mTORC2 is primarily responsive to insulin/insulin-like growth factor 1 (IGF-1) signaling, whereas mTORC1 is sensitive to insulin as well as amino acids and glucose. AMPK, Adenosine monophosphate-stimulated kinase.

dominant negative mTOR or S6K similarly increased life span (Kapahi et al. 2004). The interest surrounding the mTOR-signaling pathway increased still further when inhibition of mTOR signaling in yeast was found to increase both chronological and replicative life span (Kaeberlein et al. 2005; Powers et al. 2006). Importantly, Kaeberlein and colleagues found that calorie restriction (CR), an intervention that extends life span in yeast as well as mammals, was unable to extend the life span of long-lived *tor1Δ* yeast.

A CR diet has been the gold standard for life span interventions since its discovery in the 1930s, and extends the life span of yeast, worms, flies, mice, dogs, and nonhuman primates (reviewed in Lamming and Anderson 2014). The mechanism behind the effects of a CR diet on life span have been elusive and highly debated, and the possibility that CR might function by inhibiting mTOR pathway signaling spurred significant efforts into understanding how the mTOR-signaling pathway regulates life span. It also suggested the possibility that rapamycin, as an inhibitor of mTOR, could function as a small molecule CR mimetic and extend life span. Indeed, rapamycin was soon shown to extend life span in yeast (Powers et al. 2006; Lamming et al. 2007).

Although interest in testing rapamycin in flies and worms was intense, and has now been shown to extend life span (Bjedov et al. 2010; Robida-Stubbs et al. 2012), investigation of the effects of rapamycin on life span jumped directly to mice with the aid of the National Institute of Aging (NIA) Interventions Testing Program (ITP). The ITP was able to solve the technical challenges of delivering rapamycin to mice in chow by microencapsulating it in an enteric polymer to protect rapamycin from the acidic environment of the stomach. In 2009, the ITP published the first of several manuscripts on the effects of rapamycin on mice, showing that rapamycin could extend the life span of genetically heterogeneous HET3 mice when treatment began at 20 mo of age (Harrison et al. 2009). Subsequent studies by the ITP determined that rapamycin had a similar effect on life span when delivered starting at 9 mo of age (Fig. 2A,B), and

that the response to rapamycin was dose dependent (Miller et al. 2011b, 2014).

In addition to HET3 mice, rapamycin has now been shown to extend the life span of both male and female C57BL/6J mice (Fok et al. 2014b; Zhang et al. 2014), male C57BL/6J Rj mice (Neff et al. 2013), female 129/Sv mice (Anisimov et al. 2011), and female FVB/N HER-2/neu mice (Popovich et al. 2014). Fascinatingly, all studies that have compared the effect of rapamycin on both males and females have found that rapamycin promotes longevity in females more effectively than in males (Fig. 2C). We will discuss a possible reason for this effect below, but it is interesting to note that many genetic interventions in the insulin/IGF-1/mTOR signaling pathway also show greater benefits in females than males (Fig. 2C). This includes mice null for either *Irs1* or *S6K1* (Selman et al. 2009, 2011) and mice heterozygous for both *mTOR* and *mLST8* (Lamming et al. 2012).

HOW DOES RAPAMYCIN EXTEND LIFE SPAN, AND WHAT CAN IT TEACH US?

Because of the involvement of mTOR in so many key physiological processes, rapamycin has many different biological effects in pathways that are important in health and longevity. Interestingly, although rapamycin initially attracted attention as a CR mimetic, a microarray and metabolome study found that rapamycin and CR have surprisingly divergent effects on gene expression and metabolites in the liver (Fok et al. 2014a). As we have previously detailed (Lamming et al. 2013), the proposed mechanisms by which rapamycin extends life span include suppression of cancer, inhibition of translation, maintenance of protein quality, and effects on stem cells. We provide a brief overview of these areas with the latest research on these areas below.

Cancer

Cancer is an important cause of mortality in both mice and humans. Overall, rapamycin and derivatives such as everolimus and temsirolimus have been only modestly effective in

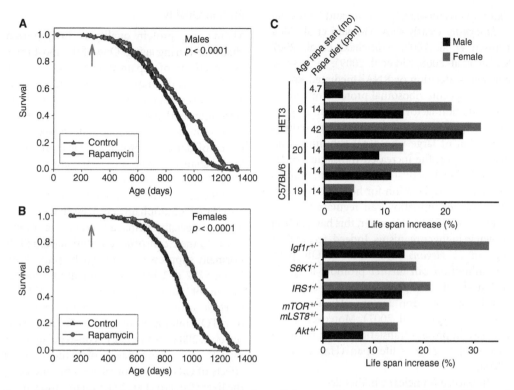

Figure 2. Sexually dimorphic impact of rapamycin and genetic interventions in the insulin/insulin-like growth factor 1 (IGF-1)/mechanistic target of rapamycin (mTOR)-signaling pathway. (*A,B*) Treatment with 14 ppm rapamycin begun at 9 mo of age extends the life span of genetically heterogeneous HET3 (*A*) males and (*B*) females. (*C*) Rapamycin and genetic interventions in the insulin/IGF-1/mTOR-signaling pathway that promote life span have a stronger effect on median female life span than on male life span (data from Holzenberger et al. 2003; Harrison et al. 2009; Selman et al. 2009, 2011; Lamming et al. 2012; Nojima et al. 2013; Fok et al. 2014b; Miller et al. 2014; Zhang et al. 2014). Mean life span is shown for HET3 mice initiated on 14 ppm rapamycin at 20 mo of age as median is not available. (Panels *A* and *B* from Miller et al. 2011a; reprinted by permission of Oxford University Press.)

humans (Miller et al. 2011b), although targeted use of rapamycin against cancers with hyperactivating mutations in the mTOR protein kinase shows significant promise (Grabiner et al. 2014; Wagle et al. 2014). Approximately 70% of HET3 mice die from lymphoma, hemangiosarcoma, and lung carcinoma, the frequency of which is not significantly shifted by rapamycin (Miller et al. 2011b), suggesting that rapamycin does not prevent cancer. Rapamycin significantly reduces the proportion of 16-mo-old mice with cancer or precancerous lesions, suggesting that rapamycin does delay cancer in mice, and it has been argued that the effect of rapamycin may be limited to delaying cancer, not aging (Neff et al. 2013). However, it is clear

that rapamycin delays many forms of age-dependent changes and preserves health span (Wilkinson et al. 2012). Although the anticancer effect of rapamycin may be important, it likely does not account for the full effects of rapamycin on the aging process.

Protein Translation

mTORC1 is an important regulator of protein translation through two distinct mechanisms: the regulation of ribosomal biogenesis via S6K1, and the regulation of mRNA translation mediated by the 4E-BPs. The importance of translation in regulating longevity in model organisms is undisputed, as experiments in *Sac-*

charomyces cerevisiae, *C. elegans*, and *Drosophila melanogaster* clearly show (Kapahi et al. 2004; Hansen et al. 2007; Syntichaki et al. 2007; Steffen et al. 2008; Zid et al. 2009). In these experiments, deletion or RNAi-mediated knockdown of specific ribosomal proteins or translation initiation factors extend life span. In yeast, the life span extension resulting from reduced expression of large ribosomal subunits is dependent upon the increased translation of the transcriptional activator GCN4, providing a mechanistic explanation for how the efficiency of translation initiation can regulate life span (Steffen et al. 2008). However, this has not been shown in higher organisms. Indeed, recent findings in *C. elegans* show that mTOR pathway inhibition can further promote longevity in long-lived *C. elegans* with RNAi-depressed translation initiation factors (Hansen et al. 2007; Syntichaki et al. 2007). Moreover, a 50% decrease in protein translation is not sufficient to extend *C. elegans* life span (Hansen et al. 2007).

In mice, it is unclear whether decreased protein translation is sufficient to extend life span. Although deletion of *S6K1* significantly extends life span (Selman et al. 2009), initial studies found that loss of *S6K1* does not impair protein translation in skeletal muscle (Mieulet et al. 2007). Although yeast lacking *Rpl22a* have extended life span, loss of *Rpl22* in mice has essentially no effect on translation because of compensatory expression of a paralog, *Rpl22l1* (O'Leary et al. 2013). A more recent study found that rapamycin does decrease skeletal muscle protein synthesis, but the amount of the change is very small, and rapamycin does not decrease protein synthesis in heart (Drake et al. 2013). A study comparing the effect of rapamycin and *S6K1* deletion in multiple tissues of mice found that although a single dose of rapamycin did decrease translation in multiple tissues, chronic treatment with rapamycin for 4 wk did not (Garelick et al. 2013). Moreover, mice lacking *S6K1* have normal translational activity and respond normally to rapamycin, suggesting that the benefits of chronic rapamycin on life span are not a result of decreased translation (Garelick et al. 2013).

Protein Quality

Maintaining protein quality is an important challenge during aging. One of the most interesting effects of rapamycin that was recently discovered is that rapamycin treatment of aged mice rejuvenates the aging heart proteome. Despite increased protein half-life, the hearts of rapamycin treated mice had a decreased abundance of damaged proteins (Dai et al. 2014). Such a change could result from increased clearance of damaged proteins.

One of the ways in which damaged proteins are cleared is autophagy, a process in which proteins, especially damaged ones, are broken down to their constituent amino acids. mTOR normally suppresses autophagy by phosphorylating S757 of Ulk1, a kinase required for initiation of autophagy (Kim et al. 2011). In *C. elegans*, autophagy is required for either CR or mTOR inhibition to extend life span (Hansen et al. 2008). Autophagy is up-regulated during CR in mice, and may mediate the beneficial effects of CR on many organ systems, including the liver (Cuervo et al. 2005; Zhang and Cuervo 2008; Kume et al. 2010; Han et al. 2012). The regulation of autophagy is likely a critical part of how rapamycin promotes life span, and it may also impact cancer, as stimulation of autophagy may be an important mechanism of tumor suppression (White et al. 2010).

A second way in which damaged proteins are cleared is proteasome activity, which has been shown to be important in yeast life span (Kruegel et al. 2011). Enhanced proteasome activity has also been found in the exceptionally long-lived naked mole rat (Rodriguez et al. 2012). It was recently found that rapamycin boosts proteasome activity in the brains of female mice treated with rapamycin (Rodriguez et al. 2014), suggesting that regulation of proteasome activity may be important for life span.

Stem Cells and Cell Senescence

Loss of stem-cell proliferative capacity, either because of a decrease in stem cell number or decreased potency, may explain many of the phenotypes of aging. Some of the first work on mTOR signaling in stem cells was performed with hematopoietic stem cells (HSCs), which show age-related declines in self-renewal and function. The function of HSCs declines during aging, and Chen et al. (2009) determined that mTOR signaling was elevated in HSCs from aged mice, and that treatment with rapamycin restored self-renewal of aged HSCs. Similarly, rapamycin treatment or CR increases the self-renewal of aged intestinal stem cells (Yilmaz et al. 2012). More recent experiments conducted in vitro have found that rapamycin can preserve mesenchymal stem-cell self-renewal and prevent epithelial stem cell senescence (Iglesias-Bartolome et al. 2012; Gharibi et al. 2014). In both cases, this appears to be largely a result of decreased damage from reactive oxygen species rather than more general protection from aging. Stem cells remain an important research area for the biology of aging, and hopefully more in vivo data will determine whether rapamycin can protect or even rejuvenate other populations of stem cells. In vivo data suggests that rapamycin may increase transcription of oxidative stress response genes in *C. elegans* and mouse liver (Robida-Stubbs et al. 2012).

WILL THE SIDE EFFECTS OF RAPAMYCIN LIMIT ITS USE AS A HUMAN ANTIAGING THERAPEUTIC?

Although rapamycin shows many beneficial effects in mice, in humans, rapamycin and rapamycin derivatives are used primarily as immunosuppressants following organ transplantation, and in the treatment of several specific types of cancer, including renal cell carcinoma, pancreatic neuroendocrine tumors, and HER2-negative breast cancer (Pusceddu et al. 2014). Serious side effects in humans include an increased incidence of viral and fungal infections including pneumonia, chronic edema, painful oral aphthous ulceration, and hair loss (Mahe et al. 2005; McCormack et al. 2011). Metabolic effects of long-term rapamycin treatment have also been observed, including decreased insulin sensitivity, glucose intolerance, and an increased risk of new-onset diabetes (Johnston et al. 2008). Finally, rapamycin treatment of mice consistently benefits females more than males,

suggesting the possibility that rapamycin treatment of humans may show a similar sexually dimorphic effect on health span and life span.

Short-term treatment with rapamycin is acceptable in the context of cancer treatment and organ transplantation, and might be acceptable for short-term treatment of specific age-related pathologies. For instance, 10 wk of rapamycin treatment reverses age-related cardiac hypertrophy and diastolic dysfunction in aged mice while rejuvenating the heart at the level of the proteome (Dai et al. 2014). However, short-term rapamycin treatment is likely to be insufficient in the case of many age-related diseases, including Alzheimer's disease. Although prophylactic dosing with rapamycin in mouse models of Alzheimer's disease significantly reduces amyloid-β, plaques, tangles, and cognitive defects, dosing older mice has no beneficial effects (Spilman et al. 2010; Majumder et al. 2012). The risks of long-term prophylactic treatment with rapamycin are therefore likely to be unacceptable.

One area in which the side effects of rapamycin treatment may be less important is in the treatment of diseases of rapid aging such as Hutchinson–Gilford progeria syndrome (HGPS). HGPS is a rare, fatal genetic disorder resulting from a mutation in *LMNA*, with no known treatment or cure. The cause of death in most cases of HGPS is progressive arterial occlusive disease, with death from heart attack or stroke occurring at an average age of 13 years (Varga et al. 2006). Treatment of human HGPS fibroblasts and mice lacking *Lmna* with rapamycin reverses HGPS phenotypes at the cellular level and promotes life span and health at the organismal level (Cao et al. 2011; Ramos et al. 2012). Although long-term treatment with rapamycin poses risks, the fatal nature of HGPS suggests that clinical trials of rapamycin in HGPS patients should be considered.

The majority of the data on the side effects of rapamycin have come from mice and from relatively sick humans, not from relatively healthy humans, and healthy humans might experience fewer serious side effects. The potential benefits of rapamycin are so large that trials in other mammals, which may be better models for humans, are getting underway. Rapamycin pharmacology studies in a nonhuman primate, the marmoset, show that rapamycin can be dosed to socially housed marmosets for more than a year without causing anemia, fibrotic lung changes, or mouth ulcers (Tardif et al. 2014). The effects on metabolism and immunity in marmosets, however, are as yet unknown. Also, these marmosets have been maintained in a relatively pathogen-free environment, not the relatively pathogen-rich environment in which humans live. To address some of these issues, a new study at the University of Washington will test the effect of rapamycin on aging phenotypes in pet dogs (Check Hayden 2014). Although these experiments have the potential to significantly advance our understanding of the real-world effects of rapamycin, they must be pursued cautiously, as negative outcomes, such as the development of diabetes in pet dogs, could taint the public perception of rapamycin as a pro-longevity intervention.

INTERMITTENT RAPAMYCIN TREATMENT: A WAY TO SIDESTEP SIDE EFFECTS

How can we use the exciting potential of rapamycin to reap the longevity dividend? Importantly, recent discoveries suggest that at least some of the negative side effects of rapamycin may be separable from its deleterious side effects. In particular, it was recently discovered that long-term treatment with rapamycin disrupts not only mTORC1, but also disrupts mTORC2 in vivo in multiple tissues, including the liver, white adipose tissue, and skeletal muscle (Lamming et al. 2012). Research from many laboratories has identified many positive roles for mTORC2 in health and longevity, and negative consequences from its disruption.

Specifically, hepatic mTORC2 is important for the regulation of gluconeogenesis, and disruption of hepatic mTORC2 by rapamycin causes hepatic insulin resistance and decreased glucose tolerance (Lamming et al. 2012, 2014a). mTORC2 is also important in the proper functioning and proliferation of β cells (Zahr et al. 2008; Yang et al. 2012). The critical role of mTORC2 in the immune system has only been

recently uncovered, with mTORC2 playing a role in the function, proliferation, and differentiation of T cells, B cells, and macrophages (Haxhinasto et al. 2008; Maier et al. 2012; Powell et al. 2012; Byles et al. 2013). A significant decrease in T-cell number and the expansion of regulatory T cells (Tregs) are the likely cause of many of the effects of rapamycin on the immune system (Powell et al. 2012), and mTORC2 activity normal suppresses the differentiation of Tregs (Haxhinasto et al. 2008). Finally, mTORC2 is extremely important in male longevity, and genetic depletion of *Rictor*, a key component of mTORC2, severely shortens male, but not female, life span (Lamming et al. 2014b).

It is possible that this male-specific effect of mTORC2 inhibition on life span explains the sexually dimorphic impact of rapamycin and genetic inhibition of insulin/IGF-1/mTOR signaling pathway on life span, but the mechanistic basis for this sexually dimorphic effect remains unknown. Many of the physiological effects of hepatic *Rictor* deletion are mediated by reduced Akt activity; however, mice heterozygous for *Rictor*, although having decreased male longevity (Fig. 3A), have normal Akt activity (Lamming et al. 2014b). A recent article examining the life span of mice heterozygous for *Akt1* found that these mice had a significant increase in life span (Fig. 3B) (Nojima et al. 2013). It is, therefore, likely that one or more additional mTORC2 substrates mediate the decreased male life span resulting from *Rictor* depletion. Although the role of the mTORC2 substrate SGK in mammalian life span is not known, recent findings in *C. elegans* suggest that SGK may play an important role in determining life span (Mizunuma et al. 2014).

Since many of the negative side effects of rapamycin are mediated by inhibition of mTORC2, drugs that specifically inhibit mTORC1 without inhibiting mTORC2 could allow us to realize the full power of rapamycin (Lamming et al. 2013). We have reported that the rapamycin analogs everolimus and temsirolimus have a decreased impact on glucose tolerance in male C57BL/6J mice, suggesting that these analogs may have a reduced impact on mTORC2, but this remains to be proven

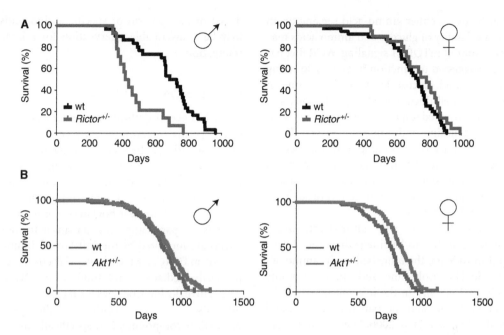

Figure 3. Haploinsufficiency of *Rictor*, but not *Akt*, significantly decreases male life span. (*A*) Kaplan–Meier plots showing life spans of male and female mice heterozygous for *Rictor*. (*B*) Kaplan–Meier survival plots showing life spans of mice heterozygous for *Akt*. (Panel *A* from Lamming et al. 2014b; reprinted, with permission, from the author. Panel *B* from Nojima et al. 2013; adapted, with permission, from the Creative Commons Attribution License, which permits unrestricted use, distribution, and reproduction in any medium.)

(Arriola Apelo et al. 2015). Unfortunately, although the scientific literature suggests several compounds specifically inhibit mTORC1, we have found that some of these results may be specific to particular cell lines or time points. Regrettably, the efforts of the pharmaceutical industry have been focused on the development of mTOR kinase inhibitors, such as Torin 1, PP242, KU63794, and WYE354 for the treatment of cancer (Liu et al. 2012). These inhibitors are extremely effective at inhibiting both mTOR complexes, and are therefore likely to have increased side effects as compared with rapamycin.

A more promising strategy is the possibility that intermittent rapamycin treatment might be sufficient to extend life span, while minimizing the time period that an individual might be immunosuppressed or at risk of diabetes. Recent findings that the effects of rapamycin on (at least) glucose tolerance and mTORC1 signaling can be washed out within a few weeks suggest this may be a feasible approach (Yang

et al. 2012; Liu et al. 2014b). A fairly intensive dosing schedule (2 wk on, 2 wk off) extends the life span of inbred female 129/Sv mice (Anisimov et al. 2011), but this dosing schedule still leads to mice spending >50% of their lives exposed to rapamycin and subject to glucose intolerance, in addition to any other metabolic and immunological impacts. We recently identified an intermittent rapamycin treatment regimen with decreased metabolic and immunological effects (Arriola Apelo et al. 2015), but it remains to be determined whether this regimen can extend life span and health span.

SUSTAINABLE DIETARY INTERVENTIONS TO INHIBIT mTORC1

An alternative approach that has not been fully explored is the possibility of inhibiting mTORC1 by altering the diet. mTORC1, but not mTORC2, is specifically sensitive to glucose and amino acid levels among other stimuli (Bar-Peled and Sabatini 2014). Interventions that focus on either amino-acid sensing or on the availability of glucose and amino acids may act to inhibit mTORC1 signaling. A CR diet has been suggested to function in part by lowering fasting blood glucose levels, which is one of the most well-documented, reproducible, and widely conserved response to a CR diet in mammals (Lamming and Anderson 2014). Treatment with acarbose, a compound that acts to slow glucose uptake from food, has been shown to extend life span (Harrison et al. 2014), and it will be interesting to learn the effect of acarbose on mTORC1 activity.

mTORC1 activity in cultured cells is extremely sensitive to leucine (Long et al. 2005), and in rodents, the branched-chain amino acids—leucine, isoleucine, and valine—promote mTORC1 activity in the liver, skeletal muscle, adipose tissue, and the pancreas (Blomstrand et al. 2006; Sans et al. 2006; Li et al. 2011; Xiao et al. 2011). Consumption of leucine also significantly affects mTORC1 activity in humans (Moberg et al. 2014). A short-term protein-free diet leads to a significant decrease in mTORC1 signaling (Harputlugil et al. 2014), and we recently found that a low protein diet can inhibit mTORC1 signaling in both the tumors and somatic tissues of a mouse xenograft model (Lamming et al. 2015).

Recent studies have clearly shown that a low-protein diet significantly extends rodent life span and is associated with reduced cancer and mortality in humans (Levine et al. 2014; Solon-Biet et al. 2014), although it is not clear whether this effect is mediated by mTOR signaling. Low-protein diets may be an attractive and more sustainable alternative to CR in humans (Fontana and Partridge 2015). A CR diet is extremely difficult for humans in the developed world to maintain, surrounded by the sights and smells of abundant food. In contrast, vegan diets may be maintainable, and it has been suggested that such plant-based diets may be particularly low in methionine (McCarty et al. 2009), which when restricted significantly extends the life span of rodents (Anthony et al. 2013). Diets restricted in specific amino acids are often used in the treatment of inborn errors of metabolism, suggesting that diets with re-

duced dietary protein or specific amino acids may be a sustainable intervention for a large population.

CONCLUSION

There has been significant excitement over the discovery that rapamycin can prolong rodent life span and may be a potent antiaging drug. Although rapamycin is very promising, its side-effect profile may limit its clinical use for the treatment of diseases of aging in humans. Although it is still unclear how mTOR inhibition extends life span, it appears that many of the side effects are mediated by the "off-target" inhibition of mTORC2, whose beneficial effects are primarily mediated by inhibition of mTORC1. Alternative dosing strategies for rapamycin that limit its effects on mTORC2, or the development of compounds that specifically inhibit mTORC1, may allow us to fully realize the health, social, and economic benefits of slowed aging. While we await these developments, consumption of a low-protein diet may promote health, perhaps in part by inhibiting mTORC1 signaling.

ACKNOWLEDGMENTS

D.W.L. is supported in part by a K99/R00 Pathway to Independence Award from the National Institutes of Health/National Institute of Aging (NIH/NIA) (AG041765). This work was supported using facilities and resources from the William S. Middleton Memorial Veterans Hospital. This work does not represent the views of the Department of Veterans Affairs or the United States Government.

REFERENCES

Anisimov VN, Zabezhinski MA, Popovich IG, Piskunova TS, Semenchenko AV, Tyndyk ML, Yurova MN, Rosenfeld SV, Blagosklonny MV. 2011. Rapamycin increases lifespan and inhibits spontaneous tumorigenesis in inbred female mice. *Cell Cycle* **10:** 4230–4236.

Anthony TG, Morrison CD, Gettys TW. 2013. Remodeling of lipid metabolism by dietary restriction of essential amino acids. *Diabetes* **62:** 2635–2644.

Arriola Apelo SI, Neuman JC, Baar EL, Syed FA, Cummings NE, Brar HK, Pumper CP, Kimple ME, Lamming DW.

2015. Alternative rapamycin treatment regimens mitigate the impact of rapamycin on glucose homeostasis and the immune system. *Aging Cell* (in press).

Bar-Peled L, Sabatini DM. 2014. Regulation of mTORC1 by amino acids. *Trends Cell Biol* **24:** 400–406.

Bar-Peled L, Chantranupong L, Cherniack AD, Chen WW, Ottina KA, Grabiner BC, Spear ED, Carter SL, Meyerson M, Sabatini DM. 2013. A tumor suppressor complex with GAP activity for the Rag GTPases that signal amino acid sufficiency to mTORC1. *Science* **340:** 1100–1106.

Betz C, Stracka D, Prescianotto-Baschong C, Frieden M, Demaurex N, Hall MN. 2013. Feature article: mTOR complex 2-Akt signaling at mitochondria-associated endoplasmic reticulum membranes (MAM) regulates mitochondrial physiology. *Proc Natl Acad Sci* **110:** 12526–12534.

Bjedov I, Toivonen JM, Kerr F, Slack C, Jacobson J, Foley A, Partridge L. 2010. Mechanisms of life span extension by rapamycin in the fruit fly *Drosophila melanogaster*. *Cell Metab* **11:** 35–46.

Blomstrand E, Eliasson J, Karlsson HK, Kohnke R. 2006. Branched-chain amino acids activate key enzymes in protein synthesis after physical exercise. *J Nutr* **136:** 269S–273S.

Byles V, Covarrubias AJ, Ben-Sahra I, Lamming DW, Sabatini DM, Manning BD, Horng T. 2013. The TSC-mTOR pathway regulates macrophage polarization. *Nat Commun* **4:** 2834.

Cao K, Graziotto JJ, Blair CD, Mazzulli JR, Erdos MR, Krainc D, Collins FS. 2011. Rapamycin reverses cellular phenotypes and enhances mutant protein clearance in Hutchinson–Gilford progeria syndrome cells. *Sci Transl Med* **3:** 89ra58.

Caron A, Richard D, Laplante M. 2015. The roles of mTOR complexes in lipid metabolism. *Annu Rev Nutr* **35:** 321–348.

Chantranupong L, Wolfson RL, Orozco JM, Saxton RA, Scaria SM, Bar-Peled L, Spooner E, Isasa M, Gygi SP, Sabatini DM. 2014. The sestrins interact with GATOR2 to negatively regulate the amino-acid-sensing pathway upstream of mTORC1. *Cell Rep* **9:** 1–8.

Check Hayden E. 2014. Pet dogs set to test anti-ageing drug. *Nature* **514:** 546.

Chen C, Liu Y, Liu Y, Zheng P. 2009. mTOR regulation and therapeutic rejuvenation of aging hematopoietic stem cells. *Sci Signal* **2:** ra75.

Cuervo AM, Bergamini E, Brunk UT, Droge W, Ffrench M, Terman A. 2005. Autophagy and aging: The importance of maintaining "clean" cells. *Autophagy* **1:** 131–140.

Dai DF, Karunadharma PP, Chiao YA, Basisty N, Crispin D, Hsieh EJ, Chen T, Gu H, Djukovic D, Raftery D, et al. 2014. Altered proteome turnover and remodeling by short-term caloric restriction or rapamycin rejuvenate the aging heart. *Aging Cell* **13:** 529–539.

Drake JC, Peelor FF III, Biela LM, Watkins MK, Miller RA, Hamilton KL, Miller BF. 2013. Assessment of mitochondrial biogenesis and mTORC1 signaling during chronic rapamycin feeding in male and female mice. *J Gerontol A Biol Sci Med Sci* **68:** 1493–1501.

Fok WC, Bokov A, Gelfond J, Yu Z, Zhang Y, Doderer M, Chen Y, Javors M, Wood WH III, Zhang Y, et al. 2014a.

Combined treatment of rapamycin and dietary restriction has a larger effect on the transcriptome and metabolome of liver. *Aging Cell* **13:** 311–319.

Fok WC, Chen Y, Bokov A, Zhang Y, Salmon AB, Diaz V, Javors M, Wood WH III, Zhang Y, Becker KG, et al. 2014b. Mice fed rapamycin have an increase in lifespan associated with major changes in the liver transcriptome. *PLoS ONE* **9:** e83988.

Fontana L, Partridge L. 2015. Promoting health and longevity through diet: From model organisms to humans. *Cell* **161:** 106–118.

Gan X, Wang J, Wang C, Sommer E, Kozasa T, Srinivasula S, Alessi D, Offermanns S, Simon MI, Wu D. 2012. PRR5L degradation promotes mTORC2-mediated PKC-δ phosphorylation and cell migration downstream of Gα$_{12}$. *Nat Cell Biol* **14:** 686–696.

Garcia-Martinez JM, Alessi DR. 2008. mTOR complex 2 (mTORC2) controls hydrophobic motif phosphorylation and activation of serum- and glucocorticoid-induced protein kinase 1 (SGK1). *Biochem J* **416:** 375–385.

Garelick MG, Mackay VL, Yanagida A, Academia EC, Schreiber KH, Ladiges WC, Kennedy BK. 2013. Chronic rapamycin treatment or lack of S6K1 does not reduce ribosome activity in vivo. *Cell Cycle* **12:** 2493–2504.

Gharibi B, Farzadi S, Ghuman M, Hughes FJ. 2014. Inhibition of Akt/mTOR attenuates age-related changes in mesenchymal stem cells. *Stem Cells* **32:** 2256–2266.

Grabiner BC, Nardi V, Birsoy K, Possemato R, Shen K, Sinha S, Jordan A, Beck AH, Sabatini DM. 2014. A diverse array of cancer-associated *MTOR* mutations are hyperactivating and can predict rapamycin sensitivity. *Cancer Discov* **4:** 554–563.

Guertin DA, Stevens DM, Thoreen CC, Burds AA, Kalaany NY, Moffat J, Brown M, Fitzgerald KJ, Sabatini DM. 2006. Ablation in mice of the mTORC components *raptor, rictor, or mLST8* reveals that mTORC2 is required for signaling to Akt-FOXO and PKC-α, but not S6K1. *Dev Cell* **11:** 859–871.

Han X, Turdi S, Hu N, Guo R, Zhang Y, Ren J. 2012. Influence of long-term caloric restriction on myocardial and cardiomyocyte contractile function and autophagy in mice. *J Nutr Biochem* **23:** 1592–1599.

Hansen M, Taubert S, Crawford D, Libina N, Lee SJ, Kenyon C. 2007. Lifespan extension by conditions that inhibit translation in *Caenorhabditis elegans*. *Aging Cell* **6:** 95–110.

Hansen M, Chandra A, Mitic LL, Onken B, Driscoll M, Kenyon C. 2008. A role for autophagy in the extension of lifespan by dietary restriction in *C. elegans*. *PLoS Genet* **4:** e24.

Harputlugil E, Hine C, Vargas D, Robertson L, Manning BD, Mitchell JR. 2014. The TSC complex is required for the benefits of dietary protein restriction on stress resistance in vivo. *Cell Rep* **8:** 1160–1170.

Harrison DE, Strong R, Sharp ZD, Nelson JF, Astle CM, Flurkey K, Nadon NL, Wilkinson JE, Frenkel K, Carter CS, et al. 2009. Rapamycin fed late in life extends lifespan in genetically heterogeneous mice. *Nature* **460:** 392–395.

Harrison DE, Strong R, Allison DB, Ames BN, Astle CM, Atamna H, Fernandez E, Flurkey K, Javors MA, Nadon NL, et al. 2014. Acarbose, 17-α-estradiol, and nordihydroguaiaretic acid extend mouse lifespan preferentially in males. *Aging Cell* 13: 273–282.

Haxhinasto S, Mathis D, Benoist C. 2008. The AKT–mTOR axis regulates de novo differentiation of CD4[+]Foxp3[+] cells. *J Exp Med* 205: 565–574.

Hernandez-Negrete I, Carretero-Ortega J, Rosenfeldt H, Hernandez-Garcia R, Calderon-Salinas JV, Reyes-Cruz G, Gutkind JS, Vazquez-Prado J. 2007. P-Rex1 links mammalian target of rapamycin signaling to Rac activation and cell migration. *J Biol Chem* 282: 23708–23715.

Holzenberger M, Dupont J, Ducos B, Leneuve P, Geloen A, Even PC, Cervera P, Le Bouc Y. 2003. IGF-1 receptor regulates lifespan and resistance to oxidative stress in mice. *Nature* 421: 182–187.

Hsu PP, Kang SA, Rameseder J, Zhang Y, Ottina KA, Lim D, Peterson TR, Choi Y, Gray NS, Yaffe MB, et al. 2011. The mTOR-regulated phosphoproteome reveals a mechanism of mTORC1-mediated inhibition of growth factor signaling. *Science* 332: 1317–1322.

Huang J, Dibble CC, Matsuzaki M, Manning BD. 2008. The TSC1-TSC2 complex is required for proper activation of mTOR complex 2. *Mol Cell Biol* 28: 4104–4115.

Iglesias-Bartolome R, Patel V, Cotrim A, Leelahavanichkul K, Molinolo AA, Mitchell JB, Gutkind JS. 2012. mTOR inhibition prevents epithelial stem cell senescence and protects from radiation-induced mucositis. *Cell Stem Cell* 11: 401–414.

Ikenoue T, Inoki K, Yang Q, Zhou X, Guan KL. 2008. Essential function of TORC2 in PKC and Akt turn motif phosphorylation, maturation and signalling. *Embo J* 27: 1919–1931.

Johnston O, Rose CL, Webster AC, Gill JS. 2008. Sirolimus is associated with new-onset diabetes in kidney transplant recipients. *J Am Soc Nephrol* 19: 1411–1418.

Kaeberlein M, Powers RW III, Steffen KK, Westman EA, Hu D, Dang N, Kerr EO, Kirkland KT, Fields S, Kennedy BK. 2005. Regulation of yeast replicative life span by TOR and Sch9 in response to nutrients. *Science* 310: 1193–1196.

Kang SA, Pacold ME, Cervantes CL, Lim D, Lou HJ, Ottina K, Gray NS, Turk BE, Yaffe MB, Sabatini DM. 2013. mTORC1 phosphorylation sites encode their sensitivity to starvation and rapamycin. *Science* 341: 1236566.

Kapahi P, Zid BM, Harper T, Koslover D, Sapin V, Benzer S. 2004. Regulation of lifespan in *Drosophila* by modulation of genes in the TOR signaling pathway. *Curr Biol* 14: 885–890.

Khanna N, Fang Y, Yoon MS, Chen J. 2013. XPLN is an endogenous inhibitor of mTORC2. *Proc Natl Acad Sci* 110: 15979–15984.

Kim J, Kundu M, Viollet B, Guan KL. 2011. AMPK and mTOR regulate autophagy through direct phosphorylation of Ulk1. *Nat Cell Biol* 13: 132–141.

Kim SJ, DeStefano MA, Oh WJ, Wu CC, Vega-Cotto NM, Finlan M, Liu D, Su B, Jacinto E. 2012. mTOR complex 2 regulates proper turnover of insulin receptor substrate-1 via the ubiquitin ligase subunit Fbw8. *Mol Cell* 48: 875–887.

Kruegel U, Robison B, Dange T, Kahlert G, Delaney JR, Kotireddy S, Tsuchiya M, Tsuchiyama S, Murakami CJ, Schleit J, et al. 2011. Elevated proteasome capacity ex-tends replicative lifespan in *Saccharomyces cerevisiae*. *PLoS Genet* 7: e1002253.

Kume S, Uzu T, Horiike K, Chin-Kanasaki M, Isshiki K, Araki S, Sugimoto T, Haneda M, Kashiwagi A, Koya D. 2010. Calorie restriction enhances cell adaptation to hypoxia through Sirt1-dependent mitochondrial autophagy in mouse aged kidney. *J Clin Invest* 120: 1043–1055.

Lamming DW, Anderson RM. 2014. Metabolic effects of caloric restriction. In *eLS*. Wiley, Chichester, NY.

Lamming DW, Medvedik O, Kim KD, Sinclair DA. 2007. Calorie restriction and TOR signaling converge on sirtuin-mediated lifespan extension in *Saccharomyces cerevisiae*. *Age* 29: 118–118.

Lamming DW, Ye L, Katajisto P, Goncalves MD, Saitoh M, Stevens DM, Davis JG, Salmon AB, Richardson A, Ahima RS, et al. 2012. Rapamycin-induced insulin resistance is mediated by mTORC2 loss and uncoupled from longevity. *Science* 335: 1638–1643.

Lamming DW, Ye L, Sabatini DM, Baur JA. 2013. Rapalogs and mTOR inhibitors as anti-aging therapeutics. *J Clin Invest* 123: 980–989.

Lamming DW, Demirkan G, Boylan JM, Mihaylova MM, Peng T, Ferreira J, Neretti N, Salomon A, Sabatini DM, Gruppuso PA. 2014a. Hepatic signaling by the mechanistic target of rapamycin complex 2 (mTORC2). *FASEB J* 28: 300–315.

Lamming DW, Mihaylova MM, Katajisto P, Baar EL, Yilmaz OH, Hutchins A, Gultekin Y, Gaither R, Sabatini DM. 2014b. Depletion of *rictor*, an essential protein component of mTORC2, decreases male lifespan. *Aging Cell* 13: 911–917.

Lamming DW, Cummings NE, Rastelli AL, Gao F, Cava E, Bertozzi B, Spelta F, Pili R, Fontana L. 2015. Restriction of dietary protein decreases mTORC1 in tumors and somatic tissues of a tumor-bearing mouse xenograft model. *Oncotarget* doi: 10.18632/oncotarget.5180.

Levine ME, Suarez JA, Brandhorst S, Balasubramanian P, Cheng CW, Madia F, Fontana L, Mirisola MG, Guevara-Aguirre J, Wan J, et al. 2014. Low protein intake is associated with a major reduction in IGF-1, cancer, and overall mortality in the 65 and younger but not older population. *Cell Metab* 19: 407–417.

Li X, Gao T. 2014. mTORC2 phosphorylates protein kinase C to regulate its stability and activity. *EMBO Rep* 15: 191–198.

Li F, Yin Y, Tan B, Kong X, Wu G. 2011. Leucine nutrition in animals and humans: mTOR signaling and beyond. *Amino Acids* 41: 1185–1193.

Liu Q, Kirubakaran S, Hur W, Niepel M, Westover K, Thoreen CC, Wang J, Ni J, Patricelli MP, Vogel K, et al. 2012. Kinome-wide selectivity profiling of ATP-competitive mammalian target of rapamycin (mTOR) inhibitors and characterization of their binding kinetics. *J Biol Chem* 287: 9742–9752.

Liu P, Wang Z, Wei W. 2014a. Phosphorylation of Akt at the C-terminal tail triggers Akt activation. *Cell Cycle* 13: 2162–2164.

Liu Y, Diaz V, Fernandez E, Strong R, Ye L, Baur JA, Lamming DW, Richardson A, Salmon AB. 2014b. Rapamycin-induced metabolic defects are reversible in both lean and obese mice. *Aging (Albany NY)* 6: 742–754.

 Cite this article as *Cold Spring Harb Perspect Med* doi: 10.1101/cshperspect.a025924

Long X, Ortiz-Vega S, Lin Y, Avruch J. 2005. Rheb binding to mammalian target of rapamycin (mTOR) is regulated by amino acid sufficiency. *J Biol Chem* **280:** 23433–23436.

Mahe E, Morelon E, Lechaton S, Sang KH, Mansouri R, Ducasse MF, Mamzer-Bruneel MF, de Prost Y, Kreis H, Bodemer C. 2005. Cutaneous adverse events in renal transplant recipients receiving sirolimus-based therapy. *Transplantation* **79:** 476–482.

Maier E, Duschl A, Horejs-Hoeck J. 2012. STAT6-dependent and -independent mechanisms in Th2 polarization. *Eur J Immunol* **42:** 2827–2833.

Majumder S, Caccamo A, Medina DX, Benavides AD, Javors MA, Kraig E, Strong R, Richardson A, Oddo S. 2012. Lifelong rapamycin administration ameliorates age-dependent cognitive deficits by reducing IL-1β and enhancing NMDA signaling. *Aging Cell* **11:** 326–335.

McCarty MF, Barroso-Aranda J, Contreras F. 2009. The low-methionine content of vegan diets may make methionine restriction feasible as a life extension strategy. *Med Hypotheses* **72:** 125–128.

McCormack FX, Inoue Y, Moss J, Singer LG, Strange C, Nakata K, Barker AF, Chapman JT, Brantly ML, Stocks JM, et al. 2011. Efficacy and safety of sirolimus in lymphangioleiomyomatosis. *N Engl J Med* **364:** 1595–1606.

Menon S, Dibble CC, Talbott G, Hoxhaj G, Valvezan AJ, Takahashi H, Cantley LC, Manning BD. 2014. Spatial control of the TSC complex integrates insulin and nutrient regulation of mTORC1 at the lysosome. *Cell* **156:** 771–785.

Mieulet V, Roceri M, Espeillac C, Sotiropoulos A, Ohanna M, Oorschot V, Klumperman J, Sandri M, Pende M. 2007. S6 kinase inactivation impairs growth and translational target phosphorylation in muscle cells maintaining proper regulation of protein turnover. *Am J Physiol Cell Physiol* **293:** C712–C722.

Miller RA, Harrison DE, Astle CM, Baur JA, Rodriguez Boyd A, de Cabo R, Fernandez E, Flurkey K, Javors MA, Nelson JF, et al. 2011a. Rapamycin, but not resveratrol or simvastatin, extends life span of genetically heterogeneous mice. *J Gerontol A Biol Sci Med Sci* **66A:** 191–201.

Miller RA, Harrison DE, Astle CM, Baur JA, Boyd AR, de Cabo R, Fernandez E, Flurkey K, Javors MA, Nelson JF, et al. 2011b. Rapamycin, but not resveratrol or simvastatin, extends life span of genetically heterogeneous mice. *J Gerontol A Biol Sci Med Sci* **66:** 191–201.

Miller RA, Harrison DE, Astle CM, Fernandez E, Flurkey K, Han M, Javors MA, Li X, Nadon NL, Nelson JF, et al. 2014. Rapamycin-mediated lifespan increase in mice is dose and sex dependent and metabolically distinct from dietary restriction. *Aging Cell* **13:** 468–477.

Mizunuma M, Neumann-Haefelin E, Moroz N, Li Y, Blackwell TK. 2014. mTORC2-SGK-1 acts in two environmentally responsive pathways with opposing effects on longevity. *Aging Cell* **13:** 869–878.

Moberg M, Apro W, Ohlsson I, Ponten M, Villanueva A, Ekblom B, Blomstrand E. 2014. Absence of leucine in an essential amino acid supplement reduces activation of mTORC1 signalling following resistance exercise in young females. *Appl Physiol Nutr Metab* **39:** 183–194.

Neff F, Flores-Dominguez D, Ryan DP, Horsch M, Schroder S, Adler T, Afonso LC, Aguilar-Pimentel JA, Becker L,

Garrett L, et al. 2013. Rapamycin extends murine lifespan but has limited effects on aging. *J Clin Invest* **123:** 3272–3291.

Nojima A, Yamashita M, Yoshida Y, Shimizu I, Ichimiya H, Kamimura N, Kobayashi Y, Ohta S, Ishii N, Minamino T. 2013. Haploinsufficiency of akt1 prolongs the lifespan of mice. *PLoS ONE* **8:** e69178.

O'Leary MN, Schreiber KH, Zhang Y, Duc AC, Rao S, Hale JS, Academia EC, Shah SR, Morton JF, Holstein CA, et al. 2013. The ribosomal protein Rpl22 controls ribosome composition by directly repressing expression of its own paralog, Rpl22l1. *PLoS Genet* **9:** e1003708.

Popovich IG, Anisimov VN, Zabezhinski MA, Semenchenko AV, Tyndyk ML, Yurova MN, Blagosklonny MV. 2014. Lifespan extension and cancer prevention in HER-2/neu transgenic mice treated with low intermittent doses of rapamycin. *Cancer Biol Therapy* **15:** 586–592.

Powell JD, Pollizzi KN, Heikamp EB, Horton MR. 2012. Regulation of immune responses by mTOR. *Annu Rev Immunol* **30:** 39–68.

Powers RW III, Kaeberlein M, Caldwell SD, Kennedy BK, Fields S. 2006. Extension of chronological life span in yeast by decreased TOR pathway signaling. *Genes Dev* **20:** 174–184.

Pusceddu S, Tessari A, Testa I, Procopio G. 2014. Everolimus in advanced solid tumors: When to start, early or late? *Tumori* **100:** 2e–3e.

Ramos FJ, Chen SC, Garelick MG, Dai DF, Liao CY, Schreiber KH, MacKay VL, An EH, Strong R, Ladiges WC, et al. 2012. Rapamycin reverses elevated mTORC1 signaling in lamin A/C-deficient mice, rescues cardiac and skeletal muscle function, and extends survival. *Sci Transl Med* **4:** 144ra103.

Robida-Stubbs S, Glover-Cutter K, Lamming DW, Mizunuma M, Narasimhan SD, Neumann-Haefelin E, Sabatini DM, Blackwell TK. 2012. TOR signaling and rapamycin influence longevity by regulating SKN-1/Nrf and DAF-16/FoxO. *Cell Metab* **15:** 713–724.

Rodriguez KA, Edrey YH, Osmulski P, Gaczynska M, Buffenstein R. 2012. Altered composition of liver proteasome assemblies contributes to enhanced proteasome activity in the exceptionally long-lived naked mole-rat. *PLoS ONE* **7:** e35890.

Rodriguez KA, Dodds SG, Strong R, Galvan V, Sharp ZD, Buffenstein R. 2014. Divergent tissue and sex effects of rapamycin on the proteasome-chaperone network of old mice. *Front Mol Neurosci* **7:** 83.

Saci A, Cantley LCt, Carpenter CL. 2011. Rac1 regulates the activity of mTORC1 and mTORC2 and controls cellular size. *Mol Cell* **42:** 50–61.

Sancak Y, Bar-Peled L, Zoncu R, Markhard AL, Nada S, Sabatini DM. 2010. Ragulator-Rag complex targets mTORC1 to the lysosomal surface and is necessary for its activation by amino acids. *Cell* **141:** 290–303.

Sans MD, Tashiro M, Vogel NL, Kimball SR, D'Alecy LG, Williams JA. 2006. Leucine activates pancreatic translational machinery in rats and mice through mTOR independently of CCK and insulin. *J Nutr* **136:** 1792–1799.

Sarbassov DD, Ali SM, Sengupta S, Sheen JH, Hsu PP, Bagley AF, Markhard AL, Sabatini DM. 2006. Prolonged rapamycin treatment inhibits mTORC2 assembly and Akt/PKB. *Mol Cell* **22:** 159–168.

Selman C, Tullet JM, Wieser D, Irvine E, Lingard SJ, Choudhury AI, Claret M, Al-Qassab H, Carmignac D, Ramadani F, et al. 2009. Ribosomal protein S6 kinase 1 signaling regulates mammalian life span. *Science* **326:** 140–144.

Selman C, Partridge L, Withers DJ. 2011. Replication of extended lifespan phenotype in mice with deletion of insulin receptor substrate 1. *PLoS ONE* **6:** e16144.

Seto B. 2012. Rapamycin and mTOR: A serendipitous discovery and implications for breast cancer. *Clin Trans Med* **1:** 29.

Settembre C, Zoncu R, Medina DL, Vetrini F, Erdin S, Erdin S, Huynh T, Ferron M, Karsenty G, Vellard MC, et al. 2012. A lysosome-to-nucleus signalling mechanism senses and regulates the lysosome via mTOR and TFEB. *EMBO J* **31:** 1095–1108.

Solon-Biet SM, McMahon AC, Ballard JW, Ruohonen K, Wu LE, Cogger VC, Warren A, Huang X, Pichaud N, Melvin RG, et al. 2014. The ratio of macronutrients, not caloric intake, dictates cardiometabolic health, aging, and longevity in ad libitum-fed mice. *Cell Metab* **19:** 418–430.

Spilman P, Podlutskaya N, Hart MJ, Debnath J, Gorostiza O, Bredesen D, Richardson A, Strong R, Galvan V. 2010. Inhibition of mTOR by rapamycin abolishes cognitive deficits and reduces amyloid-beta levels in a mouse model of Alzheimer's disease. *PLoS ONE* **5:** e9979.

Steffen KK, MacKay VL, Kerr EO, Tsuchiya M, Hu D, Fox LA, Dang N, Johnston ED, Oakes JA, Tchao BN, et al. 2008. Yeast life span extension by depletion of 60s ribosomal subunits is mediated by Gcn4. *Cell* **133:** 292–302.

Syntichaki P, Troulinaki K, Tavernarakis N. 2007. eIF4E function in somatic cells modulates ageing in *Caenorhabditis elegans*. *Nature* **445:** 922–926.

Tao R, Xiong X, Liangpunsakul S, Dong XC. 2014. Sestrin 3 protein enhances hepatic insulin sensitivity by direct activation of the mTORC2-Akt signaling. *Diabetes* **64:** 1211–1223.

Tardif S, Ross C, Bergman P, Fernandez E, Javors M, Salmon A, Spross J, Strong R, Richardson A. 2014. Testing efficacy of administration of the antiaging drug rapamycin in a nonhuman primate, the common marmoset. *J Gerontol A Biol Sci Med Sci* **70:** 577–587.

Thoreen CC, Chantranupong L, Keys HR, Wang T, Gray NS, Sabatini DM. 2012. A unifying model for mTORC1-mediated regulation of mRNA translation. *Nature* **485:** 109–113.

Varga R, Eriksson M, Erdos MR, Olive M, Harten I, Kolodgie F, Capell BC, Cheng J, Faddah D, Perkins S, et al. 2006. Progressive vascular smooth muscle cell defects in a mouse model of Hutchinson–Gilford progeria syndrome. *Proc Natl Acad Sci* **103:** 3250–3255.

Vellai T, Takacs-Vellai K, Zhang Y, Kovacs AL, Orosz L, Muller F. 2003. Genetics: Influence of TOR kinase on lifespan in *C. elegans*. *Nature* **426:** 620.

Vezina C, Kudelski A, Sehgal SN. 1975. Rapamycin (AY-22,989), a new antifungal antibiotic. I: Taxonomy of the producing streptomycete and isolation of the active principle. *J Antibiot (Tokyo)* **28:** 721–726.

Wagle N, Grabiner BC, Van Allen EM, Hodis E, Jacobus S, Supko JG, Stewart M, Choueiri TK, Gandhi L, Cleary JM, et al. 2014. Activating mTOR mutations in a patient with an extraordinary response on a phase I trial of everolimus and pazopanib. *Cancer discovery* **4:** 546–553.

White E, Karp C, Strohecker AM, Guo Y, Mathew R. 2010. Role of autophagy in suppression of inflammation and cancer. *Curr Opin Cell Biol* **22:** 212–217.

Wilkinson JE, Burmeister L, Brooks SV, Chan CC, Friedline S, Harrison DE, Hejtmancik JF, Nadon N, Strong R, Wood LK, et al. 2012. Rapamycin slows aging in mice. *Aging Cell* **11:** 675–682.

Xiao F, Huang Z, Li H, Yu J, Wang C, Chen S, Meng Q, Cheng Y, Gao X, Li J, et al. 2011. Leucine deprivation increases hepatic insulin sensitivity via GCN2/mTOR/S6K1 and AMPK pathways. *Diabetes* **60:** 746–756.

Yang SB, Lee HY, Young DM, Tien AC, Rowson-Baldwin A, Shu YY, Jan YN, Jan LY. 2012. Rapamycin induces glucose intolerance in mice by reducing islet mass, insulin content, and insulin sensitivity. *J Mol Med (Berl)* **90:** 575–585.

Yilmaz OH, Katajisto P, Lamming DW, Gultekin Y, Bauer-Rowe KE, Sengupta S, Birsoy K, Dursun A, Yilmaz VO, Selig M, et al. 2012. mTORC1 in the Paneth cell niche couples intestinal stem-cell function to calorie intake. *Nature* **486:** 490–495.

Zahr E, Molano RD, Pileggi A, Ichii H, San Jose S, Bocca N, An W, Gonzalez-Quintana J, Fraker C, Ricordi C, et al. 2008. Rapamycin impairs β-cell proliferation in vivo. *Transplant Proc* **40:** 436–437.

Zhang C, Cuervo AM. 2008. Restoration of chaperone-mediated autophagy in aging liver improves cellular maintenance and hepatic function. *Nat Med* **14:** 959–965.

Zhang Y, Bokov A, Gelfond J, Soto V, Ikeno Y, Hubbard G, Diaz V, Sloane L, Maslin K, Treaster S, et al. 2014. Rapamycin extends life and health in C57BL/6 mice. *J Gerontol A Biol Sci Med Sci* **69:** 119–130.

Zid BM, Rogers AN, Katewa SD, Vargas MA, Kolipinski MC, Lu TA, Benzer S, Kapahi P. 2009. 4E-BP extends lifespan upon dietary restriction by enhancing mitochondrial activity in *Drosophila*. *Cell* **139:** 149–160.

Zinzalla V, Stracka D, Oppliger W, Hall MN. 2011. Activation of mTORC2 by association with the ribosome. *Cell* **144:** 757–768.

Metformin: A Hopeful Promise in the Aging Research

Marta G. Novelle[1,2,3], Ahmed Ali[1], Carlos Diéguez[3], Michel Bernier[1], and Rafael de Cabo[1]

[1]Translational Gerontology Branch, National Institute on Aging, NIH, Baltimore, Maryland 21224

[2]Research Center of Molecular Medicine and Chronic Diseases (CIMUS), University of Santiago de Compostela-Instituto de Investigación Sanitaria, Santiago de Compostela 15782, Spain

[3]CIBER Fisiopatología de la Obesidad y Nutrición (CIBERobn), Santiago de Compostela 15706, Spain

Correspondence: decabora@grc.nia.nih.gov

Even though the inevitable process of aging by itself cannot be considered a disease, it is directly linked to life span and is the driving force behind all age-related diseases. It is an undisputable fact that age-associated diseases are among the leading causes of death in the world, primarily in industrialized countries. During the last several years, an intensive search of antiaging treatments has led to the discovery of a variety of drugs that promote health span and/or life extension. The biguanide compound metformin is widely used for treating people with type 2 diabetes and appears to show protection against cancer, inflammation, and age-related pathologies. Here, we summarize the recent developments about metformin use in translational aging research and discuss its role as a potential geroprotector.

Over the past two decades, metformin has emerged as the first-line treatment for people with type 2 diabetes (T2DM) and is the most widely prescribed antidiabetic drug in the world (American Diabetes Association 2014). In addition to its use in T2DM, metformin is being prescribed for the treatment of polycystic ovary syndrome, diabetic nephropathy, and gestational diabetes, and has shown early promise as a treatment for cancer. Historically, despite its well-accepted antidiabetic properties in the 1950s, and use for hyperglycemia treatment in England in 1958, metformin remained contraindicated largely because of concerns about lactic acidosis and it was not approved by the U.S. Food and Drug Administration until 1994 (Bailey and Turner 1996; Mahmood et al. 2013). We now know that the rare event of lactic acidosis occurs in >0.01 to 0.08 cases (average, 0.03) per 1000 patient-years caused by an insufficient metformin clearance by the kidneys (Bailey and Turner 1996). Therefore, the risk of side effects is relatively low in comparison to the multiple benefits of metformin.

The exact molecular mechanisms of metformin's therapeutic action still remain unknown. Metformin is a biguanide compound originally derived from a guanidine derivative found in the plant *Galega officinalis*. It acts as an insulin sensitizer and exerts its principal metabolic action on the liver. In addition to its glucoregulatory action, metformin has gained attention for its pleiotropic effects and activity in a variety of tissues, such as muscles, adipose tissue, ovary, endothelium, and brain (Diamanti-Kandarakis et al. 2010; Foretz et al. 2014). Food intake (Adeyemo et al. 2014; Pernicova and Korbonits 2014) and body weight (Glueck et al. 2001) are decreased as a result of a direct action of metformin on the hypothalamic centers regulating satiety and feeding (Stevanovic et al. 2012); it may also influence metabolic and cellular processes associated with the development of chronic conditions of aging, including inflammation, fatty liver, oxidative damage, protein glycation, cellular senescence, diminished autophagy, apoptosis, and development of several types of cancer (Isoda et al. 2006; Kita et al. 2012; Hirsch et al. 2013; Woo et al. 2014). A number of recent studies support the role of metformin in improving health span and life span in different animal models (Anisimov et al. 2011; Cabreiro et al. 2013; Martin-Montalvo et al. 2013; Anisimov 2014; De Haes et al. 2014). The possibility exists, therefore, for similar beneficial actions of metformin in human health and longevity.

The purpose of this article is to review the role of metformin as a possible geroprotector drug. We will try to summarize recent evidence for the antiaging properties of metformin, the molecular mechanisms implicated in this role, and, finally, discuss new (research opportunities) directions to better understand the translational potential of metformin.

HOW DOES METFORMIN WORK?

Metformin is excreted intact in the urine, without being metabolized by the liver or kidney. About 50%–60% of an oral dose is absorbed into the systemic circulation and distributed in most tissues at similar concentrations, although higher concentrations are found in gastrointestinal tract, liver, and kidney (Bailey and Turner 1996; Gong et al. 2012; Pawlyk et al. 2014). Age, gender, nutritional status, lifestyle, and genetic variations represent some of the factors that influence metformin's susceptibility and distribution to target tissues. For instance, membrane transporter polymorphism is a key determinant in the pharmacokinetic properties of this drug (Chen et al. 2013; Pawlyk et al. 2014). Metfor-

min exerts its therapeutic effects, through a number of mechanisms and physiological pathways that resemble those generated by caloric restriction (CR), an experimental model known to extend life span and health span in various organisms. Indeed, microarray analyses have shown that metformin induces the same gene expression profile as CR (Dhahbi et al. 2005; Spindler 2006; Martin-Montalvo et al. 2013), despite no reduction in food intake (Mercken et al. 2012; de Cabo et al. 2014).

The inhibition of hepatic gluconeogenesis and lipogenesis by metformin occurs via alterations in cellular energetics. The decrease in cellular respiration that results from metformin's inhibition of mitochondrial complex I activity (El-Mir et al. 2000; Owen et al. 2000) yields lower ATP levels. Although the interaction with mitochondrial copper ion appears essential for the metabolic effects of metformin (Logie et al. 2012), more still needs to be learned about whether the drug inhibits respiration through direct or indirect action (Fontaine 2014). The nonclassical effects of metformin on the expression of glucose transporters and glycolytic enzymes (up-regulation) result indirectly from mitochondrial respiratory chain inhibition (Owen et al. 2000). This inhibition of the electron transport chain (Batandier et al. 2006; Guarente 2008) combined with the induction of antioxidant gene expression by the SKN-1/Nrf2 transcription pathway (Onken and Driscoll 2010) provides mechanistic insights into metformin's role in lowering the production of reactive oxygen species (ROS). AMP-activated protein kinase (AMPK) is a key sensor of energy status that regulates metabolic energy balance at whole-body level (Hardie et al. 2012). The increase in AMP/ATP and ADP/ATP ratios stimulates AMPK (Stephenne et al. 2011); however, metformin can activate AMPK without eliciting detectable changes in AMP, ADP, and ATP levels (Hawley et al. 2002). It was later determined that the tumor suppressor protein LKB1 (alternatively termed SK11) was responsible for the activating phosphorylation of AMPK in response to metformin (Shaw et al. 2005). Indeed, the LKB1–AMPK pathway controls the expression of key hepatic gluconeogenic genes by regulating the transcriptional co-activator cAMP-response element-binding protein (CREB)-regulated transcription coactivator 2 (CRTC2, also known as TORC2) (Shaw et al. 2005), a key regulator of fasting glucose metabolism (Koo et al. 2005). The role of LKB1–AMPK as mediator of metformin's action on hepatic gluconeogenesis and lipogenesis (Zhou et al. 2001; Zou et al. 2004; Shaw et al. 2005) was put to test in studies using conditional *Ampk* knockout mice (Foretz et al. 2010). The observed inhibition of gluconeogenesis, independent of LKB1–AMPK signaling, was accompanied by a decrease in hepatic energy state in response to concentrations of metformin that were far higher than those reached in hepatic portal vein after standard treatment (Foretz et al. 2010). When therapeutic concentrations of metformin were tested, hepatic gluconeogenesis was suppressed via AMPK activation (Cao et al. 2014) and formation of AMPK $\alpha\beta\gamma$ complexes (Meng et al. 2014). The ability of AMPK to improve lipid metabolism helps explain the reduction in hepatic steatosis by metformin (Woo et al. 2014), which requires the inhibitory phosphorylation of acetyl-CoA carboxylase (ACC) by AMPK, an essential step toward the lipid-lowering and insulin-sensitizing effects of metformin (Fullerton et al. 2013). Moreover, metformin treatment decreases the levels of sterol regulatory element-binding protein 1 (SREBP-1), a key lipogenic transcription factor, via direct phosphorylation by AMPK (Zhou et al. 2001; Li et al. 2011). The regulation of lipid metabolism by metformin also takes place by enhancing the fatty acid β-oxidation pathway (Collier et al. 2006). New molecular mechanisms by which metformin inhibits hepatic gluconeogenesis have been proposed and include the ability of the drug to inhibit adenylate cyclase through AMP accumulation, thereby blocking the glucagon-signaling pathway (Miller et al. 2013), and direct inhibition of mitochondrial glycerophosphate dehydrogenase (mGPD) (Madiraju et al. 2014). In the latter study, metformin-mediated mGPD inhibition was accompanied by lower mitochondrial $NADH/NAD^+$ ratios, a result inconsistent with prior reports showing that complex I inhibition by metformin increased

this ratio (Owen et al. 2000). The different doses and route of administration of metformin between the two studies might explain these discrepancies (Baur and Birnbaum 2014).

Another potential mechanism through which metformin inhibits hepatic gluconeogenesis is the down-modulated expression of genes encoding for the gluconeogenic enzymes, phosphoenolpyruvate carboxykinase (PEPCK), and glucose-6-phosphatase (G6Pase), a molecular mechanism that requires AMPK-mediated upregulation of orphan nuclear receptor short heterodimer partner (SHP) expression (Kim et al. 2008). Additionally, metformin improves glucose homeostasis by promoting an increase in insulin-independent phosphorylation of insulin receptor and insulin receptor substrates (IRS)-1 and (IRS)-2, and subsequent translocation of glucose transporters GLUT4 to the plasma membrane (Gunton et al. 2003; Yuan et al. 2003). The regulation of the incretin hormone (e.g., glucagon-like peptide 1) and insulin secretory responses with metformin treatment has been reported (Cho and Kieffer 2011; Maida et al. 2011; Kim et al. 2014).

Metformin also acts as an inhibitor of mechanistic target of rapamycin complex 1 (mTORC1) through AMPK-dependent and -independent mechanisms. AMPK activation by metformin inhibits the protein kinase mTOR, thus preventing the phosphorylation of downstream targets, including S6K, rpS6, and 4E-BP1 (Dowling et al. 2007). Inhibition of the Ras-related GTP binding (Rag) GTPases (Kalender et al. 2010) and up-regulation of REDD1, a hypoxia-inducible factor 1 (HIF-1) target (Shoshani et al. 2002; Ben Sahra et al. 2011), are among the AMPK-independent mechanisms by which metformin inhibits mTORC1 signaling. Because of the many faces of mTOR in life span and metabolism, it is intriguing that metformin may act as a potential therapeutic drug for the treatment of aging and age-related diseases, such as cancer and metabolic syndrome (Johnson et al. 2013).

METFORMIN AS AN ANTI-AGING DRUG

Recent reviews have reported the geroprotective effects of biguanides, mainly metformin, because of its superior safety profile (Bulterijs 2011; Berstein 2012; Miles et al. 2014). As indicated earlier, metformin treatment enhances insulin sensitivity, induces glycolysis, and suppresses hepatic gluconeogenesis. There is some evidence that metformin may also have cardioprotective effects (Eurich et al. 2013; Hong et al. 2013) and contribute to the prevention of some forms of human cancer (Cazzaniga et al. 2013; Anisimov 2014; Laskov et al. 2014). This therapeutic profile of metformin supports its use for age-related diseases and longevity. Of significance, many studies have confirmed the positive effect of metformin on life span of worms, flies, mice, and rats. Moreover, diabetic and cardiovascular disease patients who are prescribed metformin have increased rates of survival (Scarpello 2003; Yin et al. 2013), and it was recently proposed that metformin might promote longevity by preventing frailty in older adults with T2DM (Wang et al. 2014). Chronic treatment with metformin among patients with diabetes might reduce the risk of cognitive decline and dementia (Ng et al. 2014; Patrone et al. 2014) and improve survival in several types of cancer (Greenhill 2015; Ko et al. 2015; Lin et al. 2015; Rego et al. 2015).

STUDIES IN INVERTEBRATE MODELS

Many molecular mechanisms implicated in aging and age-related diseases have been elucidated in *Caenorhabditis elegans*, an experimental model widely used for the identification of new pharmacological agents capable of delaying the aging process (Olsen et al. 2006; Lapierre and Hansen 2012). Metformin supplementation (50 mM dose) was found to increase the mean life span of *C. elegans* by about 40% without maximum life span extension. This increase in health span had CR-like features that involved activation of the LKB1–AMPK–SKN1 pathway both in wild-type worms and in mutant animals with disrupted insulin pathway (Onken and Driscoll 2010). The increase in carbohydrate levels in metformin-treated worms provides a good source of ATP to better survive 2 to 3 d of anoxia exposure through a mechanism that depends on specific AMPK subunits (LaRue

and Padilla 2011). Active bacterial metabolism is a critical nutritional requirement for *C. elegans* life span (Lenaerts et al. 2008; Cabreiro and Gems 2013). Biguanide-treated worms lived longer (~30% increase compared with their normal life span) only when cultured with a *Escherichia coli* strain sensitive to the drug, which contrasts with the pathogenic effects of drug-resistant bacteria on nematode health and aging. An alteration in microbial folate and methionine metabolism helps explain the extended longevity, which is consistent with the notion that metformin is a CR-mimetic drug. Metformin is primarily used for the management of hyperglycemia in T2DM, which led Cabreiro and colleagues to test whether high glucose adversely affected bacterial growth inhibition by metformin and, consequently, *C. elegans* longevity. The reduction in metformin-induced life span extension in response to glucose supplementation led the investigators to suggest that altering gut microbiota might represent a new therapeutic approach for delaying aging and the treatment of age-related diseases (Cabreiro et al. 2013). Conversely, glucose restriction extends *C. elegans* life span by inducing mitohormesis, a physiological process based on mitochondrial oxidative stress (Schulz et al. 2007; Zarse et al. 2012). Of significance, metformin-treated nematodes showed increased respiration and higher ROS production, consistent with the generation of a mitohormetic signal (De Haes et al. 2014). These investigators established that the mitohormetic signal was propagated by the hydrogen peroxide scavenger peroxiredoxin PRDX-2, whose expression was up-regulated after metformin treatment, and deletion of the *prdx2* gene led to decreased overall life expectancy. *C. elegans* treated with metformin also had a youthful morphology for a longer time, which contributed to their improved health span (De Haes et al. 2014).

The beneficial effects of metformin on the life span of nematodes do not appear to be evolutionarily conserved in *Drosophila*. AMPK activation increases life span in *Drosophila* (Tohyama and Yamaguchi 2010; Stenesen et al. 2013), and metformin treatment reduces lipid storage via robust activation of AMPK without promoting longevity in either male or female flies. Perturbations in intestinal homeostasis may be responsible for metformin toxicity in flies when taken in high enough doses (Slack et al. 2012) and when different antidiabetic compounds were tested for their potential antiaging properties (Jafari et al. 2007). In the latter study, metformin given in doses of 0.4, 0.8, and 1.6 mg/ml did not decrease the mortality rate in flies (Jafari et al. 2007). Even though the prolongevity effects of metformin have yet to be found, this drug can inhibit age- and oxidative-stress-induced DNA damage and delay stem cell aging in *Drosphilia* (Na et al. 2013).

STUDIES IN RODENT MODELS

C57BL/6J mice and Fisher 344 (F344) rats are the preferred strains of rodents for use in gerontological studies (Anisimov et al. 2012). The physiology of these animals, mainly at the cellular level, is very similar to humans, which allows the study of various compounds for their life extending properties and the extrapolation of these findings to human aging. The sole publication on the impact of metformin in rat longevity indicated a lack of effect of the biguanine on mean life span and mean of the last surviving 10% male F344 rats, compelling the investigators to question the claims about metformin acting as a CR-mimetic drug (Smith et al. 2010). However, this strain of rats is resistant to the health benefits of CR and, thus, may provide a partial explanation for the lack of prolongevity effects of metformin in F344 rats (Smith et al. 2010). This study was rather inconclusive and new approaches will be required to determine whether metformin can prolong life span in rats.

More studies were performed in different mouse strains using male and female animals. In general, female mice responded better to metformin vis-à-vis mean life span extension, as compared with male mice and rats. Metformin treatment (100 mg/kg in drinking water for 5 consecutive days every month) significantly increased mean (+8%) and maximal life span (+9%) of short-lived, cancer-prone female HER-2/neu transgenic mice (strain

FVB/N carrying a HER-2/neu oncogene) with significant reduction in the mean size and accumulation of mammary adenocarcinoma (Anisimov et al. 2005a,b). When combined with melatonin, metformin inhibited the growth of a HER2 mammary tumor and Ehrlich tumor growth in mice, whereas metformin treatment alone was shown to slow down the development of spontaneous mammary tumors and increase mean life span in female HER-2/neu transgenic mice (Anisimov et al. 2010a).

The geroprotective effects of metformin and its ability to suppress spontaneous tumorigenesis were also observed in other mouse strains. Long-term treatment with metformin significantly increased mean (+37.9%) and maximum life span (+10.3%) of female outbred SHR mice, and slowed down the age-associated disturbances in the estrous function without impacting on body weight or food intake (Anisimov et al. 2008). However, metformin treatment did not alter the incidence or mean latency of tumors, an unexpected finding that was attributed to the inherent genetic makeup of the SHR mouse strain. Nevertheless, this result emphasizes the fact that metformin can prolong life independently of its ability to suppress cancer (Blagosklonny and Campisi 2008). It is interesting to note that the responsiveness of female SHR mice to the prolongevity effects of metformin was dependent of the age of the animals at the onset of treatment. An increase in the mean life span was observed when metformin treatment was started at the age of 3 or 9 mo (+14.1% and +6.1%, respectively), but not at 15 mo of age (Anisimov et al. 2011). Focusing the analysis on tumor-free mice only, there was a significant increase (20.7% and 7.1%) but significant reduction (−12.8%) in mean life span when metformin administration was initiated in 3-, 9-, and 15-mo-old animals, respectively.

The possibility that metformin can improve the outcomes of two neurological disorders was investigated both in male and female mice. In the first study, different metformin doses (0, 2, or 5 mg/ml) were given in the drinking water of 5-wk-old transgenic mice with Huntington's disease (the R6/2 line with ∼150 glutamine repeats) (Ma et al. 2007). The investigators observed that metformin, only at 2 mg/ml, significantly increased mean life span (+20.1%) and decreased the duration of hind limb clasping, a phenotypic marker of motor defect, in male but not female animals. In the second study, three doses of metformin (0.5, 2, and 5 mg/ml) were given in the drinking water of male and female SOD1^{G93A} mice (transgenic model of amyotrophic lateral sclerosis [ALS]) from 35 d of age (Kaneb et al. 2011). Metformin treatment had no effect on disease onset, progression or survival in male SOD1^{G93A} mice at any dose while eliciting a dose-dependent negative neurological response in females owing to metformin's ability to inhibit estrogen production (Rice et al. 2009). Inhibition of estrogen can accelerate ALS progression and reduce life span in female SOD1^{G93A} mice (Choi et al. 2008). All treatment groups appeared to weigh less and displayed no significant differences in their life span, as compared with control mice. However, a tendency toward increased survival was observed with reduction in the dose of metformin (Kaneb et al. 2011).

The notion that the prolongevity effects of metformin depend on the developmental stage of the animal at the onset of treatment was further explored in inbred male and female 129/Sv mice (Anisimov et al. 2010b). Addition of metformin (100 mg/kg) in drinking water of 3-mo-old male animals elicited a significant decrease in mean life span (−13.4%) without affecting maximum life span. A higher incidence of chromosome aberrations was also noted in metformin-treated male mice. In females, metformin did not influence maximum life span, but it slightly increased mean and median life span by 4.4% and 7.8%, respectively, with a significant reduction in the total incidence of malignant tumors. However, an increase of benign angiogenic tumors was observed in metformin-treated female mice (Anisimov et al. 2010b). The reasons for these gender-specific differences on metformin responses are still under study and may be attributed to the fact that males and females have different mechanisms of aging. Potential gender-related variability in outcomes are exemplified by the next series of reports.

Deletion of ribosomal S6 protein kinase 1 (S6K1), a component of the mTOR pathway, significantly increased the life span of female C57BL/6 mice (+20.4%) without changes in that of male animals (Selman et al. 2009). Subcutaneous administration of metformin (100 mg/kg body weight) in 3-, 5-, and 7-d-old 129/Sv mouse pups caused an inversion of the gender response to the prolongevity effects of the drug (Anisimov et al. 2015). These investigators reported an increase in mean life span (+20%) and a slight maximum life span extension (+3.5%) in males who received metformin neonatally, while a decrease in mean and median life span (−9.1% and −13.8%, respectively) without significant differences in maximum life span was observed when female mouse pups were treated with metformin, as compared with control animals. The neonatal period is critical for the development of the hypothalamic circuits that control energy homeostasis (Contreras et al. 2013) and it has been suggested that reprogramming of these circuits, especially the mTOR-signaling pathway, may be part of the aging process (Blagosklonny 2013). Many aspects of aging are controlled by the hypothalamus, and alteration of hypothalamic pathways might allow the manifestations of aging to be modified (Zhang et al. 2013).

The long-term effects of metformin supplementation (0.1% and 1% w/w) in the food was performed in male C57BL/6 mice, starting from the age of 54 wk for the remainder of their lives (Martin-Montalvo et al. 2013). The mean life span of mice supplemented with 0.1% metformin increased by 5.83%, while that of mice on 1% metformin was significantly reduced (−14.4%), likely caused by renal failure. Diet supplementation with 0.1% metformin tended to preserve body weight with advancing age, a condition known to increase longevity in mice (Pearson et al. 2008). There were no significant differences in the number of pathologies in mice on 0.1% metformin; however, liver cancer incidence was significantly reduced with 1% metformin supplementation (3.3% vs. 26.5% in metformin- and vehicle-treated mice, respectively) (Martin-Montalvo et al. 2013). An improvement in physical performance and glucose homeostasis combined with increased insulin sensitivity, and a reduction in low-density lipoprotein and cholesterol levels occurred in 0.1% metformin-fed mice without a decrease in caloric intake. By preserving overall health span in mice, metformin may prevent the development of metabolic syndrome through significant reduction in oxidative stress and chronic inflammation (Martin-Montalvo et al. 2013). These investigators reported similar gene expression patterns in the liver (and skeletal muscle) of mice fed 40% CR and 0.1% metformin, reinforcing the role of metformin as a CR mimetic (Mercken et al. 2012; de Cabo et al. 2014). Of significance, the prolongevity effect of metformin was observed also in a second strain of male mice (hybrid B6C3F1), with a 4.15% increase in mean life span in response to 0.1% metformin supplementation in the diet (Martin-Montalvo et al. 2013).

CONCLUSIONS AND PERSPECTIVES

According to recent published data in different animal models, metformin appears to be a promising candidate as a life-extending drug (Fig. 1). This compound is generally well tolerated and its long history of clinical use makes it an even more attractive candidate. Besides, metformin is more beneficial than any other antidiabetic drug in reducing age-related diseases and improving survival in diabetic patients. Although the initial results are very hopeful, more work is needed to elucidate several aspects that still remain unclear. Many of these positive results have been obtained using doses of metformin that exceed therapeutic levels in humans (Martin-Castillo et al. 2010; Aldea et al. 2014). Moreover, the modes of administration varied among research teams, with the addition of metformin either in drinking water or to the

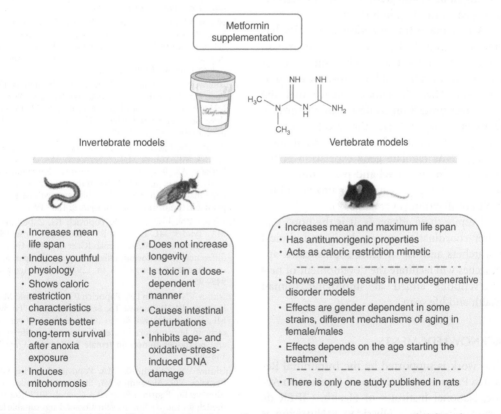

Figure 1. Summary of the effects of metformin supplementation in invertebrate (*Caenorhabditis elegans* and *Drosophila melanogaster*) and vertebrate models (rodents, mainly mice).

diet. Although female mice were initially found to show a better response to metformin supplementation, recent results from our laboratory indicated no gender or stain differences in the actions of metformin (Martin-Montalvo et al. 2013). Therefore, to establish the molecular mechanisms and pathways of aging, it is imperative to investigate potential hormone-metformin interactions in male and female animals of varying ages, as the age of starting metformin treatment determines whether an increase in mean and maximum life span occurs (Menendez et al. 2011; Anisimov et al. 2015). There are not enough studies to conclude whether there are epigenetic/genetic differences in metformin effect on aging, life span, and tumorigenesis. Because not all organisms studied seem to respond positively to metformin supplementation (e.g., flies and rats), new approaches with different protocols and experimental designs would be crucial to understanding how metformin might be a good geroprotector throughout phylogeny, including in humans.

A new interesting functional interplay has emerged during the last years that might explain some of the molecular mechanisms through which metformin could improve health and life span. There is some evidence that the anticancer protection conferred by metformin treatment may involve the modulation of miRNAs (Pulito et al. 2014). These small noncoding RNAs regulate gene expression at the posttranscriptional level and metformin modulates miRNAs that regulate apoptosis and inhibit proliferation (Li et al. 2012).

Despite these advances, it is the hope that better coordination among basic and clinical researchers and use of more sophisticated approaches will facilitate the development of new interventions aimed at improving human health and life span.

ACKNOWLEDGMENTS

This work is supported by the Intramural Research Program of the National Institute on Aging, National Institutes of Health. CIBER de Fisiopatología de la Obesidad y Nutrición is an initiative of ISCIII. The funding agency had no role in study design, data collection and analysis, decision to publish, or preparation of the manuscript.

REFERENCES

Adeyemo MA, McDuffie JR, Kozlosky M, Krakoff J, Calis KA, Brady SM, Yanovski JA. 2014. Effects of metformin on energy intake and satiety in obese children. *Diabetes Obes Metab* 17: 363–370.

Aldea M, Craciun L, Tomuleasa C, Berindan-Neagoe I, Kacso G, Florian IS, Crivii C. 2014. Repositioning metformin in cancer: Genetics, drug targets, and new ways of delivery. *Tumour Biol* 35: 5101–5110.

American Diabetes Association. 2014. Standards of medical care in diabetes—2014. *Diabetes Care* 37: S14–S80.

Anisimov VN. 2014. Do metformin a real anticarcinogen? A critical reappraisal of experimental data. *Ann Transl Med* 2: 60.

Anisimov VN, Berstein LM, Egormin PA, Piskunova TS, Popovich IG, Zabezhinski MA, Kovalenko IG, Poroshina TE, Semenchenko AV, Provinciali M, et al. 2005a. Effect of metformin on life span and on the development of spontaneous mammary tumors in HER-2/neu transgenic mice. *Exp Gerontol* 40: 685–693.

Anisimov VN, Egormin PA, Bershtein LM, Zabezhinskii MA, Piskunova TS, Popovich IG, Semenchenko AV. 2005b. Metformin decelerates aging and development of mammary tumors in HER-2/neu transgenic mice. *Bull Exp Biol Med* 139: 721–723.

Anisimov VN, Berstein LM, Egormin PA, Piskunova TS, Popovich IG, Zabezhinski MA, Tyndyk ML, Yurova MV, Kovalenko IG, Poroshina TE, et al. 2008. Metformin slows down aging and extends life span of female SHR mice. *Cell Cycle* 7: 2769–2773.

Anisimov VN, Egormin PA, Piskunova TS, Popovich IG, Tyndyk ML, Yurova MN, Zabezhinski MA, Anikin IV, Karkach AS, Romanyukha AA. 2010a. Metformin extends life span of HER-2/neu transgenic mice and in combination with melatonin inhibits growth of transplantable tumors in vivo. *Cell Cycle* 9: 188–197.

Anisimov VN, Piskunova TS, Popovich IG, Zabezhinski MA, Tyndyk ML, Egormin PA, Yurova MV, Rosenfeld SV, Semenchenko AV, Kovalenko IG, et al. 2010b. Gender differences in metformin effect on aging, life span and spontaneous tumorigenesis in 129/Sv mice. *Aging* 2: 945–958.

Anisimov VN, Berstein LM, Popovich IG, Zabezhinski MA, Egormin PA, Piskunova TS, Semenchenko AV, Tyndyk ML, Yurova MN, Kovalenko IG, et al. 2011. If started early in life, metformin treatment increases life span and postpones tumors in female SHR mice. *Aging* 3: 148–157.

Anisimov VN, Zabezhinski MA, Popovich IG, Pliss GB, Bespalov VG, Alexandrov VA, Stukov AN, Anikin IV, Alimova IN, Egormin Pcapital A C, et al. 2012. Rodent models for the preclinical evaluation of drugs suitable for pharmacological intervention in aging. *Expert Opin Drug Discov* 7: 85–95.

Cite this article as *Cold Spring Harb Perspect Med* doi: 10.1101/cshperspect.a025932

Anisimov VN, Popovich IG, Zabezhinski MA, Egormin PA, Yurova MN, Semenchenko AV, Tyndyk ML, Panchenko AV, Trashkov AP, Vasiliev AG, et al. 2015. Sex differences in aging, life span and spontaneous tumorigenesis in 129/Sv mice neonatally exposed to metformin. *Cell Cycle* **14:** 46–55.

Bailey CJ, Turner RC. 1996. Metformin. *N Engl J Med* **334:** 574–579.

Batandier C, Guigas B, Detaille D, El-Mir MY, Fontaine E, Rigoulet M, Leverve XM. 2006. The ROS production induced by a reverse-electron flux at respiratory-chain complex 1 is hampered by metformin. *J Bioenerg Biomembr* **38:** 33–42.

Baur JA, Birnbaum MJ. 2014. Control of gluconeogenesis by metformin: Does redox trump energy charge? *Cell Metab* **20:** 197–199.

Ben Sahra I, Regazzetti C, Robert G, Laurent K, Le Marchand-Brustel Y, Auberger P, Tanti JF, Giorgetti-Peraldi S, Bost F. 2011. Metformin, independent of AMPK, induces mTOR inhibition and cell-cycle arrest through REDD1. *Cancer Res* **71:** 4366–4372.

Berstein LM. 2012. Metformin in obesity, cancer and aging: Addressing controversies. *Aging* **4:** 320–329.

Blagosklonny MV. 2013. Big mice die young but large animals live longer. *Aging* **5:** 227–233.

Blagosklonny MV, Campisi J. 2008. Cancer and aging: More puzzles, more promises? *Cell Cycle* **7:** 2615–2618.

Bulterijs S. 2011. Metformin as a geroprotector. *Rejuvenation Res* **14:** 469–482.

Cabreiro F, Gems D. 2013. Worms need microbes too: Microbiota, health and aging in *Caenorhabditis elegans*. *EMBO Mol Med* **5:** 1300–1310.

Cabreiro F, Au C, Leung KY, Vergara-Irigaray N, Cocheme HM, Noori T, Weinkove D, Schuster E, Greene ND, Gems D. 2013. Metformin retards aging in *C. elegans* by altering microbial folate and methionine metabolism. *Cell* **153:** 228–239.

Cao J, Meng S, Chang E, Beckwith-Fickas K, Xiong L, Cole RN, Radovick S, Wondisford FE, He L. 2014. Low concentrations of metformin suppress glucose production in hepatocytes through AMP-activated protein kinase (AMPK). *J Biol Chem* **289:** 20435–20446.

Cazzaniga M, DeCensi A, Pruneri G, Puntoni M, Bottiglieri L, Varricchio C, Guerrieri-Gonzaga A, Gentilini OD, Pagani G, Dell'Orto P, et al. 2013. The effect of metformin on apoptosis in a breast cancer presurgical trial. *Br J Cancer* **109:** 2792–2797.

Chen S, Zhou J, Xi M, Jia Y, Wong Y, Zhao J, Ding L, Zhang J, Wen A. 2013. Pharmacogenetic variation and metformin response. *Curr Drug Metab* **14:** 1070–1082.

Cho YM, Kieffer TJ. 2011. New aspects of an old drug: Metformin as a glucagon-like peptide 1 (GLP-1) enhancer and sensitizer. *Diabetologia* **54:** 219–222.

Choi CI, Lee YD, Gwag BJ, Cho SI, Kim SS, Suh-Kim H. 2008. Effects of estrogen on lifespan and motor functions in female hSOD1 G93A transgenic mice. *J Neurol Sci* **268:** 40–47.

Collier CA, Bruce CR, Smith AC, Lopaschuk G, Dyck DJ. 2006. Metformin counters the insulin-induced suppression of fatty acid oxidation and stimulation of triacylglycerol storage in rodent skeletal muscle. *Am J Physiol Endocrinol Metab* **291:** E182–E189.

Contreras C, Novelle MG, Leis R, Dieguez C, Skrede S, Lopez M. 2013. Effects of neonatal programming on hypothalamic mechanisms controlling energy balance. *Horm Metab Res* **45:** 935–944.

de Cabo R, Carmona-Gutierrez D, Bernier M, Hall MN, Madeo F. 2014. The search for antiaging interventions: From elixirs to fasting regimens. *Cell* **157:** 1515–1526.

De Haes W, Frooninckx L, Van Assche R, Smolders A, Depuydt G, Billen J, Braeckman BP, Schoofs L, Temmerman L. 2014. Metformin promotes lifespan through mitohormesis via the peroxiredoxin PRDX-2. *Proc Natl Acad Sci* **111:** E2501–E2509.

Dhahbi JM, Mote PL, Fahy GM, Spindler SR. 2005. Identification of potential caloric restriction mimetics by microarray profiling. *Physiol Genomics* **23:** 343–350.

Diamanti-Kandarakis E, Christakou CD, Kandaraki E, Economou FN. 2010. Metformin: An old medication of new fashion: Evolving new molecular mechanisms and clinical implications in polycystic ovary syndrome. *Eur J Endocrinol* **162:** 193–212.

Dowling RJ, Zakikhani M, Fantus IG, Pollak M, Sonenberg N. 2007. Metformin inhibits mammalian target of rapamycin-dependent translation initiation in breast cancer cells. *Cancer Res* **67:** 10804–10812.

El-Mir MY, Nogueira V, Fontaine E, Averet N, Rigoulet M, Leverve X. 2000. Dimethylbiguanide inhibits cell respiration via an indirect effect targeted on the respiratory chain complex I. *J Biol Chem* **275:** 223–228.

Eurich DT, Weir DL, Majumdar SR, Tsuyuki RT, Johnson JA, Tjosvold L, Vanderloo SE, McAlister FA. 2013. Comparative safety and effectiveness of metformin in patients with diabetes mellitus and heart failure: Systematic review of observational studies involving 34,000 patients. *Circ Heart Fail* **6:** 395–402.

Fontaine E. 2014. Metformin and respiratory chain complex. I: The last piece of the puzzle? *Biochem J* **463:** e3–e5.

Foretz M, Hebrard S, Leclerc J, Zarrinpashneh E, Soty M, Mithieux G, Sakamoto K, Andreelli F, Viollet B. 2010. Metformin inhibits hepatic gluconeogenesis in mice independently of the LKB1/AMPK pathway via a decrease in hepatic energy state. *J Clin Invest* **120:** 2355–2369.

Foretz M, Guigas B, Bertrand L, Pollak M, Viollet B. 2014. Metformin: From mechanisms of action to therapies. *Cell Metab* **20:** 953–966.

Fullerton MD, Galic S, Marcinko K, Sikkema S, Pulinilkunnil T, Chen ZP, O'Neill HM, Ford RJ, Palanivel R, O'Brien M, et al. 2013. Single phosphorylation sites in Acc1 and Acc2 regulate lipid homeostasis and the insulin-sensitizing effects of metformin. *Nat Med* **19:** 1649–1654.

Glueck CJ, Fontaine RN, Wang P, Subbiah MT, Weber K, Illig E, Streicher P, Sieve-Smith L, Tracy TM, Lang JE, et al. 2001. Metformin reduces weight, centripetal obesity, insulin, leptin, and low-density lipoprotein cholesterol in nondiabetic, morbidly obese subjects with body mass index greater than 30. *Metabolism* **50:** 856–861.

Gong L, Goswami S, Giacomini KM, Altman RB, Klein TE. 2012. Metformin pathways: Pharmacokinetics and pharmacodynamics. *Pharmacogenet Genomics* **22:** 820–827.

Greenhill C. 2015. Gastric cancer: Metformin improves survival and recurrence rate in patients with diabetes and gastric cancer. *Nat Rev Gastroenterol Hepatol* **12:** 124.

Guarente L. 2008. Mitochondria—A nexus for aging, calorie restriction, and sirtuins? *Cell* **132:** 171–176.

Gunton JE, Delhanty PJ, Takahashi S, Baxter RC. 2003. Metformin rapidly increases insulin receptor activation in human liver and signals preferentially through insulin-receptor substrate-2. *J Clin Endocrinol Metab* **88:** 1323–1332.

Hardie DG, Ross FA, Hawley SA. 2012. AMPK: A nutrient and energy sensor that maintains energy homeostasis. *Nat Rev Mol Cell Biol* **13:** 251–262.

Hawley SA, Gadalla AE, Olsen GS, Hardie DG. 2002. The antidiabetic drug metformin activates the AMP-activated protein kinase cascade via an adenine nucleotide-independent mechanism. *Diabetes* **51:** 2420–2425.

Hirsch HA, Iliopoulos D, Struhl K. 2013. Metformin inhibits the inflammatory response associated with cellular transformation and cancer stem cell growth. *Proc Natl Acad Sci* **110:** 972–977.

Hong J, Zhang Y, Lai S, Lv A, Su Q, Dong Y, Zhou Z, Tang W, Zhao J, Cui L, et al. 2013. Effects of metformin versus glipizide on cardiovascular outcomes in patients with type 2 diabetes and coronary artery disease. *Diabetes Care* **36:** 1304–1311.

Isoda K, Young JL, Zirlik A, MacFarlane LA, Tsuboi N, Gerdes N, Schonbeck U, Libby P. 2006. Metformin inhibits proinflammatory responses and nuclear factor-κB in human vascular wall cells. *Arterioscler Thromb Vasc Biol* **26:** 611–617.

Jafari M, Khodayari B, Felgner J, Bussel II, Rose MR, Mueller LD. 2007. Pioglitazone: An anti-diabetic compound with anti-aging properties. *Biogerontology* **8:** 639–651.

Johnson SC, Rabinovitch PS, Kaeberlein M. 2013. mTOR is a key modulator of ageing and age-related disease. *Nature* **493:** 338–345.

Kalender A, Selvaraj A, Kim SY, Gulati P, Brule S, Viollet B, Kemp BE, Bardeesy N, Dennis P, Schlager JJ, et al. 2010. Metformin, independent of AMPK, inhibits mTORC1 in a rag GTPase-dependent manner. *Cell Metab* **11:** 390–401.

Kaneb HM, Sharp PS, Rahmani-Kondori N, Wells DJ. 2011. Metformin treatment has no beneficial effect in a dose-response survival study in the SOD1^{G93A} mouse model of ALS and is harmful in female mice. *PLoS ONE* **6:** e24189.

Kim YD, Park KG, Lee YS, Park YY, Kim DK, Nedumaran B, Jang WG, Cho WJ, Ha J, Lee IK, et al. 2008. Metformin inhibits hepatic gluconeogenesis through AMP-activated protein kinase-dependent regulation of the orphan nuclear receptor SHP. *Diabetes* **57:** 306–314.

Kim MH, Jee JH, Park S, Lee MS, Kim KW, Lee MK. 2014. Metformin enhances glucagon-like peptide 1 via cooperation between insulin and Wnt signaling. *J Endocrinol* **220:** 117–128.

Kita Y, Takamura T, Misu H, Ota T, Kurita S, Takeshita Y, Uno M, Matsuzawa-Nagata N, Kato K, Ando H, et al. 2012. Metformin prevents and reverses inflammation in a non-diabetic mouse model of nonalcoholic steatohepatitis. *PLoS ONE* **7:** e43056.

Ko EM, Sturmer T, Hong JL, Castillo WC, Bae-Jump V, Funk MJ. 2015. Metformin and the risk of endometrial cancer: A population-based cohort study. *Gynecol Oncol* **136:** 341–347.

Koo SH, Flechner L, Qi L, Zhang X, Screaton RA, Jeffries S, Hedrick S, Xu W, Boussouar F, Brindle P, et al. 2005. The CREB coactivator TORC2 is a key regulator of fasting glucose metabolism. *Nature* **437:** 1109–1111.

Lapierre LR, Hansen M. 2012. Lessons from *C. elegans*: Signaling pathways for longevity. *Trends Endocrinol Metab* **23:** 637–644.

LaRue BL, Padilla PA. 2011. Environmental and genetic preconditioning for long-term anoxia responses requires AMPK in *Caenorhabditis elegans*. *PLoS ONE* **6:** e16790.

Laskov I, Drudi L, Beauchamp MC, Yasmeen A, Ferenczy A, Pollak M, Gotlieb WH. 2014. Anti-diabetic doses of metformin decrease proliferation markers in tumors of patients with endometrial cancer. *Gynecol Oncol* **134:** 607–614.

Lenaerts I, Walker GA, Van Hoorebeke L, Gems D, Vanfleteren JR. 2008. Dietary restriction of *Caenorhabditis elegans* by axenic culture reflects nutritional requirement for constituents provided by metabolically active microbes. *J Gerontol A Biol Sci Med Sci* **63:** 242–252.

Li Y, Xu S, Mihaylova MM, Zheng B, Hou X, Jiang B, Park O, Luo Z, Lefai E, Shyy JY, et al. 2011. AMPK phosphorylates and inhibits SREBP activity to attenuate hepatic steatosis and atherosclerosis in diet-induced insulin-resistant mice. *Cell Metab* **13:** 376–388.

Li W, Yuan Y, Huang L, Qiao M, Zhang Y. 2012. Metformin alters the expression profiles of microRNAs in human pancreatic cancer cells. *Diabetes Res Clin Pract* **96:** 187–195.

Lin JJ, Gallagher EJ, Sigel K, Mhango G, Galsky MD, Smith CB, LeRoith D, Wisnivesky JP. 2015. Survival of patients with stage IV lung cancer with diabetes treated with metformin. *Am J Respir Crit Care Med* **191:** 448–454.

Logie L, Harthill J, Patel K, Bacon S, Hamilton DL, Macrae K, McDougall G, Wang HH, Xue L, Jiang H, et al. 2012. Cellular responses to the metal-binding properties of metformin. *Diabetes* **61:** 1423–1433.

Ma TC, Buescher JL, Oatis B, Funk JA, Nash AJ, Carrier RL, Hoyt KR. 2007. Metformin therapy in a transgenic mouse model of Huntington's disease. *Neurosci Lett* **411:** 98–103.

Madiraju AK, Erion DM, Rahimi Y, Zhang XM, Braddock DT, Albright RA, Prigaro BJ, Wood JL, Bhanot S, MacDonald MJ, et al. 2014. Metformin suppresses gluconeogenesis by inhibiting mitochondrial glycerophosphate dehydrogenase. *Nature* **510:** 542–546.

Mahmood K, Naeem M, Rahimnajjad NA. 2013. Metformin: The hidden chronicles of a magic drug. *Eur J Intern Med* **24:** 20–26.

Maida A, Lamont BJ, Cao X, Drucker DJ. 2011. Metformin regulates the incretin receptor axis via a pathway dependent on peroxisome proliferator-activated receptor-α in mice. *Diabetologia* **54:** 339–349.

Martin-Castillo B, Vazquez-Martin A, Oliveras-Ferraros C, Menendez JA. 2010. Metformin and cancer: Doses, mechanisms and the dandelion and hormetic phenomena. *Cell Cycle* **9:** 1057–1064.

Martin-Montalvo A, Mercken EM, Mitchell SJ, Palacios HH, Mote PL, Scheibye-Knudsen M, Gomes AP, Ward TM, Minor RK, Blouin MJ, et al. 2013. Metformin improves healthspan and lifespan in mice. *Nat Commun* 4: 2192.

Menendez JA, Cufi S, Oliveras-Ferraros C, Vellon L, Joven J, Vazquez-Martin A. 2011. Gerosuppressant metformin: Less is more. *Aging* 3: 348–362.

Meng S, Cao J, He Q, Xiong L, Chang E, Radovick S, Wondisford FE, He L. 2014. Metformin activates AMP-activated protein kinase by promoting formation of the αβγ heterotrimeric complex. *J Biol Chem* 290: 3793–3802.

Mercken EM, Carboneau BA, Krzysik-Walker SM, de Cabo R. 2012. Of mice and men: The benefits of caloric restriction, exercise, and mimetics. *Ageing Res Rev* 11: 390–398.

Miles JM, Rule AD, Borlaug BA. 2014. Use of metformin in diseases of aging. *Curr Diab Rep* 14: 490.

Miller RA, Chu Q, Xie J, Foretz M, Viollet B, Birnbaum MJ. 2013. Biguanides suppress hepatic glucagon signalling by decreasing production of cyclic AMP. *Nature* 494: 256–260.

Na HJ, Park JS, Pyo JH, Lee SH, Jeon HJ, Kim YS, Yoo MA. 2013. Mechanism of metformin: Inhibition of DNA damage and proliferative activity in *Drosophila* midgut stem cell. *Mech Ageing Dev* 134: 381–390.

Ng TP, Feng L, Yap KB, Lee TS, Tan CH, Winblad B. 2014. Long-term metformin usage and cognitive function among older adults with diabetes. *J Alzheimers Dis* 41: 61–68.

Olsen A, Vantipalli MC, Lithgow GJ. 2006. Using *Caenorhabditis elegans* as a model for aging and age-related diseases. *Ann NY Acad Sci* 1067: 120–128.

Onken B, Driscoll M. 2010. Metformin induces a dietary restriction-like state and the oxidative stress response to extend *C. elegans* healthspan via AMPK, LKB1, and SKN-1. *PLoS ONE* 5: e8758.

Owen MR, Doran E, Halestrap AP. 2000. Evidence that metformin exerts its anti-diabetic effects through inhibition of complex I of the mitochondrial respiratory chain. *Biochem J* 348: 607–614.

Patrone C, Eriksson O, Lindholm D. 2014. Diabetes drugs and neurological disorders: New views and therapeutic possibilities. *Lancet Diabetes Endocrinol* 2: 256–262.

Pawlyk AC, Giacomini KM, McKeon C, Shuldiner AR, Florez JC. 2014. Metformin pharmacogenomics: Current status and future directions. *Diabetes* 63: 2590–2599.

Pearson KJ, Baur JA, Lewis KN, Peshkin L, Price NL, Labinskyy N, Swindell WR, Kamara D, Minor RK, Perez E, et al. 2008. Resveratrol delays age-related deterioration and mimics transcriptional aspects of dietary restriction without extending life span. *Cell Metab* 8: 157–168.

Pernicova I, Korbonits M. 2014. Metformin—Mode of action and clinical implications for diabetes and cancer. *Nat Rev Endocrinol* 10: 143–156.

Pulito C, Donzelli S, Muti P, Puzzo L, Strano S, Blandino G. 2014. microRNAs and cancer metabolism reprogramming: The paradigm of metformin. *Ann Transl Med* 2: 58.

Rego DF, Pavan LM, Elias ST, De Luca Canto G, Guerra EN. 2015. Effects of metformin on head and neck cancer: A systematic review. *Oral Oncol* 51: 416–422.

Rice S, Pellatt L, Ramanathan K, Whitehead SA, Mason HD. 2009. Metformin inhibits aromatase via an extracellular signal-regulated kinase-mediated pathway. *Endocrinology* 150: 4794–4801.

Scarpello JH. 2003. Improving survival with metformin: The evidence base today. *Diabetes Metab* 29: 6S36–6S43.

Schulz TJ, Zarse K, Voigt A, Urban N, Birringer M, Ristow M. 2007. Glucose restriction extends *Caenorhabditis elegans* life span by inducing mitochondrial respiration and increasing oxidative stress. *Cell Metab* 6: 280–293.

Selman C, Tullet JM, Wieser D, Irvine E, Lingard SJ, Choudhury AI, Claret M, Al-Qassab H, Carmignac D, Ramadani F, et al. 2009. Ribosomal protein S6 kinase 1 signaling regulates mammalian life span. *Science* 326: 140–144.

Shaw RJ, Lamia KA, Vasquez D, Koo SH, Bardeesy N, Depinho RA, Montminy M, Cantley LC. 2005. The kinase LKB1 mediates glucose homeostasis in liver and therapeutic effects of metformin. *Science* 310: 1642–1646.

Shoshani T, Faerman A, Mett I, Zelin E, Tenne T, Gorodin S, Moshel Y, Elbaz S, Budanov A, Chajut A, et al. 2002. Identification of a novel hypoxia-inducible factor 1-responsive gene, *RTP801*, involved in apoptosis. *Mol Cell Biol* 22: 2283–2293.

Slack C, Foley A, Partridge L. 2012. Activation of AMPK by the putative dietary restriction mimetic metformin is insufficient to extend lifespan in *Drosophila*. *PLoS ONE* 7: e47699.

Smith DL Jr, Elam CF Jr, Mattison JA, Lane MA, Roth GS, Ingram DK, Allison DB. 2010. Metformin supplementation and life span in Fischer-344 rats. *J Gerontol A Biol Sci Med Sci* 65: 468–474.

Spindler SR. 2006. Use of microarray biomarkers to identify longevity therapeutics. *Aging Cell* 5: 39–50.

Stenesen D, Suh JM, Seo J, Yu K, Lee KS, Kim JS, Min KJ, Graff JM. 2013. Adenosine nucleotide biosynthesis and AMPK regulate adult life span and mediate the longevity benefit of caloric restriction in flies. *Cell Metab* 17: 101–112.

Stephenne X, Foretz M, Taleux N, van der Zon GC, Sokal E, Hue L, Viollet B, Guigas B. 2011. Metformin activates AMP-activated protein kinase in primary human hepatocytes by decreasing cellular energy status. *Diabetologia* 54: 3101–3110.

Stevanovic D, Janjetovic K, Misirkic M, Vucicevic L, Sumarac-Dumanovic M, Micic D, Starcevic V, Trajkovic V. 2012. Intracerebroventricular administration of metformin inhibits ghrelin-induced hypothalamic AMP-kinase signalling and food intake. *Neuroendocrinology* 96: 24–31.

Tohyama D, Yamaguchi A. 2010. A critical role of *SNF1A/ dAMPKα* (*Drosophila AMP-activated protein kinase α*) in muscle on longevity and stress resistance in *Drosophila melanogaster*. *Biochem Biophys Res Commun* 394: 112–118.

Wang CP, Lorenzo C, Espinoza SE. 2014. Frailty attenuates the impact of metformin on reducing mortality in older adults with type 2 diabetes. *J Endocrinol Diabetes Obes* 2: 1030.

Woo SL, Xu H, Li H, Zhao Y, Hu X, Zhao J, Guo X, Guo T, Botchlett R, Qi T, et al. 2014. Metformin ameliorates hepatic steatosis and inflammation without altering adipose phenotype in diet-induced obesity. *PLoS ONE* 9: e91111.

Yin M, Zhou J, Gorak EJ, Quddus F. 2013. Metformin is associated with survival benefit in cancer patients with concurrent type 2 diabetes: A systematic review and meta-analysis. *Oncologist* **18:** 1248–1255.

Yuan L, Ziegler R, Hamann A. 2003. Metformin modulates insulin post-receptor signaling transduction in chronically insulin-treated Hep G2 cells. *Acta Pharmacol Sin* **24:** 55–60.

Zarse K, Schmeisser S, Groth M, Priebe S, Beuster G, Kuhlow D, Guthke R, Platzer M, Kahn CR, Ristow M. 2012. Impaired insulin/IGF1 signaling extends life span by promoting mitochondrial L-proline catabolism to induce a transient ROS signal. *Cell Metab* **15:** 451–465.

Zhang G, Li J, Purkayastha S, Tang Y, Zhang H, Yin Y, Li B, Liu G, Cai D. 2013. Hypothalamic programming of systemic ageing involving IKK-β, NF-κB and GnRH. *Nature* **497:** 211–216.

Zhou G, Myers R, Li Y, Chen Y, Shen X, Fenyk-Melody J, Wu M, Ventre J, Doebber T, Fujii N, et al. 2001. Role of AMP-activated protein kinase in mechanism of metformin action. *J Clin Invest* **108:** 1167–1174.

Zou MH, Kirkpatrick SS, Davis BJ, Nelson JS, Wiles WGt, Schlattner U, Neumann D, Brownlee M, Freeman MB, Goldman MH. 2004. Activation of the AMP-activated protein kinase by the anti-diabetic drug metformin in vivo. Role of mitochondrial reactive nitrogen species. *J Biol Chem* **279:** 43940–43951.

Cite this article as *Cold Spring Harb Perspect Med* doi: 10.1101/cshperspect.a025932

Articulating the Case for the Longevity Dividend

S. Jay Olshansky

School of Public Health, University of Illinois at Chicago, Chicago, Illinois 60612

Correspondence: sjayo@uic.edu

The survival of large segments of human populations to advanced ages is a crowning achievement of improvements in public health and medicine. But, in the 21st century, our continued desire to extend life brings forth a unique dilemma. The risk of death from cardiovascular diseases and many forms of cancer have declined, but even if they continue to do so in the future, the resulting health benefits and enhanced longevities are likely to diminish. It is even possible that healthy life expectancy could decline in the future as major fatal diseases wane. The reason is that the longer we live, the greater is the influence of biological aging on the expression of fatal and disabling diseases. As long as the rates of aging of our bodies continue without amelioration, the progress we make on all major disease fronts must eventually face a point of diminishing returns. Research in the scientific study of aging has already shown that the aging of our bodies is inherently modifiable, and that a therapeutic intervention that slows aging in people is a plausible target for science and public health. Given the speed with which population aging is progressing and chronic fatal and disabling conditions are challenging health care costs across the globe, the case is now being made in the scientific literature that delayed aging could be one of the most efficient and promising ways to combat disease, extend healthy life, compress morbidity, and reduce health care costs. A consortium of scientists and nonprofit organizations has devised a plan to initiate an accelerated program of scientific research to develop, test for safety and efficacy, and then disseminate a therapeutic intervention to delay aging if proven to be safe and effective; this is referred to as the Longevity Dividend Initiative Consortium (LDIC). In this review, I articulate the case for the LDIC.

The rise in human longevity and the increase in our healthy lifespan are two of humanity's greatest achievements. In the developed world, and now even among growing subgroups in developing nations, increasingly larger segments of the population have gained access to one of the most precious of all commodities—the opportunity to live life into older ages. In developed nations, ∼85% of everyone born today will live to at least their 65th birthday, and >42% will live past their 85th birthday—a privilege that has been denied to most people throughout history (Human Mortality Database, www.mortality.org). However, the price to pay for common access to older ages is the opportunity to witness the aging of our bodies, and the fatal and disabling diseases that accompany extended survival.

The trade-off of chronic degenerative diseases for decades of life that was accomplished during the 20th century has undeniably been worth it, but humanity now faces a rather daunting health and economic dilemma. The combination of additional life extension with a forthcoming rapid upward shift in age structure (population aging) will lead to a dramatic increase in the prevalence of the diseases of old age—producing a major challenge to health care systems and age-entitlement programs. The National Institutes of Health, World Health Organization, United Nations, World Economic Forum, MacArthur Foundation, and other organizations have appropriately acknowledged that aging and life extension also offer an equal measure of opportunity (e.g., see Beard et al. 2011), but the rising prevalence of costly diseases is an inevitable by-product of our success.

The usual approach to combatting the diseases of old age has been to lower the behavioral risk factors that influence their expression, delay their appearance through earlier detection, and to use medical technology to extend survival for those whose bodies are already diseased. This approach has been successful in the past, but there is a growing body of evidence to suggest that continuing down this path will lead to diminishing gains in life extension and, more importantly, the possibility (perhaps likelihood) that the historic rise in health span may come to a halt—leaving future older cohorts the prospect of rising frailty and disability in later ages (Olshansky et al. 2006; Butler et al. 2008). The fact is, the longer we live, the more the biological aging of our bodies influences the fatal and disabling diseases that emerge (Miller 2012).

Recognizing the important linkage between the biological aging of our bodies and disease expression, an exciting line of scientific research has emerged that offers an opportunity to extend our healthy years further (Miller 2002; Sierra et al. 2009; Kirkland 2013). The health and economic benefits that would accrue to individuals and nations if this approach is successful has been documented (Goldman et al. 2013). As a result, a consortium of scientists has formed with the purpose of developing a new weapon to extend health span, combat the diseases of aging, compress mortality, morbidity, and disability, and help to ameliorate the economic challenges of an anticipated rising prevalence of late-onset diseases. Numerous experimental animal studies have now shown that interventions that ameliorate multiple fatal and disabling maladies of aging are possible, and will likely be transformative to human health if successfully brought to fruition. A large-scale, concerted, and coordinated effort is now underway to develop, test, and then push the translation of these findings into real-world clinical investigation and use—referred to as the Longevity Dividend Initiative Consortium (LDIC).

The LDIC aims at accelerating the pace of translation from the basic biology of aging into clinical interventions that will improve quality of life at all ages, but especially for people reaching older ages. The goals are ambitious because they address needs in several scientific domains; from basic biology, to genetics, preclinical and clinical research, as well as population-level modeling. This review and other literature provide the rationale behind the LDIC and begin outlining the various scientific pathways that researchers are pursuing to this end.

HEALTHY LIFE EXTENSION

The most precious of all commodities is life itself, and if there is one attribute most of us share, it is the desire to remain alive. The yearning for healthy life is equally important, perhaps more so—especially for those struggling to regain health that has been lost. One would think, therefore, that making the case for the development of new more effective methods of extending our healthy years would be universally accepted and easy to make, regardless of how it is achieved. Sadly, this is not the case.

In public health, examples of interventions that in the past had a profound influence on the length and quality of life include the development and dissemination of clean water, sanitation, indoor living and working environments, and refrigeration. During the last century, epidemiologists made the public aware of the life-shortening effects of smoking and other harmful risk factors, and the life-extending effects of proper diet and exercise, among others.

In the modern world of medicine and medical technology, a trip to the doctor, dentist, or other health professional is justified today as forms of primary prevention. When a health issue arises, such as a serious infection, cancer, or heart disease, it is now routine to seek out and trust modern medical treatment as the best approach to regaining one's health. In fact, a strong endorsement for the efficacy of medicine's ability to extend healthy life comes from its validation by the insurance industry.

These three pillars of healthy life extension have earned our trust, and deservedly so, but now concerns are being raised about how much more healthy life can be manufactured using these approaches. The reason is the biological aging of our bodies.

In the last half-century, a combination of public health and medicine enabled most people born in the developed world to live past age 65, and for them, a large percentage live past age 85. As appealing as this is, the problem that arises with extended survival is that a less tractable risk factor has emerged—biological aging. Public health can manufacture only so much survival time through lifestyle modification, after which medical technology has an important life-extending impact, but even these methods of life extension eventually lead the survivors to face the increased and accelerated ravages of the biological aging of our bodies.

Consider the effect of aging on the body as the same as the effect of miles on your car. Very few things go wrong with most cars during the first 3 years and 36,000 miles, and for some automobiles the warranty period has been extended to 10 years and 100,000 miles. Operate these cars beyond their warranty period, and a cluster of problems emerges. These problems are an inevitable by-product of the passage of time and the accumulation of damage that arises from operating the machine—they are not programmed to occur at a set time by the auto manufacturers. Although planned obsolescence is part of the manufacturing ethos for some manufacturers of certain products, what is

meant here is that a specific "death time" is not built into a car.

The same principles hold true for human bodies. Once we operate our bodies beyond the equivalent of their biological warranty period, a large number of health issues begin to emerge and cluster tightly into later regions of the life span. Among scientists who track these events, this is known as "competing causes," which is another way of saying that a large number of lethal and disabling conditions accumulate in aging bodies. Ameliorating any one lethal condition independent of all others leaves the person with a remaining high risk from all other remaining conditions. With time (and age), the treatments devised through medicine (that tend to focus on one disease at a time) and risk factor modification, then become progressively less effective as survivors move further into older age windows, in which aging-related diseases cluster ever more tightly together. Keep in mind that, just like automobiles, our bodies are not programmed with aging or death genes that are set off at a predetermined age. Aging is best thought of as an inadvertent by-product of fixed genetic programs that evolved under the direct force of natural selection for early life developmental events—aging is a product of evolutionary neglect, not evolutionary intent.

Recognizing the fact that competing causes place a damper on the future effectiveness of medical interventions that are disease oriented, scientists in the field of aging have proposed that the next big step in public health and healthy life extension is to attack the seeds of aging rather than just its consequences as we do now. The idea is to slow the aging of our bodies such that 1 year of clock time is matched by less than 1 year of biological time. In this way, we would retain our youthful vigor for a longer time period and, if delayed aging interventions work the way we hope they do, experience a compression of the infirmities of old age into a shorter time frame at the end of life. Delaying the biological aging of our bodies is the only viable approach to addressing the increasing importance of competing causes, and the rise of aging as an ever more important risk factor for disease.

It is at this juncture where one of the main problems occurs. The contemporary proposal to slow aging as a means to extend healthy life has historical linkages to medical deception, charlatanism, and greed (Gruman 1966). Historically, the quest for immortality was couched within a "prolongevity" message suggesting that ingesting or injecting substances with alleged "antiaging" properties could manufacture youth. One of the most famous among these is the alchemist's dream to transmute lead into gold, which was thought to confer properties of immortality to those who ingested minute quantities.

In the late 19th century, the French physiologist Charles-Édouard Brown-Séquard claimed to have discovered the secret to rejuvenation. Brown-Séquard crushed the testicles of domesticated animals, extracted "vital" substances from them, and then inoculated older people against the "aging disease." Modern versions of these ancient "antiaging" potions were described by the United States Government Accounting Office as posing the "potential for physical and economic harm" (GAO 2001).

Finally, some scientists in the field of aging have formed companies designed to attract outside investors interested in cashing in on a possible breakthrough in the field of aging (Anton 2013). Although this approach enables some aging science to occur that would not otherwise be funded, it can and has led to exaggerated claims and unproven interventions that reach the marketplace before they are fully evaluated using the tools of science. This too creates suspicion among the public who already have a difficult time distinguishing between medical fraud and genuine public health interventions.

Taken together, these historical and contemporary roadblocks to legitimacy have delayed the entrance of aging science into the realm of accepted discourse as a legitimate and, quite frankly, valuable and needed public health intervention. However, these are not the only roadblocks.

RELIGIOUS ARGUMENTS

Religious objections are sometimes posed in response to proposals to enhance public health by modulation of aging. The objection usually starts from the assertion that tampering with aging is equivalent to tampering with God's plan for us—an effort that should not be pursued. However, this argument loses its power when those proposing it admit that both they and their children have been vaccinated against lethal childhood diseases. It is hard to imagine that God's plan is to kill most children from communicable diseases before reaching the age of 10, but up until the 19th century that was humanity's fate. Most people who make this argument also admit that they would seek medical attention if they (or their loved ones) experience heart disease or cancer. Why is one form of disease prevention acceptable, whereas another is not?

POPULATION GROWTH

When delayed aging was first proposed as a public health intervention in the 1950s, rapid population growth was a concern because the growth rate (GR) in the post-World War II era was ∼3%. To place this GR into perspective, at that rate it takes the population 26 years to double in size. Thus, there was reason to be concerned about the population GR during most of the last half of the 20th century—this was alarming to both demographers and environmentalists. Although the rate of population growth has attenuated considerably since 1950, the momentum for population growth will remain with us through the middle of this century. However, environmental concerns have escalated considerably. Population growth and resource depletion should be on our minds, and these are issues that are appropriate to raise when having a discussion about healthy life extension.

The thing is, those making this argument believe that delayed aging will dramatically accelerate population growth, wipe out the reductions in the GR achieved in recent decades, further challenge resource depletion, and generate a new set of population and environmental headaches. As it turns out, none of these concerns are valid.

With regard to population growth, I have estimated how the GR would change with the hypothetical extreme scenario of immortality (i.e., no more deaths) (Olshansky 2013). Under the extreme scenario of immortality, the GR would be ∼1.5% (i.e., the GR would be defined by the birth rate because the death rate would be zero)—which is three times faster than the current GR of ∼0.5%. However, longer lives tend to be accompanied by lower fertility, so I estimate a GR under conditions of hypothetical immortality of ∼0.9%, still twice the current GR. Because immortality is not likely to happen any time soon, and because the longevity dividend associated with delayed aging would yield only marginal increases in life expectancy, the actual population GR would only rise slightly if the longevity dividend is achieved.

In fact, the population GR would also rise marginally with a hypothetical cure for cancer or heart disease. I have yet to hear anyone argue that cures for these diseases should not be pursued because success would be accompanied by accelerated population growth and resource depletion. The bottom line is that the Longevity Dividend Initiative will have a negligible effect on population growth and the environment, but it will have a dramatically positive impact on work, retirement, health care financing and costs, and physical and psychological well-being (Goldman et al. 2013).

DELAYED AGING MEANS INCREASED INFIRMITY

Perhaps the most common misconception and fear about aging science and the Longevity Dividend Initiative is the belief that delayed aging will extend the period of infirmity at the end of life—the fear that most people have as they approach older ages. There is an irony to this view because, although there may be disagreements among the scientists involved on exactly how to accomplish the goals we have set, the one thing we all have in common is the final and most important goal of extending the period of healthy life. An intervention that does not meet the test of extending the health and functionality of both body and mind together would not be pursued—in fact, such an intervention would be seen as harmful.

ARTICULATING THE CASE FOR THE LONGEVITY DIVIDEND

The case for the longevity dividend is compelling and in theory should be easy to make to funders, public health professionals, and the general public. Here is the line of reasoning:

1. Treating diseases worked well in the past to extend healthy life, but aging has emerged as the primary risk factor for the most common fatal and disabling diseases;

2. The longer we live, the greater the influence of aging on disease expression;

3. Aging science offers medicine and public health a new and potentially far more effective weapon for preventing disease, extending healthy life, and avoiding the infirmities associated with old age (Butler et al. 2008; Goldman et al. 2013);

4. Failing to take this new approach could leave people who reach older ages in the future, even more vulnerable to rising disability than they are now;

5. Aging science represents a new paradigm of primary prevention in public health that will lead to more effective methods of delaying most fatal and disabling diseases, extending healthy life, and reducing the prevalence of infirmities more commonly experienced at older ages (Sierra et al. 2009; Kirkland 2013; Tchkonia et al. 2013).

Language used to describe the longevity dividend must be unambiguous. Much like the introduction of antibiotics in the mid-20th century and the broad dissemination of basic measures of public health a century ago, humanity is once again fortunate enough to witness the rise of a new paradigm in human health. Aging science has successfully turned the spotlight on the origins of our aging bodies and minds and the fatal and disabling diseases that accompany us in our later years. What the scientific study of aging reveals shakes up a long-held assumption that aging is an inevitable and immutable by-product of the passage of time (Miller 2002), and these new discoveries fundamentally challenge the fatalist view that aging and death are nature's way of removing the old to make way for the young.

Science has now shown that aging is inherently modifiable. Furthermore, there is now reason to believe that aging science can be translated into new more effective medical and public health interventions that will be able to combat fatal and disabling diseases far more effectively than any intervention available today—yielding an extension of the period of healthy life in ways that could not even be conceived of just a few years ago.

Although people that benefit from advances in aging science will probably live longer, it is the extension of healthy life that is the primary goal along with reductions in the infirmities of old age, and increased economic value to individuals and societies, that would accrue from the extension of healthy life.

It is only a matter of time before aging science acquires the same level of prestige and confidence that medicine and public health now enjoy, and when that time comes, a new era in human health will emerge. There are an abundance of formidable obstacles standing in the way, including strongly held views of how to proceed, a history of association with dubious aging interventions, and misconceptions about the goals in mind and the impact of success on population growth and the environment. Once the air clears and aging science is translated into effective and safe interventions that can be measured and documented to extend our healthy years, the 21st century will bear witness to one of the most important new developments in the history of medicine.

ACKNOWLEDGMENTS

The MacArthur Foundation Research Network on an Aging Society supports this work. Earlier versions of this manuscript were published in *Public Policy & Aging Report* (Goldman and Olshansky 2013; Olshansky 2013).

REFERENCES

Anton T. 2013. *The longevity seekers science, business, and the fountain of youth.* University of Chicago Press, Chicago.

Beard JR, Biggs S, Bloom DE, Fried LP, Hogan P, Kalache A, Olshansky SJ, eds. 2011. *Global population ageing: Peril or promise.* World Economic Forum, Geneva.

Butler RN, Miller RA, Perry D, Carnes BA, Williams TF, Cassel C, Brody J, Bernard MA, Partridge L, Kirkwood T, et al. 2008. New model of health promotion and disease prevention for the 21st century. *BMJ* **337:** 149–150.

General Accounting Office. 2001. Health products for seniors. "Anti-aging" products pose potential for physical and economic harm, *GAO-01-1129.* United States General Accounting Office, Washington, DC.

Goldman DP, Olshansky SJ. 2013. Delayed aging versus delayed disease: A new paradigm for public health. *Public Policy Aging Rep* **23:** 16–18.

Goldman DP, Cutler D, Rowe JW, Michaud PC, Sullivan J, Peneva D, Olshansky SJ. 2013. Substantial health and economic returns from delayed aging may warrant a new focus for medical research. *Health Aff (Millwood)* **32:** 1698–1705.

Gruman G. 1966. A history of ideas about the prolongation of life. *Trans Am Phil Soc* **56:** 1–102.

Kirkland JL. 2013. Translating advances from the basic biology of aging into clinical application. *Exp Gerontol* **48:** 1–5.

Miller R. 2002. Scientific prospects and political obstacles. *Milbank Q* **80:** 155–174.

Miller R. 2012. Genes against aging. *J Gerontol A Biol Sci Med Sci* **67:** 495–502.

Olshansky SJ. 2013. Articulating the case for the longevity dividend. *Public Policy Aging Rep* **23:** 3–6.

Olshansky SJ, Perry D, Miller RA, Butler RN. 2006. In pursuit of the longevity dividend. *The Scientist* **20:** 28–36.

Sierra F, Hadley E, Suzman R, Hodes R. 2009. Prospects for life span extension. *Annu Rev Med* **60:** 457–469.

Tchkonia T, Zhu Y, van Deursen J, Campisi J, Kirkland JL. 2013. Cellular senescence and the senescent secretory phenotype: Therapeutic opportunities. *J Clin Invest* **123:** 966–972.

World Economic Forum. 2012. Global population ageing: Peril or promise? Global Agenda Council on an Ageing Society, Geneva, Switzerland.

Cite this article as *Cold Spring Harb Perspect Med* doi: 10.1101/cshperspect.a025940

The Economic Promise of Delayed Aging

Dana Goldman

USC Schaeffer Center for Health Policy and Economics, University of Southern California, Los Angeles, California 90089

Correspondence: dpgoldma@usc.edu

Biomedicine has made enormous progress in the last half century in treating common diseases. However, we are becoming victims of our own success. Causes of death strongly associated with biological aging, such as heart disease, cancer, Alzheimer's disease, and stroke, cluster within individuals as they grow older. These conditions increase frailty and limit the benefits of continued, disease-specific improvements. Here, we show that a "delayed-aging" scenario, modeled on the biological benefits observed in the most promising animal models, could solve this problem of competing risks. The economic value of delayed aging is estimated to be $7.1 trillion over 50 years. Total government costs, including Social Security, rise substantially with delayed aging—mainly caused by longevity increases—but we show that these can be offset by modest policy changes. Expanded biomedical research to delay aging appears to be a highly efficient way to forestall disease and extend healthy life.

"Moonshot" was the word Google founder and chief executive officer Larry Page used to describe Calico—a biotech start-up focused on the challenge of aging and age-associated pathology. In fact, the chances of success may be much greater.

Through rapid advances in biological research and medical innovation, the biomedical industry has made enormous strides in improving the health and lengthening the lives of Americans as a whole. Life expectancy at birth has risen from age 70 in 1960 to ~79 in 2015. The innovations in health care over the course of the century have led to a rapid drop in infant mortality. Fewer are dying from heart disease. Cancer survival rates are increasing. On the horizon, we see advances in medicine, genetics, and biomedical technology that may take us to new heights in slowing aging, preventing inherited diseases, and solving the health riddles that currently stump us.

These advances will leave the nation with an older society as more people live into their 80s, 90s, and even the 100s, and as the baby boomers enter old age. The potential costs of an aging society are well known, including rising health care costs at a time of shrinking government budgets. The solvency of Medicare is a perennial issue. Yet, thanks to health care and other advances, the health of the elderly population has also improved. In but one example, the nursing home population has declined along with disability rates. In 1985, 5.4% of the elderly were in nursing homes. By 1995, the share was down to 4.6%, on an age-adjusted basis (Cutler 2001).

Cite this article as *Cold Spring Harb Perspect Med* doi: 10.1101/cshperspect.a025072

If advances in science and health care continue, people may continue to reach older ages as healthier and more active individuals (Lowsky et al. 2013). As a result, they may work longer and use fewer health care resources, while continuing to give back to society in numerous ways.

THE ROLE OF TECHNOLOGY

The specter of an aging society regularly ignites debate in health policy circles on costs versus benefits. Is it right, some wonder, that older adults in their last years of life consume 24% of Medicare's expenditures? Or is it cost-effective to pay for a cancer treatment drug that will prolong an inevitable death for only a few months? In Britain, for example, the National Institute for Health and Clinical Excellence determines what therapies the National Health Service will cover. It generally recommends against paying for a therapy that costs more than $31,000 to $47,000 for each year of life gained, adjusted for quality.

Yet medical advances, although costly, are proving worth it in many instances, according to a vast body of research on cost effectiveness of medical care. There is good evidence that the benefits of health improvements dominate any additional costs for most new technologies. For example, life expectancy increased ~7 yr from 1960 to 2000 in most developed nations. During that span, the increases in medical spending have provided reasonable value, with the exception of spending increases in medical care for the elderly since 1980.

Even in some of the most difficult diseases, we have made good progress (Goldman et al. 2005; Shekelle et al. 2005). Cancer is a salient example, as the policy debates about cancer costs are frequently most pronounced (Experts in Chronic Myeloid Leukemia 2013). Many say we are losing the war on cancer that Richard Nixon first declared in 1971. After all, cancer is the second leading cause of death, and accounts for approximately one-fourth of all deaths in a year. However, the reality is much more positive. Today, cancer patients live longer, healthier, and happier lives than those in prior decades.

Survival rates for all cancers increased by almost 4 years from 1988 to 2000, creating 23 million additional life-years and generating $1.9 trillion in additional value to society, once the health gains are tallied (Lakdawalla et al. 2010; Sun et al. 2010; Goldman and Philipson 2014; Stevens et al. 2015). Survival rates have continued to improve in recent years. Compared with research and development spending—both private and public—one can easily see a substantial return on investment. Furthermore, progress is being made in dealing with the extreme toxicity of chemotherapy and radiation regimens (Hsu et al. 2013). So, although cancer still remains a pernicious disease, there is hope that it can eventually be managed as a chronic illness with modest side effects.

THE CHALLENGES OF THE SINGLE-DISEASE MODEL

In some ways though, we are becoming victims of our own success. Increased disability rates are now accompanying increases in life expectancy, leaving the length of a healthy life span unchanged (Bhattacharya et al. 2004; Lakdawalla et al. 2004, 2010; Crimmins and Beltran-Sanchez 2011) or even shorter than in the past (Hulsegge et al. 2014). As people age, they are now much less likely to fall victim to a single isolated disease than was previously the case. Instead, competing causes of death more directly associated with biological aging (e.g., heart disease, cancer, stroke, Alzheimer's, etc.) cluster within individuals as they reach older ages. These conditions elevate mortality risk, as well as create the frailty and disabilities that can accompany old age.

Fortunately, new research is emerging that has the potential to extend life while reducing the prevalence of comorbidities over the entire lifetime (Kirkland 2013; Tchkonia et al. 2013). Scientists have been asking whether we can decelerate the process by which the cluster of conditions described above arises, making people healthier at older ages and even lowering spending on health care (Fries 1980; Fries et al. 1993; Miller 2002; Martin et al. 2007; Butler et al. 2008; Sierra et al. 2009; Tchkonia et al. 2013).

 Cite this article as *Cold Spring Harb Perspect Med* doi: 10.1101/cshperspect.a025072

Simply put, can we age more slowly—thereby delaying the onset and progression of all fatal and disabling diseases simultaneously?

At the practical level, delayed aging means having the body and mind of someone who is years younger than the majority of today's population at one's chronological age and spending a larger proportion of one's life in good health and free from frailty and disability (Fries 1980; Vergara et al. 2004; Butler et al. 2008). Experimental studies involving animal models have already succeeded in accomplishing this in the laboratory (Miller 2002), and this collection is filled with the latest research developments in aging science, suggesting that a therapeutic intervention to delay aging is on the horizon. In addition, there is evidence that centenarians (whose longevity is at least partially heritable) often have delayed onset of age-related diseases and disabilities, which suggests that they senesce (grow old biologically) more slowly than the rest of the population (Lipton et al. 2010).

By manipulating genes, altering reproduction, reducing caloric intake, modulating the levels of hormones that affect growth and maturation, and altering insulin-signaling pathways, it has been possible to extend the life span—and the healthy life span—of invertebrates and mammals (Tatar et al. 2003; Sebastiani and Perls 2012; Kirkland 2013). These specific manipulations are unlikely to be directly applicable in humans, but they may lead scientists in the right direction.

In addition, clinical interventions to delay aging have been proposed that involve interfering with chronic inflammation. In mice, the selective removal of senescent cells has been documented to lead to significant improvements in health, an intervention that many researchers believe could be clinically effective in people (Tchkonia et al. 2013). Some scientists contend that such interventions are sufficiently close to fruition that people alive today will benefit from them (Miller 2002; Martin et al. 2007; Butler et al. 2008; Sierra et al. 2009; Kirkland 2013; Tchkonia et al. 2013). Should we continue on this path of discovery?

In deciding whether and how much society should continue to invest in delayed aging, two specific questions arise. First, what are the social returns—in terms of health and spending—on continued investments in a "disease model" versus the returns on investments in delayed aging? Second, can society afford to invest in the accelerated development of interventions that extend healthy life, given fiscal uncertainties? In this article, we compare the future health and economic benefits—as well as the costs—of continuing to prioritize the "disease model" with the benefits and costs of placing a new emphasis on delayed aging.

ECONOMIC SCENARIOS OF DELAYED AGING

We examined the economic benefits and costs of delayed aging, with a focus on the fiscal impact, in an article published in *Health Affairs* (Goldman et al. 2013). In what follows, we summarize these results. We specifically looked at the costs of major entitlement programs, specifically, Federal and State spending for Medicare and Medicaid, and Federal income support through old age, survivors, and disability insurance and supplemental security income. Economic outputs were aggregated into fiscally relevant variables using benefit rules for particular programs. Annual costs are given in constant 2010 dollars. All cumulative costs are discounted using a 3% annual discount rate (Gold et al. 1996).

We developed four scenarios (one representing the status quo or baseline) and compared the health and medical spending they would involve. For each scenario, we conducted the simulation 50 times and averaged the outcomes. We assumed that all changes were accomplished at no additional cost relative to baseline, to allow us to focus on population benefits. Each scenario assumed that changes in mortality and disease processes occurred in the period 2010–2030. The scenarios also assumed that progress ceased after 2030, but that the effects of earlier changes continued to play out.

Two disease-specific scenarios were meant to represent optimistic developments in medical research, disease treatment, and improvements in behavioral risk factors. In other words, these scenarios assumed that by attacking diseases individually through treatments or systemically through behavior modification, the incidence of disease, and the impact of cases of disease would be reduced.

The fourth scenario (assuming delayed aging) is a hypothetical assessment of a successful effort to translate research on the biology of aging into therapeutic interventions that would reduce and compress both morbidity and mortality into a shorter period of time at the end of life (Olshansky et al. 2009). Unlike the delayed disease interventions in the two disease-specific scenarios—which face diminishing returns because of competing causes of sickness and death in aging populations—the delayed aging scenario assumed that all fatal and disabling diseases were influenced simultaneously. Thus, this scenario represents what might best be thought of as a superefficient method of attacking the fatal and disabling diseases that are most prevalent at older ages—a form of primary prevention that would simultaneously influence all fatal and disabling diseases at once.

Status Quo

In the status quo (or baseline) scenario, we used the mortality forecasts for all-cause mortality in the intermediate projections of the Social Security Administration (The Board of Trustees of Federal Old-Age and Survivors Insurance and Federal Disability Insurance Trust Funds, see www.socialsecurity.gov/oact/TR/2011/tr2011.pdf). We did not change the incidence of disease. Heuristically, in this scenario, mortality improvements can be seen as the result of improved treatments for people with disease.

Delayed Cancer

In the first disease-specific scenario, we modified the status quo scenario by reducing the incidence of cancer over time. From 2010 to 2030, we phased in a linear 25% reduction in cancer incidence. We assumed that this change was accomplished at no additional Medicare cost relative to the baseline. Historical evidence suggests that there was a reduction of 1.3% in overall cancer incidence rates for men per year from

2000 to 2006 and a reduction of 0.5% for women per year from 1998 to 2006 (Jemal et al. 2010). Averaged over 20 years, these trends yield a range of reductions in cancer incidence of 10% to 26%. Thus, our assumptions in this scenario were within the bounds of the observed trends. We assumed that the reduced incidence rate continued until the end of the simulation. To account for improvements in health before age 51, the prevalence of chronic conditions in the incoming cohorts of 51-yr-olds was adjusted to match the prevalence for 44-yr-olds in the target year, as measured in the National Health Interview Survey.

Delayed Heart Disease

We modified the status quo scenario by reducing the incidence of heart disease over time. As was the case with cancer, we assumed a linear reduction in the incidence of 25% between 2010 and 2030, and no change thereafter (Kubo et al. 2003). And, again as in the delayed cancer scenario, we assumed that there was no additional Medicare cost, and we adjusted the prevalence of chronic conditions in the incoming cohorts.

Delayed Aging

We assumed that improvements in mortality and health started earlier in life than they did in the disease-specific scenarios. We assumed that the slope of the intrinsic mortality curve— that is, mortality from factors such as age, as opposed to exposure to external risks such as trauma or smoking—observed in 2000 for both men and women ages 15 to 50 would decline by 20% by 2050. These hypothesized changes are consistent with research on the biology of aging, which suggests that the health benefits of delayed aging would begin at puberty—the time when mortality begins rising exponentially (de Magalhaes et al. 2005; Edlin and Stiglitz 2012).

Delayed Aging with an Eligibility Fix

We modeled a variant of the delayed aging scenario that included an adjustment to the eligibility age for Medicare and the normal retirement age for Social Security. Social Security provides a strong precedent for such a policy fix. The statutory full retirement age was raised in 1983 from 65 to 66, and the age will increase to 67 for people born in 1960 and later. Our "eligibility fix" consisted of a gradual increase in the eligibility age for Medicare from 35 to 38, and for Social Security from 67 to 68 (extending the Social Security amendments of 1983, which mandated gradual increases in the retirement over a 22-yr period starting in 2000, for ~10 yr). In this scenario, people enrolled in Medicare Part A as soon as they were eligible to do so. The delayed aging scenario with the eligibility fix—because of the later official statutory retirement age—would result in more taxes collected during working years than in the original delayed aging scenario without an eligibility change, and less lifetime benefits paid because of the later start of retirement.

The impact of the changing rates of transition of disease and functional status can be seen in the change in average life-cycle characteristics. Life expectancy at age 51 in 2030 was 35.8 yr in the status quo scenario, based on current Social Security Administration projections (see www.socialsecurity.gov/oact/TR/2011/tr 2011.pdf). It improved by ~1 yr in both the delayed cancer (36.9 yr) and delayed heart disease (36.6 yr) scenarios. In the delayed aging scenario, however, it increased to 38.0 yr, an improvement of 2.2 yr (Fig. 1) (Sullivan et al. 2013).

Under the status quo, the number of elderly people—those age 65 or older—in the United States more than doubled, increasing from 43 million in 2010 to 106 million in 2060. The scenarios of delayed cancer and delayed heart disease diverged little from the first scenario, leading to only 0.8% and 2.0% more elderly people in 2060, respectively. In contrast, the delayed aging scenario added 6.9% more elderly people. These demographic gains would occur quickly, with 6.1% more elderly Americans than in the status quo scenario after only 20 yr.

Of course, it matters whether these survivors would be healthy or disabled. In the status quo scenario, 31.0 million people age 65 or older were not disabled in 2010; the number was 75.5 million in 2060 (Fig. 2) (Manton et al. 1997; Freedman et al. 2004). In the dis-

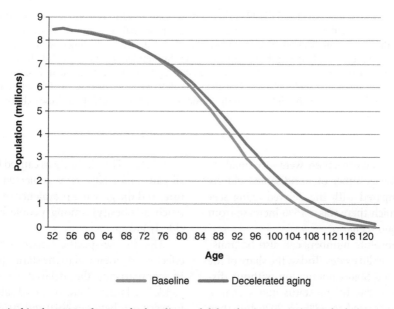

Figure 1. Survival in the 2030 cohort under baseline and delayed aging scenarios. Calculations are derived from the Future Elderly Model (see www.rand.org/pubs/research_briefs/RB9324/index1.html).

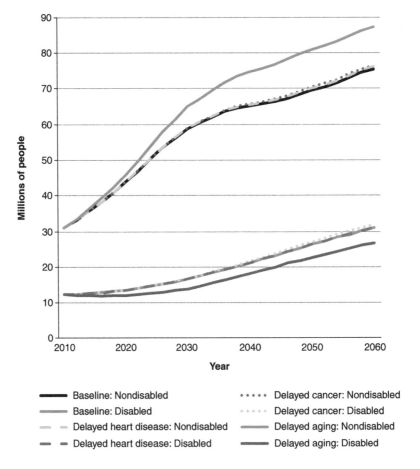

Figure 2. Millions of nondisabled and disabled elderly Americans in various scenarios, 2010–2060. The figure shows the number of elderly Americans (age 65 or older) projected to be either nondisabled or disabled according to the various medical progress scenarios. Disabled is defined as having one or more limitations in instrumental activities of daily living, having one or more limitations in activities of daily living, living in a nursing home, or a combination of the three. The delayed aging scenario resulted in a substantially higher percentage and number of nondisabled people than the delayed heart disease or delayed cancer scenario. Calculations are derived from the Future Elderly Model (see www.rand.org/pubs/research_briefs/RB9324/index1.html).

ease-specific scenarios, there were very small increases in the number of nondisabled elderly people compared with the delayed aging scenario, in which there was a 15% increase from the status quo scenario.

These absolute numbers can also be translated into disability rates. Today, the share of the elderly United States population without disabilities is ~72%. In the status quo scenario, this share increased to 78% in 2026 but then declined to 71% in 2060 (Fig. 3) (Sullivan

et al. 2013). This decline was caused by the lower all-cause mortality rates projected for the future, and the growing prevalence of health risks (such as obesity) among people entering the elderly group.

The disease-specific scenarios both had an effect nearly identical to the status quo scenario. In comparison, the delayed aging scenario yielded a larger share of nondisabled seniors in every year between 2010 and 2026, compared with the status quo scenario. Although the size

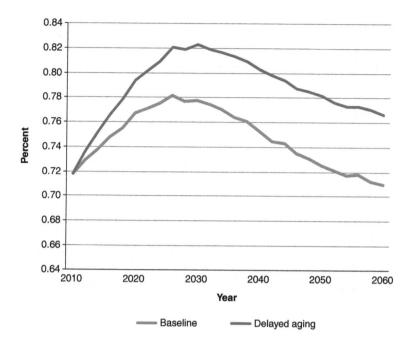

Figure 3. Fraction of 65 and older population without disability. Calculations are derived from the Future Elderly Model (see www.rand.org/pubs/research_briefs/RB9324/index1.html).

of the difference declined from 2030 to 2060, during that 30-yr period, an additional 5% of elderly people were nondisabled in the delayed aging scenario. Per capita Medicare spending was also lower in the delayed aging scenario than in the status quo scenario.

On the population level, the aggregate costs show the fiscal strain imposed by delayed aging (Fig. 4). In that scenario, more elderly people were alive. Consequently, more people qualified for entitlement programs, and costs were higher. In 2060, spending in the delayed aging scenario was $295 billion more than in the status quo scenario. In contrast, the delayed cancer scenario led to only a modest increase, and the delayed heart disease scenario brought spending below the level in the status quo scenario.

The gap in income support was also considerable. Spending beyond that in the status quo scenario was relatively low in the disease-specific scenarios (Fig. 5). In comparison, it climbed to around $125 billion in the delayed aging scenario by 2055. Delayed aging would add nearly $420 billion to the entitlement deficit in the

status quo scenario in 2060, 70% of which would come from increased outlays for Medicare and Medicaid.

Figure 6 shows the fiscal effects of the four main scenarios as well as the effect of delayed aging with the eligibility fix to Medicare and Social Security described above. The eligibility fix would more than offset the additional costs of delayed aging relative to the costs of the status quo scenario.

THE VALUE OF DELAYED AGING

Our results show that shifting the focus of medical investment to delayed aging would lead to a set of desirable, but economically challenging, circumstances. The potential gains are significant. Although the disease model has reduced mortality from lethal conditions dramatically in the past century, its influence is now waning because of competing risks. As people live longer, they are more likely to fall victim to multiple diseases. Our simulations of reduced incidence of heart disease and cancer suggested incrementally smaller gains in longevity going forward.

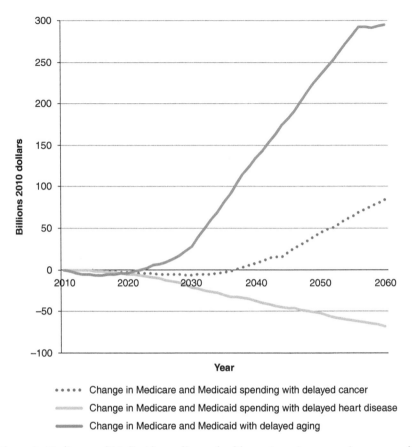

Figure 4. Change in Medicare and Medicaid spending on health care in various scenarios compared with status quo, billions of dollars, 2010–2060. All spending is in 2010 dollars. The figure shows per period (nondiscounted) projected spending on Medicare and Medicaid under various medical progress scenarios, relative to the status quo scenario for Americans aged 51 or older. Spending is much higher in the delayed aging scenario because of the larger increase in the total population, even though per period costs for Medicare are lower. Calculations are derived from the Future Elderly Model (see www.rand.org/pubs/research_briefs/RB9324/index1.html).

The medical costs of treating these diseases independently would rise but, for example, would produce only a 3.2-yr increase in life expectancy for 65-yr-olds from 2010 to 2060 (The Board of Trustees of Federal Old-Age and Survivors Insurance and Federal Disability Insurance Trust Funds, see www.socialsecurity.gov/oact/TR/2011/tr2011.pdf).

Recent research has shown that the decades-long improvement in the functional status of older Americans halted in 2002 (Lakdawalla et al. 2004; Crimmins and Beltran-Sanchez 2011). This suggests that many of the historical drivers of better health in the elderly may no longer work. Declining disability buttresses the case for research on slowing aging by compressing morbidity and extending healthy life, which would provide an adequate workforce for producing the goods and services that the future aging society would use, and would yield direct benefits to those older people who remain socially engaged.

Still, the fact remains that longer lives would mean greater fiscal burdens for Social Security and other income support programs and increased Medicare and Medicaid expenditures, even as per capita medical costs declined. An unequivocal answer to the question of whether

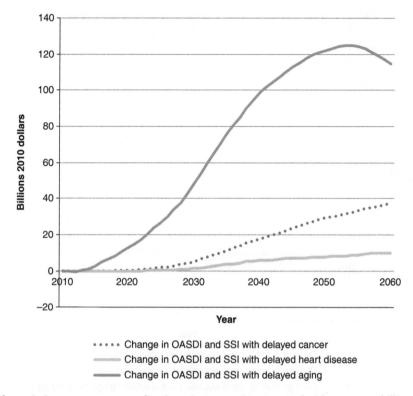

Figure 5. Change in income support spending in various scenarios compared with status quo, billions of dollars, 2010–2060. All spending is in 2010 dollars. The figure shows per period (nondiscounted) projected spending on income support under various medical progress scenarios, relative to the status quo scenario. Income support includes all Old-Age and Survivors Insurance, Social Security Disability Insurance (OASDI) and Supplemental Security Income (SSI) spending on Americans age 51 or older. Spending is much higher under the delayed aging scenario because of the larger increase in the total population. Calculations are derived from the Future Elderly Model (see www.rand.org/pubs/research_briefs/RB9324/index1.html).

the current focus of medical research and investment should be shifted from the disease model to delayed aging depends on whether the potential gains could be realized and the adverse consequences allayed. One way to reflect on the future gains is to look at the presented discounted value of all the additional, quality-adjusted life-years that arise from delayed aging relative to the status quo. These can then be valued using a conservative metric such as $100,000 per life-year. Doing so yields a social benefit related to delayed aging of ~$7.1 trillion, without even considering the cognitive benefits that could arise from these interventions (Christensen et al. 2013).

Given the large social return, the question then becomes how we accommodate these changes fiscally. Several policy measures might achieve fiscal balance; we show one involving eligibility changes, but a full evaluation of the options is beyond the scope of this research. However, we note here one benefit of delayed aging that might enlarge the set of possibilities. With people staying healthy until a much later age, it might be more feasible to justify raising the eligibility age for public programs for seniors. Arguments against doing so often note that life expectancy increases in lower socioeconomic groups have lagged far behind those in better-off groups (Ketcham and Simon 2008; Kindig and Cheng 2013). A future in which delayed aging increased the health of all socioeconomic groups would make these increases in eligibility ages more palatable.

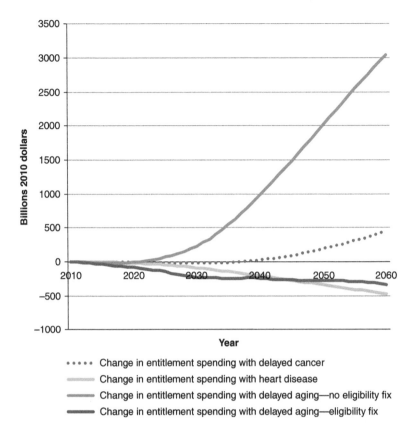

Figure 6. Change in major entitlement spending in various scenarios compared with status quo, billions of dollars, 2010–2060. All spending is in 2010 dollars. The figure shows the cumulative fiscal impact of spending on Medicare, Medicaid, Old-Age and Survivors Insurance, Social Security Disability Insurance (OASDI), and Supplemental Security Income (SSI) (discounted at 3%) of various medical progress scenarios relative to the status quo scenario. The fix is a gradual increase in the eligibility ages for Medicare and Social Security. The inclusion of the eligibility fix would result in no additional entitlement spending relative to that in the status quo scenario, despite much larger increases in the older population. Calculations are derived from the Future Elderly Model (see www.rand.org/pubs/research_briefs/RB9324/index1.html).

CONCLUSION

It is clear that competing health risks limit the impact of major clinical breakthroughs for specific diseases. In other words, making progress against one disease means that another one will eventually emerge in its place. However, evidence suggests that if aging is delayed, all fatal and disabling disease risks would be lowered simultaneously.

The potential economic benefits of delayed aging are enormous. We find that realizing the promise of the current biological models might net society >$7 trillion in net benefits. (This

windfall may explain why the private sector has become increasingly keen to research these mechanisms.) The major challenges of delayed aging appear to be of a fiscal nature, but they are manageable. The benefits to society of delayed aging would accrue rapidly and would extend to all future generations. Investing in research to delay aging should become a priority.

More generally, innovations to prevent disease have enormous economic salience, even at older ages. However, our current health system rewards treatment at the expense of prevention. This has likely led to dramatic underinvestment in forestalling disease onset. Clearly, there is a

 Cite this article as *Cold Spring Harb Perspect Med* doi: 10.1101/cshperspect.a025072

public role to play in tilting the playing field back from treatment to prevention, and as evidence presented here indicates, delayed aging could very well become the most efficient method of primary prevention available.

REFERENCES

Bhattacharya J, Cutler DM, Goldman DP, Hurd MD, Joyce GF, Lakdawalla DN, Panis CW, Shang B. 2004. Disability forecasts and future Medicare costs. *Front Health Policy Res* **7**: 75–94.

Butler RN, Miller RA, Perry D, Carnes BA, Williams TF, Cassel C, Brody J, Bernard MA, Partridge L, Kirkwood T, et al. 2008. New model of health promotion and disease prevention for the 21st century. *BMJ* **337**: a399.

Christensen K, Thinggaard M, Oksuzyan A, Steenstrup T, Andersen-Ranberg K, Jeune B, McGue M, Vaupel JW. 2013. Physical and cognitive functioning of people older than 90 years: A comparison of two Danish cohorts born 10 years apart. *Lancet* **382**: 1507–1513.

Crimmins EM, Beltran-Sanchez H. 2011. Mortality and morbidity trends: Is there compression of morbidity? *J Gerontol B Psychol Sci Soc Sci* **66**: 75–86.

Cutler DM. 2001. Declining disability among the elderly. *Health Affairs* **20**: 11–27.

Experts in Chronic Myeloid Leukemia. 2013. The price of drugs for chronic myeloid leukemia (CML) is a reflection of the unsustainable prices of cancer drugs: From the perspective of a large group of CML experts. *Blood* **121**: 4439–4442.

Freedman VA, Crimmins E, Schoeni RF, Spillman BC, Aykan H, Kramarow E, Land K, Lubitz J, Manton L, Martin LG, et al. 2004. Resolving inconsistencies in trends in old-age disability: Report from a technical working group. *Demography* **41**: 417–441.

Fries JF. 1980. Aging, natural death, and the compression of morbidity. *N Engl J Med* **303**: 130–135.

Fries JF, Koop CE, Beadle CE, Cooper PP, England MJ, Greaves RF, Sokolov JJ, Wright D. 1993. Reducing health care costs by reducing the need and demand for medical services. *N Engl J Med* **329**: 321–325.

Gold M, Siegel J, Russell L, Weinstein M. 1996. *Cost-effectiveness in health and medicine.* Oxford University Press, New York.

Goldman DP, Philipson T. 2014. Five myths about cancer care in America. *Health Aff (Millwood)* **33**: 1801–1804.

Goldman DP, Shang B, Bhattacharya J, Garber AM, Hurd M, Joyce GF, Lakdawalla DN, Panis C, Shekelle PG. 2005. Consequences of health trends and medical innovation for the future elderly. *Health Aff (Millwood)* **24**: W5-R5–W5-R17.

Goldman DP, Cutler D, Rowe JW, Michaud PC, Sullivan J, Peneva D, Olshansky SJ. 2013. Substantial health and economic returns from delayed aging may warrant a new focus for medical research. *Health Aff (Millwood)* **32**: 1698–1705.

Hsu T, Ennis M, Hood N, Graham M, Goodwin PJ. 2013. Quality of life in long-term breast cancer survivors. *J Clin Oncol* **31**: 3540–3548.

Hulsegge G, Picavet HS, Blokstra A, Nooyens AC, Spijkerman AM, van der Schouw YT, Smit HA, Verschuren WM. 2014. Today's adult generations are less healthy than their predecessors: Generation shifts in metabolic risk factors: The Doetinchem cohort study. *Eur J Prev Cardiol* **21**: 1134–1144.

Jemal A, Siegel R, Xu J, Ward E. 2010. Cancer statistics, 2010. *CA Cancer J Clin* **60**: 277–300.

Ketcham JD, Simon KI. 2008. Medicare Part D's effects on elderly patients' drug costs and utilization. *Am J Manag Care* **14**: SP14–SP22.

Kindig DA, Cheng ER. 2013. Even as mortality fell in most US counties, female mortality nonetheless rose in 42.8 percent of counties from 1992 to 2006. *Health Aff (Millwood)* **32**: 451–458.

Kirkland JL. 2013. Translating advances from the basic biology of aging into clinical application. *Exp Gerontol* **48**: 1–5.

Kubo M, Kiyohara Y, Kato I, Tanizaki Y, Arima H, Tanaka K, Nakamura H, Okubo K, Iida M. 2003. Trends in the incidence, mortality, and survival rate of cardiovascular disease in a Japanese community: The Hisayama study. *Stroke* **34**: 2349–2354.

Lakdawalla DN, Bhattacharya J, Goldman DP. 2004. Are the young becoming more disabled? *Health Aff (Millwood)* **23**: 168–176.

Lakdawalla DN, Sun EC, Jena AB, Reyes CM, Goldman DP, Philipson TJ. 2010. An economic evaluation of the war on cancer. *J Health Econ* **29**: 333–346.

Lipton RB, Hirsch J, Katz MJ, Wang C, Sanders AE, Verghese J, Barzilai N, Derby CA. 2010. Exceptional parental longevity associated with lower risk of Alzheimer's disease and memory decline. *J Am Geriatr Soc* **58**: 1043–1049.

Lowsky DJ, Olshansky SJ, Bhattacharya J, Goldman DP. 2013. Heterogeneity in healthy aging. *J Gerontol A Biol Sci Med Sci* **69**: 640–649.

Manton KG, Corder L, Stallard E. 1997. Chronic disability trends in elderly United States populations: 1982–1994. *Proc Natl Acad Sci* **94**: 2593–2598.

Martin GM, Bergman A, Barzilai N. 2007. Genetic determinants of human health span and life span: Progress and new opportunities. *PLoS Genet* **3**: e125.

Miller RA. 2002. Extending life: Scientific prospects and political obstacles. *Milbank Q* **80**: 155–174.

Olshansky SJ, Goldman DP, Zheng Y, Rowe JW. 2009. Aging in America in the twenty-first century: Demographic forecasts from the MacArthur Foundation Research Network on an Aging Society. *Milbank Q* **87**: 842–862.

Sebastiani P, Perls TT. 2012. The genetics of extreme longevity: Lessons from the New England Centenarian Study. *Front Genet* **3**: 277.

Shekelle PG, Ortiz E, Newberry SJ, Rich MW, Rhodes SL, Brook RH, Goldman DP. 2005. Identifying potential health care innovations for the future elderly. *Health Aff (Millwood)* **24**: W5-R67–W5-R76.

Sierra F, Hadley E, Suzman R, Hodes R. 2009. Prospects for life span extension. *Annu Rev Med* **60**: 457–469.

Stevens W, Philipson TJ, Khan ZM, MacEwan JP, Linthicum MT, Goldman DP. 2015. Cancer mortality reductions were greatest among countries where cancer care spending rose the most, 1995–2007. *Health Aff (Millwood)* **34:** 562–570.

Sullivan J, Goldman D, Michaud PC, Peneva D. 2013. Online supplement for "Attacking diseases by slowing aging: Health and economic implications." University of Southern California, Los Angeles.

Sun E, Jena AB, Lakdawalla D, Reyes CM, Philipson T, Goldman D. 2010. The contributions of improved therapy and earlier detection to cancer survival gains, 1988–2000. *Forum Health Econ Policy* **13:** 1–20.

Tatar M, Bartke A, Antebi A. 2003. The endocrine regulation of aging by insulin-like signals. *Science* **299:** 1346–1351.

Tchkonia T, Zhu Y, van Deursen J, Campisi J, Kirkland JL. 2013. Cellular senescence and the senescent secretory phenotype: Therapeutic opportunities. *J Clin Invest* **123:** 966–972.

Vergara M, Smith-Wheelock M, Harper JM, Sigler R, Miller RA. 2004. Hormone-treated snell dwarf mice regain fertility but remain long lived and disease resistant. *J Gerontol A Biol Sci Med Sci* **59:** 1244–1250.

Cite this article as *Cold Spring Harb Perspect Med* doi: 10.1101/cshperspect.a025072

Past, Present, and Future of Healthy Life Expectancy

Hiram Beltrán-Sánchez[1], Samir Soneji[2], and Eileen M. Crimmins[3]

[1]Center for Demography of Health and Aging, University of Wisconsin-Madison, Madison, Wisconsin 53706

[2]Dartmouth Institute for Health Policy & Clinical Practice, Geisel School of Medicine Dartmouth College, Lebanon, New Hampshire 03756

[3]Davis School of Gerontology, University of Southern California, Los Angeles, California 90089-0191

Correspondence: beltrans@ssc.wisc.edu

The success of the current biomedical paradigm based on a "disease model" may be limited in the future because of large number of comorbidities inflicting older people. In recent years, there has been growing empirical evidence, based on animal models, suggesting that the aging process could be delayed and that this process may lead to increases in life expectancy accompanied by improvements in health at older ages. In this review, we explore past, present, and future prospects of healthy life expectancy and examine whether increases in average length of life associated with delayed aging link with additional years lived disability-free at older ages. Trends in healthy life expectancy suggest improvements among older people in the United States, although younger cohorts appear to be reaching old age with increasing levels of frailty and disability. Trends in health risk factors, such as obesity and smoking, show worrisome signs of negative impacts on adult health and mortality in the near future. However, results based on a simulation model of delayed aging in humans indicate that it has the potential to increase not only the length of life but also the fraction and number of years spent disability-free at older ages. Delayed aging would likely come with additional aggregate costs. These costs could be offset if delayed aging is widely applied and people are willing to convert their greater healthiness into more years of work.

How long we live and what proportion of that life is spent in good health have important implications for individuals and societies. The implications for individuals span a wide range of possibilities including potential social burden of caregiving from surviving family members, valuing life insurance premiums, and adequacy of retirement benefits and savings. The societal effects include a changing dependency ratio (the ratio of dependent [older] to independent [younger] people), which has major consequences on the fiscal viability of social transfer programs, such as Social Security and Medicare, and the size and demographic composition of the workforce.

Average years of life (life expectancy) have continuously increased in most countries over the last century with no apparent plateau (Vaupel 2010). In low-mortality countries, most of the recent rise in life expectancy has been attrib-

Cite this article as *Cold Spring Harb Perspect Med* doi: 10.1101/cshperspect.a025957

uted to declining mortality rates at older ages (Rau et al. 2008). Whether additional years of life are also accompanied by years in good health has become a subject of intense interest. Many disciplines contribute answers to this question, and several frameworks for assessing healthy aging have emerged (Gruenberg 1977; Fries 1980; Manton 1982). Recent developments in the biology of aging suggest that the aging process could be delayed (Kirkwood and Austad 2000; Sierra et al. 2009; Miller 2012) and that this process may lead to faster increases in life expectancy accompanied by improvements in health at older ages (Goldman et al. 2013). Unlike current medical and health care policy approaches that typically focus on reducing progression and lethality of major chronic diseases one by one, delayed aging focuses on postponing age-dependent deterioration in dividing cells, nondividing cells, cell parts, and extracellular materials (Miller 2012). As a result, delayed aging has the potential to postpone both physiological deterioration and comorbidities over the life cycle, and extend healthy years of life (Goldman et al. 2013). If delayed aging occurs among populations as posited, the study of healthy aging may require either revamping standard theories or formulating new ones.

In this article, we provide a general overview on trends in healthy life expectancy in the United States and other high-income countries and further elaborate on the implications of delayed aging for the future of healthy life expectancy. In particular, we examine whether increases in average length of life associated with delayed aging link with additional years lived disability-free at older ages. The review is structured as follows: We describe empirical evidence on healthy life expectancy to assess trends in the past and present. Next, we evaluate the implications of delayed aging for the future of healthy life expectancy and, finally, we discuss prospects of healthy life expectancy in the near future and conclude.

HEALTHY LIFE EXPECTANCY IN THE PAST AND PRESENT

Mortality trends in high-income countries between 1900 and 1950 showed a clear age-pattern shift. Mortality at young ages and from infectious conditions was rapidly receding, whereas mortality at older ages and from chronic conditions began to dominate (Omran 1971; Preston 1976). By the 1960s, major medical improvements in cardiovascular survival led to an increasing prevalence of heart disease at older ages. These developments focused attention on the morbidity as well as the mortality of the increasing older population. By the late 1970s and early 1980s, researchers had devised theoretical frameworks as well as markers of morbidity for assessing healthy aging. We briefly review three of these frameworks—failure of success, compression of morbidity, and dynamic equilibrium—that have guided significant amounts of research on healthy life expectancy in the last decades.

The first framework proposed by Gruenberg (1977) argued that declines in mortality from chronic disease would invariably lead to increase disease prevalence, which he termed "the failure of success." In his view, mortality declines would arise from higher survival of individuals with health problems thereby increasing disease prevalence and morbid life in the population. Others noted that the interaction of mortality declines with disease incidence (Fries 1980) and that disease progression and its severity (Manton 1982) had an important role for shaping the length of morbid life. The second framework developed by Fries (1980) introduced the idea of "compression of morbidity" in which he argued that the same forces that resulted in decreased mortality would be linked to lower incidence of chronic disease and higher age of onset of chronic disease resulting in a shortening of the length of morbid life. The third framework developed by Manton (1982) introduced the idea of "dynamic equilibrium" to highlight the link with disease progression and its severity. He argued that the severity and progression of chronic disease would change at the same pace as mortality improvements so that the progression of disease would be stopped at early stages, resulting in potentially more disease in the population but disease with decreased consequences.

Cite this article as *Cold Spring Harb Perspect Med* doi: 10.1101/cshperspect.a025957

Measuring Health at Older Adult Ages and Its Connection with Length of Life

Fries' hypothesis guided most empirical research since the 1980s. In his framework, he advocated the study of disability and functional mobility indicators as proxy markers to test compression of morbidity under the assumption that these indicators "represented" the morbidity status of the population. These indicators were initially developed in the 1970s by Nagi (1979), of which activities of daily living (ADLs)—eating, bathing, walking, toileting, and dressing by oneself—are the most commonly used, and adopted internationally by the World Health Organization (World Health Organization 1980) in the 1980s. They were further elaborated by Verbrugge and Jette (1994) under the framework of the "disablement process" in the early 1990s. Under this framework, disability is thought to be influenced by the interaction of physical ability (intraindividual) and environmental challenge (extraindividual), and the focus is on how chronic and acute conditions affect critical physical functions, such as ADLs (Verbrugge and Jette 1994). In addition to these internal and external factors, disability levels are also affected by the social roles used to define disability, and the environment in which it is measured. That is, disability is considered to be the end result of a pathologic process, and so the framework's goal is to assess the trajectory of functional consequences over time and the factors that may affect it. This approach has guided the majority of research on healthy life expectancy since the mid-1990s.

Since the 1980s, many national health surveys have begun collecting biological markers that appeared to be better suited for assessing underlying physiological damage as precursors of overt morbidity. These markers are thought to represent a latent trait of functioning of major organ systems and their physiological processes. The use of these markers in social science research is quite recent (since the early 2000s), and has opened a new gate of possibilities for studying healthy aging, and it is becoming the guide for current and future research on linking mortality improvements with health (Goldman et al. 2006, 2009; Crimmins et al. 2009, 2010). Most of these biomarkers were initially developed to assess individuals' risk of cardiovascular events (e.g., Framingham Risk Score) and cardiometabolic status (e.g., metabolic syndrome). Additional composite indexes have been created to incorporate a broader array of physiological factors linked with highly prevalent health outcomes at older ages, such as markers of stress and disease accumulation (e.g., allostatic load) (Seeman et al. 1997) and markers of inflammation. For instance, recent evidence suggests that inflammatory markers related to cardiovascular and metabolic disease, such as interleukin-6 (IL-6) and soluble intercellular adhesion molecule 1 (sICAM-1), are linked with survival among middle-aged and older adults (Crimmins et al. 2010; Glei et al. 2014).

Measuring morbidity at older ages requires data on individuals at multiple points in time to assess changes in health (e.g., transition probabilities) as they reach older ages. This data is difficult to come by at the national level except for a handful of longitudinal studies, such as the Health and Retirement Study (HRS) in the United States. Thus, most evidence on morbidity indicators comes from cross-sectional surveys in the form of prevalence rates, with a few exceptions (for example, see Cai and Lubitz 2007; Crimmins et al. 2009; Cai et al. 2010). The typical approach for estimating healthy life expectancy is the Sullivan method (Sullivan 1971)—a technique that allocates years of life into years lived with and without morbidity based on prevalence rates. Thus, trends in healthy life expectancy are largely driven by prevalence rates in a given morbidity indicator. In the following sections, we provide a brief overview on past and current trends of the most commonly used morbidity indicators when they are measured by traditional disability and functional mobility as well as those related to chronic disease and biomarkers of health.

Functional Mobility-Based Indicators

Empirical evidence on past and current trends in disability and functional mobility indicators is mixed. In the United States, there is evidence that ADL disability prevalence declined among those older than age 65 until the 1990s (Freedman et al. 2013), with a decline in the severity of ADL disability among people aged 65 and older between 1992 and 2002, a decline in the prevalence of people unable to complete at least three ADLs, but no significant change in moderate disability (disabled in one or two ADLs) (Cai and Lubitz 2007). Data from Medicare Current Beneficiary Survey (Cutler et al. 2013) and from the Health and Retirement Study (Smith et al. 2013) show that ADL disability is increasingly compressed within the last 2 years before death. Other research, however, does not indicate declines in disability. Data from early 2000s show stagnation and even deterioration in mobility functioning and disability (Crimmins and Beltrán-Sánchez 2011; Freedman et al. 2013). Additional evidence indicates increasing disability rates in recent years among younger American adults aged 40–64 years (Seeman et al. 2010). Similarly, results from two major studies of aging, the Longitudinal Studies of Aging and the Medicare Current Beneficiary Survey, show no changes in age at onset of disability between 1984 and 1994 (Crimmins et al. 1994) and between 1992 and 2002 (Cai and Lubitz 2007). In Europe, trends in disability[4] at age 65 are similarly mixed with some studies indicating increases in disability rates in nine out of 13 European countries,[5] with reductions in two countries (Austria and Italy) and stable rates in two countries (Belgium and Spain) (European Health Expectancy Monitoring Unit 2009; Bowling 2011). Finally, a large-scale study of 187 countries from the global burden of disease indicates that as life expectancy has increased between 1990 and 2010, the number of healthy years lost to disability has also increased (Salomon et al. 2012).

---

[4]Disability is estimated from the question: "Are you hampered in your daily activities by any physical or mental health problem, illness or disability"?

[5]Countries included: Austria, Belgium, Denmark, Finland, France, Germany, Greece, Ireland, Italy, the Netherlands, Portugal, Spain, and the United Kingdom.

Chronic Disease Indicators and Health Risk Factors

Arthritis is the leading cause of disability in the United States, especially common among adults with multiple chronic conditions (Barbour et al. 2013). In 2005, 19% of adults who reported disabilities indicated arthritis or rheumatism as the main cause of their disability (Brault et al. 2013). Among the 53 million adults who reported doctor-diagnosed arthritis, 23 million also reported arthritis-attributable activity limitation. Arthritis is an especially common chronic condition among adults with heart disease, diabetes, and obesity. For example, 49% of adults with heart disease also had arthritis between 2010 and 2012 (Centers for Disease Control and Prevention 2013).

In addition to arthritis, major chronic conditions and cognitive impairments appear to be on the rise among U.S. older adults. Nearly half of all Medicare beneficiaries during the 1990s and early 2000s received care for at least one of the following: cancer, chronic kidney disease, chronic obstructive pulmonary disease, depression, diabetes, or heart failure (Schneider et al. 2009). Moreover, it is estimated that the prevalence of dementia in the early 2000s at ages 71+ was ~14% and ~10% for Alzheimer's disease specifically (Plassman et al. 2007), although some substantial portion of the increase is likely attributable to better diagnosis. Finally, lack of access to health insurance may exacerbate the future consequences on the health status of older adults. A substantial proportion of working-age adults with chronic conditions—who are not yet old enough to receive Medicare benefits—are uninsured (Wilper et al. 2008). These uninsured and chronically ill adults were less likely to visit a health professional and have a standard site of care (other than the emergency department) compared with their insured counterparts.

These patterns are not unique to the United States. Globally, chronic diseases are the leading cause of mortality and morbidity (World Health Organization 2002; Yach et al. 2004). The global prevalence of chronic conditions is projected to increase substantially over the next decade as the leading causes of death and dis-

Cite this article as *Cold Spring Harb Perspect Med* doi: 10.1101/cshperspect.a025957

ability shift from communicable disease to noncommunicable chronic diseases as described by the "epidemiological transition." Cardiovascular disease, cancer, and diabetes are among the chronic conditions projected to increase the most (Yach et al. 2004). Behavioral risk factors including alcohol abuse, tobacco use, and obesity contribute substantially to disability (Salomon et al. 2012). For example, 58% of diabetes, 21% of ischemic heart disease, and between 8% and 42% of cancers were attributable to obesity (body mass index ≥ 21 kg/m^2).

Physiological Status Indicators

Adverse levels in biomarkers of health slowly develop into chronic conditions over the individual's life cycle. There is little evidence from recent trends in markers of cardiometabolic risk of improvements in health as people approach old age. Trends in physiological indicators representing average functioning of multiple bodily systems indicate a deterioration in recent years in some markers of inflammation and glucose levels (diabetes indicator), but improvements in average lipid levels and markers of cardiovascular health (e.g., hypertension) (Crimmins et al. 2010; Beltrán-Sánchez et al. 2013). From the late 1980s to about 2005, time trends were stable for C-reactive protein (CRP), a marker of inflammation, and for glycosylated hemoglobin, a marker for diabetes, among people aged 65 and older (Crimmins et al. 2010). In the same time period, there were also reductions in the prevalence of high-risk cholesterol level and hypertension (Crimmins et al. 2010). Among younger adults aged 40 to 64, some evidence indicates increasing prevalence of CRP among males and higher levels of glycosylated hemoglobin for females between 1999 and 2006 (Martin et al. 2010). Importantly, declines in lipid levels and hypertension appeared to be driven by increased use and efficacy of medications, rather than reductions in the incidence of these conditions (Beltrán-Sánchez et al. 2013). For example, among the adult U.S. population aged 20 and older, the use of lipid-modifying agents almost doubled between 1999 and 2010, whereas the use of antihypertensive medications reached ~28% in 2010 (Beltrán-Sánchez et al. 2013).

Socioeconomic and Racial Differences in Healthy Life Expectancy in the United States

Period-based evidence consistently shows disparities in healthy life expectancy by sex, race/ethnicity, and socioeconomic status. Generally, the proportion of remaining life spent in good health is higher for men compared with women, for whites compared with racial and ethnic minorities, and for the most educated compared with the least educated (Crimmins et al. 1989; Manton and Stallard 1991; Guralnik et al. 1993; Hayward and Heron 1999; Crimmins and Saito 2001; Molla et al. 2004; Solé-Auró et al. 2015). Period-based studies have also found evidence of widening racial and sex disparities in healthy life expectancy over time, although these studies are based on tenuous assumptions of constant age-specific mortality and disability rates over time (Dowd and Bengtson 1978; Chappell and Havens 1980; Carreon and Noymer 2011).

Cohort studies, which are not subject to these assumptions, produce mixed results about whether racial and sex gaps in healthy life expectancy narrow, persist, or expand over age and time. For example, Kelley-Moore and Ferraro (2004) found evidence of persistent racial and sex disparities in disability, and Ferraro and Farmer (1996) found similar patterns of persistent disparities in subjective health. In contrast, Ferraro et al. (1997) found widening racial disparities in self-assessed health. Soneji (2006) concluded that cohort patterns in racial disparities in healthy life expectancy may depend on the severity level of disability, which is consistent with Manton's hypothesis of dynamic equilibrium.

Three theories in aging may help to explain the varying results from cohort studies. First, the "age as leveler" theory rests on selective survival and posits that earlier gaps in healthy life will narrow in advanced age (Kent 1971; Dowd and Bengtson 1978). Indeed, convergence in racial and sex gaps has been observed among the oldest old in chronic disease prevalence and physical disability, as well as functional health (Gibson 1991; Clark et al. 1993; Johnson 2000;

Manton and Gu 2001). Second, the theory of "persistent inequality" asserts that sex and racial gaps in earlier life will continue throughout life. Stable sex and racial gaps have been found in physical disability (Ferraro and Farmer 1996; Kelley-Moore and Ferraro 2004). Finally, the "cumulative disadvantage theory" argues that the gap in healthy life expectancy experienced by racial and ethnic minorities and women will widen in advanced age. Such widening gaps have been noted in disability and institutionalization but not in mortality (Clark 1997; Liao et al. 1999).

OUTLOOKS FOR THE FUTURE: DELAYED AGING

An important tenet of delayed aging is that all fatal and disabling disease risks are lowered simultaneously, thereby leading to a postponement of the age at onset of these conditions. To provide a sense of the possible impact of delayed aging on future healthy life expectancy in the United States, we use projections of mortality and ADL disability prevalence from 2010 to 2060 by Goldman et al. (2013) based on a delayed aging model. Briefly, Goldman et al. created population projections using a micro-

simulation model, called the Future of Elderly Model (FEM), that takes into account time trends in disability, improved prevention of diseases, and impact of new medical technologies using data from cross-sectional national health surveys (e.g., National Health Interview Survey) and the largest longitudinal studies of aging in the United States (Health and Retirement Study and the Medicare Current Beneficiary Survey). Additionally, Goldman and colleagues also projected population counts for a baseline scenario using mortality projections from the Social Security Administration (for further details, see Goldman et al. 2013).

Implications of Delayed Aging for Disability

Figure 1A shows the proportion of people aged 65 or older projected to be disability free between 2010 and 2060 in the delayed aging scenario and in the baseline scenario. Disability is defined as having one or more limitations in instrumental activities of daily living (IADLs), having one or more limitations in ADLs, living in a nursing home, or a combination of the three (Goldman et al. 2013). Results clearly show that under the delayed aging scenario there would be higher prevalence of disability-free older adults

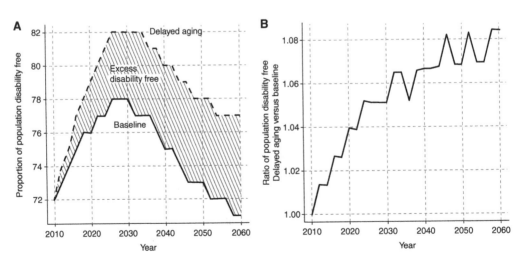

Figure 1. Effect of delayed aging on future population disability free. Proportion of people aged 65 or older projected to be disability free (*A*), and excess disability free (*B*), between 2010 and 2060 in the delayed aging scenario and in the baseline scenario, respectively. (From Goldman et al. 2013; with permission, from the authors.)

in every year between 2010 and 2060 with a peak in 2030, relative to the baseline scenario. Importantly, the magnitude of the difference between scenarios, that is, excess disability-free, slightly increases from 2030 to 2060 resulting from a lower annual rate of decline in disability-free prevalence in the delayed aging scenario (slope $= -0.0020$ in delaying aging vs. slope $= -0.0025$ in the baseline). Thus, there is an upward trend in excess of disability-free older adults in the delayed aging scenario (Fig.1B).

To highlight the impact of delayed aging on life expectancy, we estimate healthy life expectancy measured by years of life disability free at age 65 in 2030, the peak year of disability-free prevalence (Table 1). We use Sullivan's method (Sullivan 1971) to split remaining years of life at age 65 into years lived with disability and disability free using survival probabilities and disability prevalence from Goldman et al. (2013). According to the projections, delayed aging would lead to ~9% (2.4 yr) higher life expectancy at age 65 in 2030 relative to the baseline scenario, from ~25.5 yr in the baseline scenario to ~27.9 yr in delayed aging. Estimates of healthy life expectancy also show higher number of reaming years of life disability free under the delayed aging scenario, ~22.9 yr out of 27.9, relative to the baseline scenario. More importantly, however, is whether the additional 2.4 years of life under delayed aging are also accompanied by more years disability free. We use a simple and well-known decomposition approach (Kitagawa 1955) to assess how much of the additional 2.4 years are because of changes in prevalence of disability free and disability between scenarios. Our results indicate that ~80% of the extra years of life (1.9 yr) under delayed aging would be disability free. This exercise highlights that delayed aging has the potential for increasing not only the length of life but also the fraction and number of years spend disability free at older ages.

Implications of Delayed Aging for Chronic Disease Morbidity

With a few exceptions (for example, see Goldman et al. 2013), most of the evidence of delayed aging is based on animal models and results show promising prospects for postponing the age at onset of chronic disease and disability accompanied by improved physiological status (Harrison et al. 2009). There is a growing literature of empirical studies showing viable interventions to slow aging and extend life including caloric restriction (Anderson and Weindruch 2012), single-gene mutations (Bartke 2011), inhibitors of the target of rapamycin pathways (Miller et al. 2010), senescent cell elimination (Baker et al. 2011), and transplants of stem cells from young to old mice (Conboy et al. 2005), to name a few.

The common theme in recent studies is the consistency of findings suggesting that these interventions improve both lifespan and healthspan in animal subjects. For instance, caloric restriction is thought to change the regulation metabolism, which in turn activates pathways leading to increase disease resistance with delays in the onset of chronic disease (Anderson and Weindruch 2012). Single-gene mutation that affects signaling of growth hormones (e.g., Pit1 and Prop-1 genes) and insulin (e.g., insulin-like growth factor [IGF]-1) has been shown to delay age-related diseases, such as oxidative stress resistance, cardiac and ocular pathology, and atherosclerosis (Bartke 2011). Results using rapamycin indicate that this inhibitor may play a major role in the target of rapamycin pathways in control of aging in mammals and in the pathogenesis of late-life illnesses (Harrison et al. 2009; Laplante and Sabatini 2009; Miller et al. 2010). Some studies show that rapamycin delays the age at onset of conditions, such as cancer (Blagosklonny 2008) and Alzheimer's disease (Caccamo et al. 2010), and reduces atherosclerotic plaque progression (Pakala et al. 2005). Additionally, there is evidence in humans that incidence of type 2 diabetes in older adults can be delayed through medication (metformin) (Knowler et al. 2002). Although there is limited empirical evidence that these interventions may have the same health benefits in humans, there is the potential that delayed aging through pharmacotherapy may lead to health improvements at older ages.

CONCLUSION

As we look into the future of healthy life expectancy, there are some concerns regarding trends in health risk factors, such as obesity and smoking, as well as large socioeconomic differences in health. In the United States, for example, a report of the National Academy of Sciences indicates that Americans have much higher rates of unhealthy behaviors (e.g., smoking and obesity) than their counterparts in high-income countries (Woolf et al. 2013). Some research predicts a slow down of life expectancy as a result of obesity (Olshansky et al. 2005; Stewart et al. 2009), whereas there is compelling evidence that smoking has had a great toll on adult mortality and will likely continue to do so in the near future, at least for females (Preston et al. 2010). Although recent evidence shows improvements in healthy life expectancy among recent cohorts of older people in the United States, there appears to be increasing levels of frailty and disability among younger generations, leading some researchers to believe that future cohorts of older people will likely show declining health expectancy (Martin et al. 2010). However, trends on obesity from 1999 to 2006 suggests a leveling off in the prevalence in adult females and children, with similar trends among adult males from 2003 to 2006. (Ogden et al. 2007, 2008). Educational attainment among older adults is likely to increase in the next decades as new cohorts of younger adults entering old age appear to have high average educational levels (Martin et al. 2010). Because education is positively linked with health and functioning through many mechanisms (e.g., healthier lifestyles), increasing educational levels among new cohorts of older adults is likely to lead to improvements in late-life health, including disability and functional mobility.

Continuing on the current biomedical paradigm based on a "disease model" is also likely to lead to health improvements, albeit with diminishing returns because of a large number of comorbidities inflicting older people. Results from Goldman et al. (2013) simulating two scenarios (cancer and heart disease) representing optimistic developments in medical research, disease treatment, and improvements in behavioral risk factors show a slight increase in older adults from 2010 to 2060 (0.8% and 2% more people in 2060 for cancer and heart disease, respectively, relative to a baseline scenario) with one-fourth (25%) of them having disability over the period when either cancer or heart disease are arrested. These values are ~20% higher than those under the delayed aging scenario. If these projections produce an accurate rendition of the near future, the current biomedical paradigm could potentially lead to an increasing fraction of the population with disability.

On the other hand, delayed aging appears to have important consequences on health and functioning of older mammals based on animal studies, although there is limited evidence on humans. As shown in the static exercise, delayed aging has the potential for increasing not only the length of life but also the fraction and number of years spent disability free at older ages. Although disability is only one dimension of health, results from animal models suggest that delayed aging could have far-reaching health benefits by delaying the age at onset of underlying physiological processes, reducing disease progression, or both. Nonetheless, delayed aging also poses important challenges. In high-income countries, population aging is already occurring or will inevitably occur in the next decades; this process could be exacerbated under a delayed aging scenario. Even if biomedical breakthroughs eventually provide means of slowing the rate of aging, they may not be applied on a wide scale. They may prove to be exceptionally expensive, so that only a small minority may benefit from them. But even if they are inexpensive to use on a personal level, the social costs may be prohibitive. As shown by Goldman and colleagues (2013), achieving delayed aging is likely to put pressure on public transfer programs (e.g., Social Security and Medicare) with additional aggregate costs resulting from a large number of people surviving to older ages. These costs could potentially be offset by changing the eligibility age for Medicare and the normal retirement age for Social Security (Goldman et al 2013). This is not an easy task. Changing eligibility of transfer pro-

Cite this article as *Cold Spring Harb Perspect Med* doi: 10.1101/cshperspect.a025957

grams is a core source of the current financial and political turmoil in Europe (e.g., France). The United States is likely to follow similar turmoil unless the benefits of delayed aging are applied on a wide scale and people are willing to convert their greater healthiness into more years of work.

ACKNOWLEDGMENTS

We thank Dana P. Goldman for kindly providing us with results from the FEM model used in the section, Outlooks for the Future: Delayed Aging. We also thank S. Jay Olshansky for helpful comments and suggestions. The authors acknowledge financial support from Grants R24 HD047873, P30AG017266 (H.B.-S.), KL2TR 001088 (S.S.), and P30AG17265 (E.M.C.).

REFERENCES

Anderson RM, Weindruch R. 2012. The caloric restriction paradigm: Implications for healthy human aging. *Am J Hum Biol* **24:** 101–106.

Baker DJ, Wijshake T, Tchkonia T, LeBrasseur NK, Childs BG, van de Sluis B, Kirkland JL, van Deursen JM. 2011. Clearance of p16^{Ink4a}-positive senescent cells delays aging-associated disorders. *Nature* **479:** 232–236.

Barbour KE, Helmick CG, Theis KA, Murphy LB, Hootman JM, Brady TJ, Cheng YLJ. 2013. Prevalence of doctor-diagnosed arthritis and arthritis-attributable activity limitation—United States, 2010–2012. *MMWR Morb Mortal Wkly Rep* **62:** 869–873.

Bartke A. 2011. Single-gene mutations and healthy ageing in mammals. *Philos Trans R Soc B* **366:** 28–34.

Beltrán-Sánchez H, Harhay MO, Harhay MM, McElligott S. 2013. Prevalence and trends of Metabolic Syndrome in the adult US population, 1999–2010. *J Am Coll Cardiol* **62:** 697–703.

Blagosklonny MV. 2008. Prevention of cancer by inhibiting aging. *Cancer Biol Ther* **7:** 1520–1524.

Bowling A. 2011. Commentary: Trends in activity limitation. *Int J Epidemiol* **40:** 1068–1070.

Brault M, Hootman JM, Helmick CG, Theis KA, Armour B. 2013. Prevalence and most common causes of disability among adults—United States, 2005. *MMWR Morb Mortal Wkly Rep* **58:** 421–426.

Caccamo A, Majumder S, Richardson A, Strong R, Oddo S. 2010. Molecular interplay between mammalian target of rapamycin (mTOR), amyloid-β, and tau effects on cognitive impairments. *J Biol Chem* **285:** 13107–13120.

Cai LM, Lubitz J. 2007. Was there compression of disability for older Americans from 1992 to 2003? *Demography* **44:** 479–495.

Cai LM, Hayward MD, Saito Y, Lubitz J, Hagedorn A, Crimmins E. 2010. Estimation of multi-state life table functions and their variability from complex survey data using the SPACE program. *Demogr Res* **22:** 129–157.

Carreon D, Noymer A. 2011. Health-related quality of life in older adults: Testing the double jeopardy hypothesis. *J Aging Stud* **25:** 371–379.

Centers for Disease Control and Prevention. 2013. Prevalence of doctor-diagnosed arthritis and arthritis-attributable activity limitation—United States, 2010–2012. *MMWR Morbid Mortal Wkly Rep* **62:** 869–873.

Chappell NL, Havens B. 1980. Old and female: Testing the double jeopardy hypothesis. *Sociol Quart* **21:** 157–171.

Clark DO. 1997. U.S. trends in disability and institutionalization among older blacks and whites. *Am J Public Health* **87:** 438–440.

Clark DO, Maddox GL, Steinhauser K. 1993. Race, aging, and functional health. *J Aging Health* **5:** 536–553.

Conboy IM, Conboy MJ, Wagers AJ, Girma ER, Weissman IL, Rando TA. 2005. Rejuvenation of aged progenitor cells by exposure to a young systemic environment. *Nature* **433:** 760–764.

Crimmins EM, Beltrán-Sánchez H. 2011. Mortality and morbidity trends: Is there compression of morbidity? *J Gerontol B Psychol Sci Soc Sci* **66B:** 75–86.

Crimmins EM, Saito Y. 2001. Trends in healthy life expectancy in the United States, 1970–1990: Gender, racial, and educational differences. *Soc Sci Med* **52:** 1629–1641.

Crimmins EM, Saito Y, Ingegneri D. 1989. Changes in life expectancy and disability-free life expectancy in the United States. *Popul Dev Rev* **15:** 235–267.

Crimmins EM, Hayward MD, Saito Y. 1994. Changing mortality and morbidity rates and the health status and life expectancy of the older population. *Demography* **31:** 159–175.

Crimmins EM, Kim JK, Seeman TE. 2009. Poverty and biological risk: The earlier "aging" of the poor. *J Gerontol A Biol Sci Med Scis* **64A:** 286–292.

Crimmins EM, Kim JK, Vasunilashorn S. 2010. Biodemography: New approaches to understanding trends and differences in population health and mortality. *Demography* **47:** S41–S64.

Cutler DM, Ghosh K, Landrum MB. 2013. Evidence for significant compression of morbidity in the elderly U.S. population. Working Paper Series No. 19268. National Bureau of Economic Research, Cambridge, MA.

Dowd JJ, Bengtson VL. 1978. Aging in minority populations. An examination of the double jeopardy hypothesis. *J Gerontol* **33:** 427–436.

European Health Expectancy Monitoring Unit. 2009. Trends in disability-free life expectancy at age 65 in the European Union 1995–2001: A comparison of 13 EU countries. Technical report 2009_5.1. European Health Expectancy Monitoring Unit, Montpellier Cedex, France.

Ferraro KF, Farmer MM. 1996. Double jeopardy, aging as leveler, or persistent health inequality? A longitudinal analysis of White and Black Americans. *J Gerontol B Psychol Sci Soc Sci* **51:** S319–S328.

Ferraro KF, Farmer MM, Wybraniec JA. 1997. Health trajectories: Long-term dynamics among Black and White adults. *J Health Soc Behav* **38:** 38–54.

Freedman VA, Spillman BC, Andreski PM, Cornman JC, Crimmins EM, Kramarow E, Lubitz J, Martin LG, Merkin SS, Schoeni RF, et al. 2013. Trends in late-life activity limitations in the United States: An update from five national surveys. *Demography* **50**: 661–671.

Fries JF. 1980. Aging, natural death, and the compression of morbidity. *N Engl J Med* **303**: 130–135.

Gibson RC. 1991. Age-by-race differences in the health and functioning of elderly persons. *J Aging Health* **3**: 335–351.

Glei DA, Goldman N, Rodriguez G, Weinstein M. 2014. Beyond self-reports: Changes in biomarkers as predictors of mortality. *Popul Dev Rev* **40**: 331–360.

Goldman N, Turra CM, Glei DA, Seplaki CL, Lin YH, Weinstein M. 2006. Predicting mortality from clinical and nonclinical biomarkers. *J Gerontol A Biol Sci Med Sci* **61**: 1070–1074.

Goldman N, Glei DA, Lin YH, Weinstein M. 2009. Improving mortality prediction using biosocial surveys. *Am J Epidemiol* **169**: 769–779.

Goldman DP, Cutler D, Rowe JW, Michaud PC, Sullivan J, Peneva D, Olshansky SJ. 2013. Substantial health and economic returns from delayed aging may warrant a new focus for medical research. *Health Aff (Millwood)* **32**: 1698–1705.

Gruenberg EF. 1977. The failures of success. *Milbank Mem Fund Q Health Soc* **55**: 3–24.

Guralnik JM, Land KC, Blazer D, Fillenbaum GG, Branch LG. 1993. Educational status and active life expectancy among older blacks and whites. *N Engl J Med* **329**: 110–116.

Harrison DE, Strong R, Sharp ZD, Nelson JF, Astle CM, Flurkey K, Nadon NL, Wilkinson JE, Frenkel K, Carter CS, et al. 2009. Rapamycin fed late in life extends lifespan in genetically heterogeneous mice. *Nature* **460**: 392–395.

Hayward MD, Heron M. 1999. Racial inequality in active life among adult Americans. *Demography* **36**: 77–91.

Johnson NE. 2000. The racial crossover in comorbidity, disability, and mortality. *Demography* **37**: 267–283.

Kelley-Moore JA, Ferraro KF. 2004. The Black/White disability gap: Persistent inequality in later life? *J Gerontol B Psychol Sci Soc Sci* **59**: S34–43.

Kent DP. 1971. The Negro aged. *Gerontologist* **11**: 48–51.

Kirkwood TBL, Austad SN. 2000. Why do we age? *Nature* **408**: 233–238.

Kitagawa E. 1955. Components of a difference between two rates. *J Am Stat Assoc* **50**: 1168–1194.

Knowler WC, Barrett-Connor E, Fowler SE, Hamman RF, Lachin JM, Walker EA, Nathan DM; Diabetes Prevention Program Research Group. 2002. Reduction in the incidence of type 2 diabetes with lifestyle intervention or metformin. *N Engl J Med* **346**: 393–403.

Laplante M, Sabatini DM. 2009. mTOR signaling at a glance. *J Cell Sci* **122**: 3589–3594.

Liao YL, McGee DL, Cao GC, Cooper RS. 1999. Black–White differences in disability and morbidity in the last years of life. *Am J Epidemiol* **149**: 1097–1103.

Manton KG. 1982. Changing concepts of morbidity and mortality in the elderly population. *Milbank Mem Fund Q Health Soc* **60**: 183–244.

Manton KG, Gu XL. 2001. Changes in the prevalence of chronic disability in the United States black and nonblack population above age 65 from 1982 to 1999. *Proc Natl Acad Sci* **98**: 6354–6359.

Manton KG, Stallard E. 1991. Cross-sectional estimates of active life expectancy for the U.S. elderly and oldest-old populations. *J Gerontol* **46**: S170–182.

Martin LG, Schoeni RF, Andreski PM. 2010. Trends in health of older adults in the United States: Past, present, future. *Demography* **47**: S17–S40.

Miller RA. 2012. Genes against aging. *J Gerontol A Biol Sci Med Sci* **67**: 495–502.

Miller RA, Harrison DE, Astle C, Baur JA, Boyd AR, de Cabo R, Fernandez E, Flurkey K, Javors MA, Nelson JF. 2010. Rapamycin, but not resveratrol or simvastatin, extends life span of genetically heterogeneous mice. *J Gerontol A Biol Sci Med Sci* **66**: 191–201.

Molla MT, Madans JH, Wagener DK. 2004. Differentials in adult mortality and activity limitation by years of education in the United States at the end of the 1990s. *Popul Dev Rev* **30**: 625–646.

Nagi SZ. 1979. The concept and measurement of disability. In *Disability policies and government programs* (ed. Berkowitz ED), pp. 1–15. Praeger, New York.

Ogden CL, McDowell MA, Carroll MD, Flegal KM, National Center for Health Statistics (U.S.). 2007. Obesity among adults in the United States no statistical change since 2003–2004. NCHS Data Brief No 1. NCHS, Hyattsville, MD.

Ogden CL, Carroll MD, Flegal KM. 2008. High body mass index for age among U.S. children and adolescents. *JAMA* **299**: 2401–2405.

Olshansky SJ, Passaro DJ, Hershow RC, Layden J, Carnes BA, Brody J, Hayflick L, Butler RN, Allison DB, Ludwig DS. 2005. A potential decline in life expectancy in the United States in the 21st century. *N Engl J Med* **352**: 1138–1145.

Omran AR. 1971. Epidemiologic transition: A theory of the epidemiology of population change. In *International encyclopedia of population*, pp. 172–183. The Free Press, New York.

Pakala R, Stabile E, Jang GJ, Clavijo L, Waksman R. 2005. Rapamycin attenuates atherosclerotic plaque progression in apolipoprotein E knockout mice: Inhibitory effect on monocyte chemotaxis. *J Cardiovasc Pharmacol* **46**: 481–486.

Plassman BL, Langa KM, Fisher GG, Heeringa SG, Weir DR, Ofstedal MB, Burke JR, Hurd MD, Potter GG, Rodgers WL, et al. 2007. Prevalence of dementia in the United States: The aging, demographics, and memory study. *Neuroepidemiology* **29**: 125–132.

Preston SH. 1976. *Mortality patterns in national populations: With special reference to recorded causes of death*. Academic, New York.

Preston SH, Glei DA, Wilmoth JR. 2010. Contribution of smoking to international differences in life expectancy. In *International differences in mortality at older ages: Dimensions and sources* (ed. Crimmins EM, Preston SH, Cohen B), pp. 105–131. National Academies, Washington, DC.

Rau R, Soroko E, Jasilionis D, Vaupel JW. 2008. Continued reductions in mortality at advanced ages. *Popul Dev Rev* **34:** 747–768.

Salomon JA, Wang H, Freeman MK, Vos T, Flaxman AD, Lopez AD, Murray CJ. 2012. Healthy life expectancy for 187 countries, 1990–2010: A systematic analysis for the Global Burden Disease Study 2010. *Lancet* **380:** 2144–2162.

Schneider KM, O'Donnell BE, Dean D. 2009. Prevalence of multiple chronic conditions in the United States' Medicare population. *Health Qual Life Outcomes* **7:** 82.

Seeman TE, Singer BH, Rowe JW, Horwitz RI, McEwen BS. 1997. Price of adaptation—Allostatic load and its health consequences. MacArthur studies of successful aging. *Arch Intern Med* **157:** 2259–2268.

Seeman TE, Merkin SS, Crimmins EM, Karlamangla AS. 2010. Disability trends among older Americans: National Health and Nutrition Examination Surveys, 1988–1994 and 1999–2004. *Am J Public Health* **100:** 100–107.

Sierra F, Hadley E, Suzman R, Hodes R. 2009. Prospects for life span extension. *Annu Rev Med* **60:** 457–469.

Smith AK, Walter LC, Miao Y, Boscardin W, Covinsky KE. 2013. Disability during the last two years of life. *JAMA Intern Med* **173:** 1506–1513.

Solé-Auró A, Beltrán-Sánchez H, Crimmins EM. 2015. Are differences in disability-free life expectancy by gender, race, and education widening at older ages? *Popul Res Policy Rev* **34:** 1–18.

Soneji S. 2006. Disparities in disability life expectancy in us birth cohorts: The influence of sex and race. *Biodemography Soc Biol* **53:** 152–171.

Stewart ST, Cutler DM, Rosen AB. 2009. Forecasting the effects of obesity and smoking on US life expectancy. *N Engl J Med* **361:** 2252–2260.

Sullivan DF. 1971. A single index of mortality and morbidity. *HSMHA Health Rep* **86:** 347–354.

Vaupel JW. 2010. Biodemography of human ageing. *Nature* **464:** 536–542.

Verbrugge LM, Jette AM. 1994. The disablement process. *Soc Sci Med* **38:** 1–14.

Wilper AP, Woolhandler S, Lasser KE, McCormick D, Bor DH, Himmelstein DU. 2008. A national study of chronic disease prevalence and access to care in uninsured U.S. adults. *Ann Intern Med* **149:** 170–176.

Woolf SH, Aron LY; National Academies (U.S.). 2013. Panel on understanding cross-national health differences among high-income countries, Institute of Medicine (U.S.). Board on Population Health and Public Health Practice. In *US health in international perspective: Shorter lives, poorer health.* The National Academies, Washington, DC.

World Health Organization. 1980. *International classification of impairments, Disabilities and handicaps: A manual of classification relating to the consequences of disease.* World Health Organization, Geneva.

World Health Organization. 2002. *The world health report 2002: Reducing risks, promoting healthy life.* World Health Organization, Geneva.

Yach D, Hawkes C, Gould CL, Hofman KJ. 2004. The global burden of chronic diseases: Overcoming impediments to prevention and control. *JAMA* **291:** 2616–2622.

Has the Rate of Human Aging Already Been Modified?

S. Jay Olshansky

School of Public Health, University of Illinois at Chicago, Chicago, Illinois 60612

Correspondence: sjayo@uic.edu

In recent years, three hypotheses have been set forth positing variations on a common question—Has the rate of human aging already been modified? There is no disputing that people now live longer than ever before in history, and considerable variation in duration of life persists as a fundamental attribute of human longevity, but are these events caused by a measurable and verifiable difference in the rate at which people age, or are there other reasons why they occur? In this article, I explore the historical record involving changes in survival and life expectancy at older ages dating back to 1900, and examine what factors will likely contribute to changes in longevity in the United States through 2040. Evidence suggests that despite the absence of verifiable metrics of biological age, delayed aging is unlikely to be a cause of secular increases in life expectancy, but it could explain variation in survival among population subgroups, and it is the most likely explanation for why exceptionally long-lived people experience less disease and live longer than the rest of the population. If genetic heterogeneity explains any significant part of current variation in longevity, this opens the door to the development of therapeutic interventions that confer these advantages to the rest of the population.

The rise in human longevity is one of humanity's crowning achievements. There has been more survival time manufactured by public health and medical technology in the last 115 years than in all of human history combined. It has been well established that ~80% of the rise in longevity in the first half of the 20th century in the United States (and most other developed nations) was because of declining early age mortality—mostly from reductions in death rates from communicable diseases. Since 1950, ~80% of the gain in longevity was because of declining death rates at middle and older ages owing to progress made against chronic fatal diseases such as cardiovascular diseases (CVDs) (Kinsella 1992). Death rates at younger and middle ages have declined dramatically. Now ~85% of all babies born today in developed nations will survive to at least age 65, and most deaths are concentrated in the 30-yr time period between ages 65 and 95 (see the Human Mortality Database, www.mortality.org).

Although the proximate causes of declining middle and old age mortality are not in dispute (e.g., a combination of advances in public health, improved behavioral and social risk factors, and medical technology), some researchers have speculated that improved mortality statistics alone are sufficient reason to suspect that people in the modern era may be aging more slowly relative to the generations that preceded them. For example, after observing declines in death rates in Sweden from 1900 to 1990, Vaupel and Lundstrom concluded that ". . . one inter-. pretation of these shifts [declining old age mortality across time] is that the process of aging has been slowed or delayed in Sweden such that elderly Swedish men are effectively three or four years 'younger' than they used to be . . ." (Vaupel and Lundstrom 1994, p. 90). This view was repeated by Vaupel when he stated ". . . before the discovery that senescence could be postponed [a conclusion based on the observation that death rates at older ages declined], geriatric medicine was viewed as a laudable but rather futile effort to palliate the misery of those in the process of dying" (Vaupel 2010, p. 536). Levine and Crimmins (2014) documented morbidity and mortality differentials and measured biomarkers among Blacks and Whites in the United States using the National Health and Nutrition Examination Survey, concluding that premature declines in health among Blacks observed today may be indicative of an acceleration of the biological processes of aging. The study of centenarians and their offspring has led Atzmon et al. (2005) and Perls et al. (2002) to speculate that this subgroup of the population lives so long, or at least has the potential for exceptionally long lives, because they are biologically aging at a slower pace relative to the rest of the population. That is, for long-lived people, it is hypothesized that 1 yr of clock time is matched by less than 1 yr of biological time.

The preceding paragraph exemplifies the difficulty in understanding the nuances of aging science and how scientists in other disciplines interpret it. One group of researchers observed secular declines in death rates at older ages and concluded from this that more recent cohorts passing through the same age window relative to previous cohorts are "aging more slowly" (referred to here for simplicity as the "secular aging rate hypothesis"). A second set of researchers observed mortality differentials among population subgroups at a single moment in time, and speculated that the shorter-

lived subgroup is "aging more rapidly" because of variation in behavioral or environmental risk factors (referred to here as the "comparative aging rate hypothesis"). The third set of researchers focused on the genetics of long-lived people and their offspring, and concluded that these people live so long because they possess "protective genes" that enable them to escape or delay the ravages of time and "age more slowly" than the rest of the population (referred to here as the "heterogeneity aging rate hypothesis"). Can all three of these variations in aging rate be operating?

Let us begin by defining terms, because it is important to know what is meant by "rate of aging" and what it does not mean. According to Olshansky et al. (2008), a defensible definition of aging is:

. . . the accumulation of random damage to the building blocks of life—especially to DNA, certain proteins, carbohydrates, and lipids (fats)—that begins early in life and eventually exceeds the body's self-repair capabilities. This damage gradually impairs the functioning of cells, tissues, organs, and organ systems, thereby increasing vulnerability to disease and giving rise to the characteristic manifestations of aging, such as a loss of muscle and bone mass, a decline in reaction time, compromised hearing and vision, and reduced elasticity of the skin. This accretion of molecular damage comes from many sources, including, ironically, the life-sustaining processes involved in converting the food we eat into usable energy. . . . Aging, in our view, makes us ever more susceptible to such ills as heart disease, Alzheimer's disease, stroke and cancer, but these age-related conditions are superimposed on aging, not equivalent to it. Therefore, even if science could eliminate today's leading killers of older individuals, aging would continue to occur, ensuring that different maladies would take their place. In addition, it would guarantee that one crucial body component or another—say, the cardiovascular system—would eventually experience a catastrophic failure. It is an inescapable biological reality that once the engine of life switches on, the body inevitably sows the seeds of its own destruction.

Aging is not defined by the diseases that kill us or which make us more frail and disabled, because most (but not all) of these conditions are heri-

Cite this article as *Cold Spring Harb Perspect Med* doi: 10.1101/cshperspect.a025965

table on the one hand and, more importantly, inherently modifiable through risk factor modification. It was suggested long ago that biological age can be measured by actuarial analysis of large populations, through assessment of overall morbidity, or observation of chronic degenerative changes (Ludwig and Smoke 1980). But this confusion of biological aging with measurable health and mortality outcomes expressed at later ages is perhaps the source of the problem. The fact is, biological aging itself cannot as yet be reliably measured by any single or multiple "biomarkers of aging" (Baker and Sprott 1988; Kirkwood and Austad 2000), although there may come a time in the future when trustworthy biomarkers of aging are discovered. So, if we cannot measure or directly observe biological aging, let us begin by exploring the documented reasons behind the rise in longevity in the 20th century, with a particular focus on later ages, and what factors what might influence these trends going forward. By revealing what is known about mortality and survival across time and between population subgroups, it is then possible to assess whether any of the three hypotheses set forth are plausible.

DEMOGRAPHIC COMPARISON OF THREE OLDER COHORTS (1965, 2015, and 2040)

Let us begin with some basic historical longevity statistics. In 1965, the life expectancy of the U.S. population was 70.2 yr. In 1900, life expectancy was 47.7 yr. The probability of a baby born in 1965 reaching the age of 65 was 71.3%. In 1900, it was only 39.1%. Among those that had already celebrated their 65th birthday in 1965, 28.1% would survive to their 85th birthday. In 1900, only 14% of those reaching age 65 survived to age 85 (see Life Tables for the United States, www .ssa.gov/oact/NOTES/as120/LifeTables_Tbl_ 6.html). Statistics like these appear repeatedly throughout the literature describing historical trends in longevity in the United States.

There is no doubt that, based on these simple statistics alone, longevity and the enriched health conditions required to generate such increases in survival, improved dramatically in the 20th century. Many more Americans now survive to ages 65+ than at any time in history. We survive longer once having reached this age, and based on an inevitable rapid shift in America's age structure, it is certain that the prevalence of people surviving up to and beyond age 65 will rise dramatically by midcentury. However, the picture these statistics convey is only part of the story of our past and present (perhaps not even the most important elements). There is a unique history behind secular trends in mortality and survival statistics that, once understood, will provide a full three-dimensional view of the forces that influenced past and current health and longevity attributes of the U.S. population aged 65+, and the factors most likely to influence this age cohort in the future. More importantly, they provide important insights regarding speculation that people are living longer today because they are aging more slowly relative to their age-matched counterparts who lived years earlier.

Population Distribution by Race

In 1965, about 84% of the total population was White, Blacks comprised 12%, Hispanics 3.5%, and all other races represented the remaining 0.5% (Table 1). The United States experienced a rapid transformation in its racial composition to the present day, where Whites now comprise 62.6%, Blacks 12.4%, Hispanics 17.1%, and all other races 7.9% of the total. The most notable change in the U.S. population distribution by race in the year 2040 will be the rise of the Hispanic population—which is expected to comprise 28% of the total population, Blacks will be 14%, Whites 49%, and all other races will be 9% of the total. Hispanics will also occupy an increasing share of the 65+ population in 2040. The relative distribution of the Hispanic population into first-, second-, and third-generational components will be important in understanding their impact on the health and longevity of the 65+ population in 2040.

Population Distribution by Age

One of the most notable demographic changes in the history of the United States and other

Table 1. Demographic and health characteristics of the total United States and 65+ populations

	1965	2010	2040
Birth era (aged 65–110)[a]	1855–1900	1905–1950	1930–1975
Total U.S. population (in millions)[b]	194.3	316.1	380.0
Population distribution by race[c]	B 12.0%	B 12.4%	B 14.0%
	W 84.0%	W 62.6%	W 49.0%
	H 3.6%	H 17.1%	H 28.0%
	O 0.5%	O 7.9%	O 9.0%
Population aged 65+ (in millions)[d]	18.5	47.7	79.7
Population aged 85+ (in millions)	1.1	6.3	14.1
Population aged 100+	4000	78,000	230,000
Proportion of population aged 65+	9.5%	14.8%	21.0%
Proportion of population aged 85+	0.6%	2.0%	3.7%

B, Black; W, White; H, Hispanic; O, other.

[a]These are the years during which members of the 65 and older cohort were born. The entire cohort is contained within the age group 65–110, but the majority of these cohorts were born during the last two decades of each era.

[b]Source of population statistics for 2014/15 and 2040: U.S. Census Bureau, Table 6, Percent of the Population by Race and Hispanic Origin for the United States: 2015 to 2060, 2012 National Population Projections (U.S. Census Bureau, Population Division, Washington, DC, 2012); www.census.gov/population/projections/data/national/2012.html (accessed October 15, 2014); www.census.gov/population/estimates/nation/popclockest.txt.

[c]The "White" population is defined as the "non-Hispanic White" population. Percentages may not add to 100 because of rounding error. The data for 1965 and 2040 are interpolated from the following source that provided estimates only for 1960 and 2050—the percentages are not expected to change much in this short time frame (see thesocietypages.org/socimages/2012/11/14/u-s-racialethnic-demographics-1960-today-and-2050).

[d]1965 estimated from www.census.gov/prod/99pubs/p23-199.pdf; current and projected from www.census.gov/prod/1/pop/p23-190/p23-190.pdf. 1965, Past; 2010, present; and 2040, future.

developed nations is the absolute increase in the size of the older population. The population aged 65+ in 1965 was 18.5 million, increasing to 47.7 million today, and its projected to rise dramatically to about 80 million by 2040. The primary factor influencing the rise in the population aged 65+ is a shift in the age distribution—that is, larger birth cohorts moving through the age structure. Once the age distribution settles down and the historical blips pass through, the size of the older population is likely to stabilize or slightly increase past 2040. There will also be notable increases in the population aged 85+, rising from 1.1 million in 1965, to 6.3 million today, with a projection of 14.1 million by 2040. The number of centenarians increased from 4000 in 1965 to 78,000 today (Table 1), and the Census Bureau projects there will be about 230,000 centenarians in 2040. All of these estimates for 2040 could rise more rapidly than projected if death rates decline faster than predicted and the reverse is also true—if health conditions worsen for these cohorts, these estimates could be too high.

Life Expectancy

In 1965, life expectancy at birth was only 70.2 yr (Table 2). It rose to 78.8 today, and is projected by the Social Security Administration to rise to 80.5 by 2040. This means the annual rate of change was +0.17 yr annually between 1965 and 2015, and the projected rate of change is expected to decelerate to less than half of that to a +0.07 annual increase between now and 2040. The rise in life expectancy at age 65 is more relevant for this discussion, and here we saw a rise from 14.7 yr in 1965 to 19.3 yr today. The projection for 2040 is only 19.6 yr, which means the U.S. Social Security Administration does not anticipate much of an improvement in duration of life lived at older ages over the next 25 yr.

Survival

The story of survival up to and beyond age 65 is similar to that of life expectancy. In 1965, about 71% of all babies born in that year were expect-

Table 2. Life expectancy (in years) at birth and at age 65, and conditional survival to ages 65 and from ages 65 to 85 (United States: 1965, 2010, 2040)

	1965[a]	2010[a]	2040[b]
Life expectancy at birth	70.2	78.8	80.5
Males	66.8	76.4	78.5
Females	73.9	81.2	82.5
Life expectancy at age 65	14.7	19.3	19.6
Males	12.9	17.9	18.4
Females	16.4	20.5	20.9
Probability of surviving to age 65 from birth	71.3%	84.4%	87.5%
Males	73.9	80.7	85.4
Females	78.9	88.1	90.2
Probability of surviving to age 85 conditional on survival to age 65	28.1%	50.1%	50.9
Males	20.0	43.4	46.0
Females	35.1	55.7	57.6

[a]www.mortality.org.

[b]www.ssa.gov/oact/NOTES/as120/LifeTables_Tbl_6.html.

ed to reach age 65; by 2015, this rose dramatically to over 84% (Table 2). The primary reason for this improvement is a large reduction in death rates at younger ages. By 2040, survival to age 65 is expected to rise to 87.5%, which means the rate of improvement in survival is expected to decelerate rapidly. The reason is the phenomena known as entropy in the life table and competing risks (Olshansky et al. 1990) (discussed in more detail below). Perhaps, more importantly, survival between the ages of 65 and 85 did in fact increase dramatically from 28% in 1965 to 50% today. The marginal expected improvement in survival to age 85 projected for 2040 indicates that, just like life expectancy, the U.S. Social Security Administration does not anticipate much further improvement in survival for the population aged 65+ beyond that already achieved.

Health

Crimmins and Saito (2001) have shown that, from 1970 through 1990, there were varying patterns of healthy life expectancy (HLE) as a function of race and level of completed education. Overall patterns of HLE improved consistently only for those with 13+ years of education; there were some improvements for those who completed 9–12 years of schooling, but among the least educated there was a decline in HLE. More importantly, for the population aged 65 and older, there were vast differences in HLE as a function of completed education—with the most highly educated living several more healthy years relative to the least educated.

LIFE HISTORY OF THE POPULATION AGED 65+ IN 1965

Perhaps the most important thing to keep in mind when discussing the population aged 65+ in different eras is that the people who occupied this age window in the past, those in this age window now, and the people reaching that age window by 2040, are each a by-product of a very unique set of life history characteristics (e.g., cohort events). For example, almost the entire population aged 65 and older in 1965, was born in the 19th century, and most of them were born between 1880 and 1900. When they were children, they had scarce access to anything resembling a vaccination, and most vaccines designed to prevent fatal and disabling diseases, as we know them today, did not exist. Antibiotics had not yet been invented, malnutrition and undernutrition were common, infant mortality rates were as high as 35% in some parts of the country, the primary causes of death were communicable diseases (CVD and cancer were common among only the few who were fortunate enough to survive to older ages), some behavioral risk factors now known to be harmful (e.g., smoking) were considered by some in the medical community as "healthy," and the foundations of public health (clean water, sanitation, refrigeration, etc.) were just beginning to emerge, and they were inequitably distributed where they did exist (Riley 2001).

For those fortunate enough to survive childhood, the early 20th century brought the hazards of World War I and the 1918 influenza pandemic, and for the survivors of these harsh events, malnutrition and difficult living and working environments plagued this cohort in their adult years. The bottom line is that unforgiving environmental conditions led to early deaths for many children born before 1900 (the oldest members of the population aged 65+ in 1965 were the survivors), and extremely harsh environmental, living and working conditions persisted for this cohort in their first few decades of life.

What are the health and longevity implications of these harsh early life developmental events on the 65 and older population in 1965? The irony is that the relatively few members of the late 19th century birth cohorts fortunate to be alive in 1965 represented the hardiest members of their era. This does not mean they did not get sick, nor does it mean they lived a particularly long life after reaching age 65, it just means that they faced harsh living conditions throughout their lives and they were lucky to have survived. Time and the environment acted as a sieve through which members of their birth cohort passed, so although the population aged 65+ in 1965 still died out at rates that are faster than observed today, that generation should be thought of as a very unique population, the likes of which will probably never be seen again.

LIFE HISTORY OF THE POPULATION AGED 65+ TODAY

The current population aged 65+ was born between 1905 and 1950, with the vast majority born during the decades of the 1930s and 1940s into an entirely different set of environmental and health conditions relative to the 1965 cohort of people aged 65+. They carried with them the advantage of major public health advances developed in the early 20th century, but they were also accompanied by harmful behavioral health practices that elevated their risk of chronic fatal and disabling conditions (e.g., the increased use of tobacco). The rise of Alzheimer's disease along with cancer, CVD, and sensory impairments, was the price that recent older generations had to pay for their extended survival. However, some subgroups of the current population aged 65+ are also showing signs of improved health in the form of declining frailty and disability.

Almost the entire population aged 65+ today was born between the milestone events of World War I and the first few years of the post-World War II baby boom. On the positive side of the health and longevity equation, these were the first U.S. generations to have broad access to vaccines for yellow fever, typhus, influenza, and polio (among others). Shortly thereafter, younger members of this generation were vaccinated against measles, mumps, and rubella, although it is certain that many among this generation exhibited these diseases before the vaccinations became available. Death and disability often resulted from many of these communicable diseases throughout human history, but in the modern era, survival rates after acquiring some of these diseases improved dramatically. It is unclear at this point whether exhibiting these full-blown communicable diseases early in life conferred any protection from chronic/fatal conditions later in life, or whether they raised the risk of these conditions when older. In the cases of yellow fever, polio, and typhus, we now know that vaccines for these diseases saved millions of lives of people who otherwise would not have lived to ages 65 and older (World Health Organization 2009). It is most certainly the case that a substantial proportion of today's population aged 65+ would not even be alive had they been raised in the same environmental conditions present during the lives of the 1965 cohort aged 65+.

The infant mortality rate for this generation declined dramatically across the United States relative to previous cohorts as most births occurred in a hospital. After the depression there was widespread availability of food. The education of children spread rapidly and those acquiring a college education increased and the basic principles of public health had already become common, enabling most members of this generation access to clean water, sanitation, refrigeration, and indoor living and working environments. Medical technology advanced rapidly, and, by midcentury, maternal mortality was low as most births took place in the far safer environment of a hospital. The bottom line is that life expectancy rose rapidly during this era as the high force of extrinsic mortality (primarily from communicable diseases) waned in the face of advances in public health and medicine.

However, there were also distinctively harmful conditions that arose during this era that set this generation apart from their predecessors. The 1918 influenza pandemic had a profound (albeit short-term, 2-yr) negative effect on the entire U.S. population, essentially wiping out large segments of both the young and old. Few human generations throughout recorded history went through such a lethal pandemic of this magnitude. A handful of survivors from this pandemic are still alive today (they are now 96+ yr of age).

The current population aged 65+ was also characterized by a high and rapidly growing prevalence of some hazardous behavioral risk factors (e.g., smoking). Although women entering the labor force opened up opportunities for half of this generation, it did not come without risks. The rising prevalence of CVD hit both younger men and women alike among the older members of this cohort, as did the rising prevalence of cancer. However, the younger members of this cohort benefitted from a number of advances in medical diagnosis and care for CVD and cancer that occurred during the last quarter of the 20th century (when CVD mortality declined dramatically and cancer case fatality rates improved).

Overall, today's population aged 65+ faced a much milder set of environmental conditions throughout their lives relative to the 1965 cohort, and they benefitted from advances in public health and the revolution in medicine in the 20th century. This was the first generation in human history to live through and benefit from the first longevity revolution brought forth by public health and modern medical technology.

The good news in this message is that people 65 and older today live longer than any generation in history, and they experience more healthy years than ever before. The bad news is that this is also the first generation to participate in a dangerous Faustian trade—added years of life (decades for many) in exchange for an increase in the amount of survival time spent in a state of frail health accompanied by a suite of chronic fatal and disabling conditions (Alz-

heimer's disease, cancer, CVD, sensory impairments, osteoporosis, osteoarthritis, etc.). Older Americans are most certainly living longer today relative to any time in history, and the absolute number of years spent in a state of good health has expanded with it, but the Faustian trade of an extension of the period of old age must be acknowledged as a by-product of our success in extending life.

The bottom line is that the generation now passing into the age window of 65+ is, overall, much less "selected" out for early death by harsh environmental conditions relative to the generation that preceded them in 1965. Although known for acquiring harmful behavioral risk factors early on, they are also known for abandoning them in favor of improved lifestyles. Today's generation of people aged 65+ is the most diverse of all older cohorts ever seen in the United States, and this diversity includes not just ethnic diversity and its unique impact on health and duration of life, but, more importantly, the positive extremes of healthy life brought forth by favorable behavioral risk factors (e.g., reductions in the prevalence of smoking and increased education), and the negative extremes of unhealthy life brought forth by harmful behavioral risk factors (e.g., obesity and drug use) and the ability of modern medical technology to postpone death and prolong the period of life at the end where frailty, disability, and health care costs are the highest. It is safe to say that a population of people like this, all of whom have been influenced by a unique set of historical environmental hazards and cohort-specific risk factors that cannot be repeated for cohorts reaching aged 65+ in the future, will never be seen again.

LIFE HISTORY OF THE POPULATION AGED 65+ IN 2040

The characteristics of the populations aged 65+ in 1965 and today each offer unique birth era and lifetime experiences (e.g., cohort effects) that set them apart from any previous generation. As far apart as these generations are, the differences between their birth and life experiences will pale in comparison to how far different the population aged 65+ in 2040 is likely to be relative to today. The discussion will begin with an examination of how the population aged 65+ in 2040 started their lives, but the majority of the discussion to follow will address what conditions could be like in 2040.[1]

The population aged 65+ in 2040 was born at the dawn of the depression era. They lived through World War II, the Korean War, and the Vietnam War, and the vast majority was born between 1955 and 1975. Most of this cohort (excluding the centenarians and supercentenarians of 2040 born in the 1930s and 1940s), was vaccinated early in life against all major communicable diseases. They benefitted in mid-life from unprecedented reductions in CVD in the last quarter of the 20th century brought forth by advances in medical technology; they benefitted in midlife from reductions in cancer case fatality rates; each successive cohort was more highly educated than the last, making them the most highly educated birth cohort in American history; and a sizeable proportion of the original cohort adhered to improvements in some behavioral risk factors (e.g., although the older members of the cohort picked up smoking early in life, later members of the cohort took up smoking with a lower frequency and many decided to quit smoking once the scientific evidence on its harm became compelling). There will be a sizable number of "former smokers" among the population aged 65+ of 2040, and the largest group (both in terms of numbers and percentages) of nonsmokers and people rarely exposed to second-hand tobacco, in history (the latter a by-product of progressive public health efforts across many parts of the nation to ban smoking from indoor environments). This is a very powerful health advantage to this cohort, as it will significantly reduce their risk of lung cancer and related complications associated with exposure to tobacco.

The one major health blemish in the birth cohort that will reach ages 65+ in 2040 is that this generation of adults succumbed to the many forces that contributed to the rise of adult-onset obesity. Unlike smoking, where quitting the habit is possible, obesity is a far more challenging public health problem because people cannot stop eating—we have to form a new relationship with food to successfully combat obesity. This is the reason why most behavioral efforts to combat obesity do not work. Unless this problem is fixed, there is a high probability that levels of frailty and disability associated with obesity and its many complications (especially diabetes, CVD, cancer, joint problems, and depression) will dominate up to 50% or more of older cohorts in 2040 and for many subsequent years. Because medical technology is also likely to become increasingly more efficient in treating diabetes and its complications over the next quarter century (Flegal et al. 2010), the magnitude and duration of frailty and disability among future cohorts of people aged 65+ (in 2040 and beyond) are likely to skyrocket.

Finally, there is one critical demographic fact to keep in mind. Everyone who will fall within the age range of 65+ in 2040 is already alive, and existing survival data (see the Human Mortality Database, www.mortality.org) indicate that most people born in that historical era will survive to at least age 65 (i.e., ~85% survival). This means the combination of life extension brought forth by declining middle and old age mortality and a rapidly shifting and much older age structure, will produce a dramatic increase in the absolute size of the population aged 65+ in 2040 (to at least 80 million people relative to the estimated 48 million people of that age in the United States today). As a result, genetic heterogeneity will increase among cohorts reaching older ages in the coming decades.

WILL THE POPULATION AGED 65+ IN 2040 LIVE MUCH LONGER THAN TODAY'S GENERATION?

The question of how high life expectancy in the United States can rise (at birth and at older ages) has been the subject of debate for decades. To

---

[1] The difference between a projection and a forecast is that the former is a hypothetical scenario of what might be, and the latter is a scenario that the investigator actually believes will transpire. Because the population aged 65+ in 2040 has already been born and much is known about their early life conditions, the following discussion of health and longevity conditions for this cohort should be viewed as a forecast.

Cite this article as *Cold Spring Harb Perspect Med* doi: 10.1101/cshperspect.a025965

get a sense of what life expectancy and survival past age 65 will be like in 2040 in the United States, it is important to place our current longevity within the context of history. The rapid increase in life expectancy in the early 20th century was initially because of advances in public health that saved the lives of the young. When the lives of young people are extended, life expectancy rises rapidly because a large number of person-years of life are added to the population—a phenomenon that can only occur once. Once reductions in early age mortality were achieved, future gains in life expectancy must then be a product of reductions in death rates in other (middle and older) regions of the life span. That is exactly what happened.

The United States is now in a position where the only way to significantly increase life expectancy in the future is to generate large reductions in death rates at the oldest ages, and simultaneously push out the envelope of survival into outer regions of the life span (ages 120+) in which only a handful of people have ever lived. That is, large increases in life expectancy at birth in the future require not only large declines in death rates for people aged 65 and older that would have to occur faster and be of a larger magnitude than what was observed in the past for young and middle aged populations (Carnes et al. 2012). It also requires that most people either routinely live past the age of 120, or that a significant segment of the population begin surviving well past the age of 130. Some researchers contend this is exactly what will happen. Vaupel (2010) suggests that advances in biomedical technology yet to be identified or developed, will continue to yield gains in life expectancy in the future on par with patterns of change observed in the past—thus, the rationale for using linear extrapolation to project the future course of mortality and life expectancy.

Other researchers have suggested that there are two reasons why linear extrapolation of life expectancy is inappropriate in long-lived populations. First, it is critical to acknowledge that the biological processes of aging represent the most important risk factor for fatal diseases expressed at older ages (Harman 1991). Although declines in death rates at older ages are possible

and even likely in the coming decades, it has been suggested that this approach will yield diminishing returns with regard to gains in life expectancy—a product of competing risks at older ages that accumulate in aging bodies as life expectancy increases. Because it is not yet possible to alter the biological processes of aging, nor is there evidence to suggest that aging has ever been modified in the past, there is no reason to suspect that dramatic declines in death rates among the extreme elderly are plausible. The U.S. Social Security Administration acknowledges this both implicitly and explicitly by forecasting decelerating rates of increase in life expectancy by midcentury (see www.ssa.gov/oact/NOTES/as120/LifeTables_Tbl_6.html).

Second, the age distribution of death in long-lived populations like the United States has already shifted to later ages, but this shift has been characterized by a compression of death into a fairly narrow region between the ages of 65 and 95. There is no evidence that the prospects for surviving past the age of 120 are improving. There is no reason to expect people will routinely live beyond the age of 130 where no human has been documented to live in history; nor is there reason to believe that, in a genetically heterogeneous population, everyone has the potential to live as long as the longest-lived member of the population.

Currently, life expectancy at age 65 in the United States is 17.9 yr for men and 20.5 yr for women (Table 2). The current probability of surviving to age 85 (conditional on having survived to age 65) in 2010 is 43.4% for men and 55.7% for women. According to middle-range forecasts made by the U.S. Social Security Administration, by 2040, life expectancy at age 65 will be 18.4 for men and 20.9 for women. These represent fairly optimistic improvements because it requires large reductions in death rates at older ages to generate just a 1-yr increase in life expectancy at age 65. Conditional survival to age 85 from age 65 in 2040 is expected to rise to 46.0% for men and 57.6% for women (see www.ssa.gov/oact/NOTES/as120/LifeTables_Tbl_6.html)—a very small improvement over the 2010 cohort. Thus, for every 100 people aged 65 in the United States today, about 50 will live to age 85; for every 100 people aged 65 in the United States in 2040, about 51 are expected to live to age 85.

Overall, there is reason to expect that life expectancy and conditional survival in the United States will rise marginally in the coming decades; the rise will be at a much slower pace than that observed in previous decades, and there is no evidence to support any radical increase in life expectancy or survival in the population aged 65+ in the next quarter century. The implications of these observations are straightforward—mortality compression is the most likely scenario going forward, and this, in turn, is expected to be accompanied by a decelerating increase in life expectancy.

Thus, with regard to the "secular aging rate hypothesis," there are decades worth of definitive research documenting the fact that reductions in death rates at middle and older ages are a direct by-product of improved behavioral risk factors, increasingly more favorable environmental conditions that reduced death rates at older ages, and advances in medical technology that "manufactured" survival time by diagnosing and treating fatal conditions with greater efficiency. There is no direct or indirect evidence favoring the hypothesis that death rates have declined across time because more recent cohorts are aging at a slower pace.

THE ROLE OF DISPARITIES

At one level, there is reason to be optimistic about the future of health and longevity in the United States during the next 25 years. After all, advances in medicine and biomedical technology are occurring rapidly. Genetic engineering offers the promise of curing or controlling some inherited diseases, the prevalence of smoking has declined, educational attainment is increasing, aging science offers the prospect of an extension of youthful vigor—but this is not yet possible, and recent efforts to attack childhood and adult obesity are encouraging. Having said that, it is not possible to ignore some worrisome and potentially disastrous trends in health that have emerged in recent years, such as the rise of childhood obesity, diabetes, and their compli-

cations. The latent effects of these trends will not be seen for decades, which means that by 2040 we could very well witness the long-term negative effect of these trends on the population reaching older ages of the acquisition of harmful behavioral risk factors that were acquired in childhood. Childhood and adult-onset obesity represent the most worrisome health trends at the moment.

There is a tendency in medicine and public health to focus only on what is happening at the national level, and this is understandable given the need to summarize health statistics in a succinct way. However, beneath the surface of national vital statistics is an alarming persistent divide among subgroups within the United States that has been known for decades (Kitagawa and Hauser 1973), and which are invisible when the focus is just on life expectancy, survival, and health for the nation as a whole. These disparities persist even when some major causes of differential survival among the subgroups are eliminated (Lantz et al. 1998; Link and Phelan 2005). Health disparities reflect racial and ethnic differences in addition to variation in education and income that are linked directly and indirectly to behavioral risk factors (Adler et al. 1999), and it has even been proposed that social conditions can be "fundamental causes" of health inequalities, which is why interventions based exclusively on modifying biomedical risk factors have not been, and are not likely to be, successful in substantially reducing health disparities (Link and Phelan 2005). This concept of "fundamental causes" can be interpreted, in part, as variations in the rate at which biological aging is experienced, but in the absence of reliable biomarkers of aging, it is not yet possible to know with certainty whether variation in biological aging causes disparities.

Measures of health disparities as we gauge them today, did not exist in 1965, but Kitagawa and Hauser (1973) documented shortly thereafter that income, education, and occupation (among other socioeconomic status variables) modulate the dramatic variation in longevity and health that existed within the United States at that time. To understand the magnitude and importance of disparities, consider the fact that, in 2008, a White female that acquired a college education could expect to live 10 yr longer than a White female with less than 12 yr of education. To gain an understanding of the magnitude of this difference, relative to someone without a high school degree, a college education is equivalent to taking a pill that cures all CVD and cancer (Beltrán-Sánchez et al. 2008).

A difference in life expectancy of 1 yr is large; a difference of 10 yr means the person with the lower life expectancy is living in the equivalent of the early 20th century. This does not mean people in the modern era are aging more slowly, it just means that in 1900 there were many more risk factors for fatal diseases. Data from the MacArthur Research Network on an Aging Society has shown that these disparities in longevity also apply to the population aged 65 and older (Olshansky et al. 2012)—a particularly notable finding that has important implications for cohorts reaching older ages in the future. This means close attention should be paid to the physical and mental health status and racial composition of younger generations now moving through the age structure because it is these generations that will enter the 65+ age window by 2040.

Furthermore, it is worth noting that, in 2008, the effect of education on the proportion of the population that survived to age 65 (conditional on having survived to age 25) was dramatic for all race–sex groups with the exception of Hispanic females (Table 3). For example, note that among White males with <12 yr of education, only 61.3% of those reaching the age of 25 are expected to survive to age 65. By comparison, 91% of White males with a college education are expected to reach age 65. Education currently has less of an effect on Hispanic men and woman (the longest-lived subgroup in the United States today) because of a phenomenon known as the Hispanic paradox (discussed below).

SHIFTING DEMOGRAPHICS—THE HISPANIC PARADOX AND OLD AGE MORTALITY IN THE FUTURE

When rapid increases in longevity combined with declining fertility in the latter part of the

Table 3. Proportion surviving to age 65 conditional on having survived to age 25 by sex, race, and level of completed education in the United States in 2008

| | Years of school completed | | | |
	<12	12	13–15	16+
White males	61.3	73.2	86.4	91.0
White females	72.8	84.1	91.9	93.8
Black males	59.8	63.6	80.5	84.5
Black females	73.1	76.9	86.6	88.4
Hispanic males	83.5	81.8	89.6	92.2
Hispanic females	90.6	90.2	94.0	95.0

Unpublished data from research conducted for the MacArthur Research Network on an Aging Society. (Related data from the same life tables were published in Olshansky et al. 2012 and reprinted here, with permission, from the author.)

20th century, the U.S. age structure began shifting to a more rectilinear form. By 2040, the age structure of the United States and all other developed nations will be in the shape of a square or nearly so, with at least as many people alive at older ages as there are at younger ages. This new shape to the U.S. age structure is likely to be a permanent feature of our population for the foreseeable future. However, beneath the surface of a visibly shifting age structure are forthcoming changes to our demographics that will alter the course of American health and longevity by 2040, especially among future cohorts of older people. Three major events are unfolding now.

First, there is evidence to suggest that subgroups of the U.S. population are experiencing significantly different health trajectories (Olshansky et al. 2012). Although the least educated among us are a slowly shrinking segment of the population, being less educated today is far more lethal now than it was just two decades ago. This trend will not have a profound influence on national vital statistics because the proportion of the total population that falls into this category is relatively small and shrinking, but it will be a health challenge nonetheless for a number of people reaching older ages in the years 2040 and beyond.

A second factor that will influence the age structure in this century are advances in public health and the biomedical sciences that are likely to yield improvements in health and longevity. Included among them are continued efforts to reduce smoking prevalence, greater success in the treatment of complications associated with obesity, traction beginning in the battle against the rise of childhood obesity, and anticipated advances in aging science that could yield an extension of healthy life by 2040. It is unclear exactly how these positive developments will manifest over the next 25 years, but the most likely scenario is that the one subgroup likely to benefit first will be those with the highest education, highest income, and overall highest socioeconomic status.

Finally, one of the more interesting developments in shifting American demographics is the anticipated dramatic increase in the Hispanic population, and the unique impact this will have on health and longevity over the next few decades (Table 1). Details of this event are addressed by Hummer and Hayward (2015), but for now it is important to recognize that the proportion of the total U.S. population that is Hispanic will rise from 17.1% today to 28% by 2040. More important, Hispanics now represent only 7% of the population aged 65+, but this will rise to 18% by 2040. Neither of these demographic events would ordinarily be notable, except for the fact that Hispanics represent perhaps one of the more interesting anomalies in American demographics.

Hispanics currently have the highest life expectancy among the main population subgroups in the United States today. Hummer and Hayward (2015) have shown that this is because of the fact that the Hispanic population in the United States today is currently dominated by first-generation immigrants who are known to have healthier lifestyles than either their country of origin or the general U.S. population. This has led to what is commonly known as the Hispanic paradox (Markides et al. 2015)—the unexpected observation that Hispanic immigrants currently live longer than the resident population.

What makes the Hispanic impact on American demographics even more interesting is the likelihood that the health and longevity of this subgroup is on a trajectory to worsen in the coming decades. Why? Because evidence has emerged to indicate that second- and third-generation Hispanics are experiencing notable declines in health caused by the acquisition of increasingly more harmful behavioral risk factors such as smoking and obesity (Hayward et al. 2014). Thus, because Hispanics are about to noticeably increase their presence as part of American demographics, and their future health and longevity trajectory is spiraling downward, there is reason to believe that this will have a notable negative impact on the life and health expectancy of all Medicare-eligible cohorts between now and 2040.

COMPETING RISKS

One of the most important concepts in the field of aging science, which is also either unknown or misunderstood by many, is referred to as "competing risks," and it will have a profound influence on the prospects for both life expectancy and healthy life span for the older population in this century (current and future cohorts). Imagine a cohort of 100,000 female babies born in selected years in the United States, say, in 1900 and 2010. If one applied the observed death rates in those years to these hypothetical babies and plotted out the ages they would all die based on those death rates, one would observe the "age distribution of death" (see Fig. 1).

Note that, in 1900, the death distribution was skewed toward younger ages—this reflects the high infant, child, and maternal mortality observed in that year. However, even in 1900, once individuals survived past the first few years of life, most deaths were then distributed between the ages of 50 and 85 (solid black line). During the course of the 20th century, death rates declined dramatically at younger ages, enabling people who in 1900 would ordinarily have died early in life, an opportunity to survive up to and beyond the age of 65 with regularity. The dashed line illustrates what the distribution of death looks like today. Based on past, current, and projected period life tables, we now know that a female baby born in 1965 had a 78.9% chance of reaching their 65th birthday; today that probability is over 88%, and by 2040 it

Cite this article as *Cold Spring Harb Perspect Med* doi: 10.1101/cshperspect.a025965

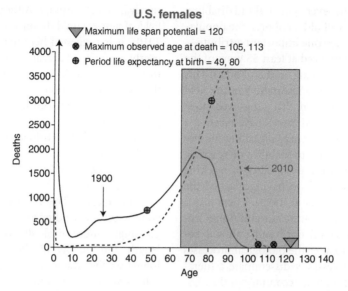

Figure 1. Age distribution of death, period life expectancy at birth, maximum observed age at death, and maximum life span potential for U.S. females in 1900 and 2010. The red shaded box is the "red zone"—an age window in which it becomes increasingly more difficult to influence both length and quality of life. (Image prepared by author based on survival data from the $d(x)$ column of life tables kindly provided to the author by Dr. Steven Goss, Chief Actuary of the U.S. Social Security Administration.)

has been estimated by the U.S. Social Security Administration that >90% will survive to their 65th birthday. However, because death rates are now so low in the United States before age 65, there is not much room left for improvement in survival to this age. This is why the rise in the proportion of the population expected to reach older ages in the future is not going to be far different from what is observed today—it is difficult to improve on a mortality profile that is already so favorable before age 65.

So what exactly are competing risks and why are they important? When death occurred at younger ages in the early 20th century, it was largely a product of communicable diseases. Now that most deaths have been redistributed to the 65+ age window, the price to be paid for extended survival is a dramatic rise in the prevalence of diseases of aging—CVD, cancer, Alzheimer's disease, dementia, and a suite of nonfatal disabling conditions (sensory impairments, osteoporosis, osteoarthritis, etc.). No one would dispute that the trade-off of chronic aging-related diseases for a longer life was worth it, but nonetheless, it was a Faustian trade that

people in the modern era have to live with from now on.

Here is the dilemma. Now that most deaths occur past the age of 65, and given that it is not currently possible to modulate or even measure the rate of biological aging, what is happening in the United States and in other long-lived populations is that older people are now routinely accumulating a broad range of fatal and disabling diseases in their bodies. The longer we live, the more diseases we accumulate. It is as if fatal aging-related diseases are "competing" for our lives in some sort of perverse death battle. If medical technology heroically saves the life of an older person, which happens with regularity, that person lives a bit longer, only to face the next cause of death that is competing for their life.

Because death is a zero sum game, when any one cause of death declines (for whatever reason, including either advances in medical technology or improved behavioral risk factors, such as reductions in smoking or the adoption of a healthy lifestyle), the risk of some other fatal condition must eventually rise! Although there is variability in the severity of fatal and disabling conditions across all older cohorts, the population aged 65+ share one important attribute—they have all experienced at least 65 years of the ravages of biological aging on their bodies. This is the reason why some researchers in the field of aging have suggested that long-lived populations like the United States have reached a point of diminishing returns regarding the rise in life expectancy, the probability of reaching the age of 65, and the probability of reaching older ages conditional on having reached the age of 65 (Olshansky et al. 1990). There just is not much room for significant additional improvement in survival until and unless it becomes possible to alter the biological processes of aging and successfully disseminate a therapeutic intervention that accomplishes this. Using a running analogy, although a number of people have broken the 4-minute mile since this record was first breached in 1954, improvements in running speed have decelerated dramatically since then because of limitations imposed by body design—in the absence of genetically fixed programs that could preclude faster running times.

Thus, there are two reasons why the concept of competing causes is so important to the future of survival and the prospect that the aging rate has already been modified. First and foremost, there is no reason to expect any dramatic improvement in survival to the population aged 65+ between now and 2040. Even the Social Security Administration anticipates no more than a 3.1% improvement in survival to age 65 between now and 2040, and that is with a fairly optimistic set of assumptions about anticipated future declines in death rates before age 65 over the next 25 years. Conditional survival to age 85 from age 65 is also not expected to change much by 2040 for the same reason. This does not mean the absolute number of people reaching ages 65 and 85 will not rise rapidly—that is an inevitable by-product of a shifting age structure.

Once one understands the basic demographics of population shifts and entropy in the life table, you can put to rest the idea that any dramatic improvement in life expectancy or

survival is forthcoming. Although some in the field of aging would disagree with this (Vaupel 2010), it is suggested here that these views are overly optimistic (unrealistic to some) because they fail to take into account anything more than the extrapolation of historical trends into the future, including our basic biology—which precludes extended survival for most people (Carnes et al. 2012; Olshansky 2013).

The second reason why competing causes is so important is perhaps even more intriguing. Note the red shaded box in the age window of 65+ in Figure 1. This is referred to as the "red zone"—an age window in which it becomes increasingly more difficult to influence both length and quality of life. Importantly, the longer we live, the more difficult it becomes to intervene in any positive way. In fact, to the contrary, I contend that if we continue with current efforts to extend life (even marginally) by combatting only major fatal diseases (consider CVD and cancer as our primary targets), we will probably achieve some additional measure of success in reducing case fatality rates for these competing causes, but the price to be paid for success may very well be an expansion of morbidity and a rise in other competing causes that yield more years of frailty and disability. In other words, the current medical model that is designed to forestall death from fatal conditions may yield marginal additional survival time, but it is distinctly possible that the trade-off will be an increase in the number and severity of unhealthy years. Why? Because treatments for major fatal diseases leave unaltered the basic biological processes of aging that give rise to most diseases competing for our lives in the red zone.

To be clear, biological aging gives rise to an increased risk of fatal diseases because aging is the primary risk factor for competing causes, but medicine and current medical technology does not "treat" aging, it addresses only its consequences. This is the reason why a large number of researchers in the field of aging are now suggesting that the time has arrived to attack biological aging itself rather than just its disease manifestations (Miller 2002; Olshansky et al. 2006; Butler et al. 2008; Sierra et al. 2009; Gold-

 Cite this article as *Cold Spring Harb Perspect Med* doi: 10.1101/cshperspect.a025965

man et al. 2013; Kirkland 2013). Unless successful in this effort, the United States and other long-lived populations run the risk that, by midcentury, the population reaching ages 65+ in the coming decades could (overall) experience higher levels of frailty and disability associated with Alzheimer's disease and nonfatal disabling conditions associated with aging. The irony is that this unwelcome scenario would be a product of success, not failure, in combatting major fatal diseases.

CONCLUSIONS

Is there empirical evidence to support the "secular aging rate hypothesis"? That is, have successive generations passing through the same age window experienced declining mortality and extended lives because they are aging more slowly? The absence of definitive biomarkers of aging makes it impossible to empirically test this hypothesis. But evidence presented here shows that death rates have been declining and life expectancy has been rising among older people because of a combination of improved behavioral risk factors (e.g., declines in smoking), advances in medical technology that detect and postpone the onset and age progression of major fatal diseases more efficiently, and because of vastly improved environmental conditions under which people were born and lived their lives. There is no empirical evidence anywhere in the scientific literature to suggest or even intimate that the cause of secular declines in death rates (and rising life expectancy) at middle and older ages are a product of improvements in the rate of biological aging among successive generations across time. It is important to emphasize again that the observed death rate in a population is not a measure of biological aging.

Is there empirical evidence to support the "comparative aging rate hypothesis," suggesting that subgroups of the population age at different rates in a single time period? This is somewhat more complicated. In the absence of a definitive measure of the rate at which biological aging is occurring in individuals, despite alleged biomarkers of aging being measured from Na-

tional Health and Nutrition Examination Survey (NHANES) data, this is currently unknowable. However, there is evidence to suggest that the lifelong accumulation of stress associated with social disadvantage, leads to a measurable accelerated shortening of telomeres (Epel et al. 2004)—one of the potential biomarkers of human aging that may be associated with rate of biological aging. Given that survival rates and the risk of major fatal conditions occur routinely at younger ages for disadvantaged subgroups of the population, the hypothesis by Levine and Crimmins (2014) that shorter-lived subgroups may experience higher death rates because they are aging more rapidly is plausible but currently immeasurable. It is important to recognize that such a hypothesis does not mean that population subgroups are destined to die earlier or live longer than others, it just means that it appears likely that behavioral and environmental risk factors have the potential to accelerate biological aging at the individual level. There are no documented methods of decelerating biological aging as yet, although researchers may be closing in on therapeutic interventions that will do just this (Kirkland 2013), but there are likely to be a large number of aging accelerators (exposure to ionizing radiation, pollutants, possible obesity, etc.). This is the reason why the "comparative aging rate hypothesis" seems plausible, whereas the "secular aging rate hypothesis" does not.

Is there empirical evidence to support the "heterogeneity aging rate hypothesis" that long-lived people and their offspring experience lower mortality and greater longevity because they possess "protective genes" that enable them to escape or delay the ravages of time more efficiently than the rest of the population? Once again, it is not yet possible for anyone to measure the rate of biological aging in individuals, but evidence from empirical research has documented the powerful effect of genetics on survival and mortality (Kirkwood and Austad 2000; Perls 2003). Given the often harmful behavioral habits practiced by many documented centenarians and supercentenarians (Perls and Silver 2008), there is reason to believe that something is protecting them from harmful risk factors that kill most other people at younger ages. It would be difficult to posit anything other than a strong genetic influence on mortality risk, so the "heterogeneity aging rate hypothesis" seems likely.

To summarize, empirical evidence shows that, during the last 115 years, people in general are living longer and healthier lives than at any time in history. The majority of the rise in longevity in the last century is a by-product of declining early age mortality brought forth by public health advances that reduced the risk of death from communicable diseases—a by-product of forces that were external to human bodies. Declines in death rates at middle and older ages since 1965 are largely a product of improved behavior risk factors and advances in medical technology that successfully manufactured survival time that would not otherwise have been possible. There is no evidence to support the hypothesis that people are living longer today because they are aging more slowly than earlier generations. The fact that genetic heterogeneity influences variation in human longevity lends credence to the "comparative aging rate hypothesis" and the "heterogeneity aging rate hypothesis" because their premise is that genetic variation drives mortality risk differentials—a fact well documented in the literature. This conclusion also opens up the door to promising therapeutic interventions designed to slow aging because such interventions would be fundamentally based on leveraging the genetic advantages for health and longevity that already exist among population subgroups.

REFERENCES

Adler NE, Marmot M, McEwen BS, Stewart J. eds. 1999. *Socioeconomic status and health in industrial nations: Social, psychological, and biological pathways.* New York Academy of Sciences, New York.

Atzmon G, Rincon M, Rabizadeh P, Barzilai N. 2005. Biological evidence for inheritance of exceptional longevity. *Mech Ageing Dev* 126: 341–345.

Baker GT, Sprott RL. 1988. Biomarkers of aging. *Exp Gerontol* 23: 223–239.

Beltrán-Sánchez H, Preston SH, Canudas-Romo V. 2008. An integrated approach to cause-of-death analysis: Cause-deleted life tables and decompositions of life expectancy. *Demogr Res* 19: 1323.

Butler RN, Miller RA, Perry D, Carnes BA, Williams TF, Cassel C, Brody J, Bernard MA, Partridge L, Kirkwood T, et al. 2008. New model of health promotion and disease prevention for the 21st century. *BMJ* 337: a399.

Carnes BA, Olshansky SJ, Hayflick L. 2012. Can human biology allow most of us to become centenarians? *J Gerontol A Biol Sci Med Sci* 68: 136–142.

Crimmins E, Saito Y. 2001. Trends in healthy life expectancy in the United States, 1970–1990: Gender, racial, and educational differences. *Soc Sci Med* 52: 1629–1641.

Epel ES, Blackburn EH, Lin J, Dhabhar FS, Adler NA, Morrow JD, Cawthon RM. 2004. Accelerated telomere shortening in response to life stress. *Proc Natl Acad Sci* 101: 17312–17315.

Flegal KM, Ogden CL, Yanovski JA, Freedman DS, Shepherd JA, Graubard BI, Borrud LG. 2010. High adiposity and high body mass index-for-age in US children and adolescents overall and by race-ethnic group. *Am J Clin Nutr* 91: 1020–1026.

Goldman DP, Cutler D, Rowe JW, Michaud PC, Sullivan J, Peneva D, Olshansky SJ. 2013. Substantial health and economic returns from delayed aging may warrant a new focus for medical research. *Health Aff (Millwood)* 32: 1698–1705.

Harman D. 1991. The aging process: Major risk factor for disease and death. *Proc Natl Acad Sci* 88: 5360–5363.

Hayward MD, Hummer RA, Chiu C, Gonzalez-Gonzalez C, Wong R. 2014. Does the Hispanic paradox in U.S. adult mortality extend to disability? *Popul Res Policy Rev* 33: 81–96.

Hummer RA, Hayward MD. 2015. Hispanic older adult health and longevity in the United States: Current patterns and concerns for the future. *Daedalus* 144: 20–30.

Kinsella KG. 1992. Changes in life expectancy 1900–1990. *Am J Clin Nutr* 55: 1196S–1202S.

Kirkland JL. 2013. Translating advances from the basic biology of aging into clinical application. *Exp Gerontol* 48: 1–5.

Kirkwood TB, Austad SN. 2000. Why do we age? *Nature* 408: 233–238.

Kitagawa EM, Hauser PM. 1973. *Differential mortality in the United States: A study in socioeconomic epidemiology.* Harvard University Press, Cambridge, MA.

Lantz PM, House JS, Lepowski JM, Williams DR, Mero RP, Chen J, 1998. Socioeconomic factors, health behaviors, and mortality: Results from a nationally representative prospective study of US adults. *JAMA* 279: 1703–1708.

Levine ME, Crimmins E. 2014. Evidence of accelerated aging among African Americans and its implications for mortality. *Soc Sci Med* 118: 27–32.

Link BG, Phelan JC. 2005. Fundamental sources of health inequalities. In *Policy challenges in modern health care* (ed. Mechanic D, et al.), pp. 71–84. Rutgers University Press, New Brunswick, NJ.

Ludwig FC, Smoke ME. 1980. The measurement of biological age. *Exp Aging Res* 6: 497–522.

Markides KS, Samper-Terment R, Al Snih S. 2015. Aging and health in Mexican Americans: Selected findings from the Hispanic EPESE. In *Race and social problems*, pp. 171–186. Springer, New York.

Miller RA. 2002. Extending life: Scientific prospects and political obstacles. *Milbank Q* **80:** 155–174.

Olshansky SJ. 2013. Can a lot more people live to one hundred and what if they did? *Accid Anal Prev* **61:** 141–145.

Olshansky SJ, Carnes BA, Cassel C. 1990. In search of Methuselah: Estimating the upper limits to human longevity. *Science* **250:** 634–640.

Olshansky SJ, Perry D, Miller RA, Butler RN. 2006. In pursuit of the longevity dividend. *The Scientist* **20:** 28–36.

Olshansky SJ, Hayflick L, Carnes BA. 2008. No truth to the fountain of youth. *Sci Am* **14:** 98–102.

Olshansky SJ, Antonucci T, Berkman L, Binstock RH, Boersch-Supan A, Cacioppo JT, Carnes BA, Carstensen LL, Fried LP, Goldman DP, et al. 2012. Differences in life expectancy due to race and educational differences are widening, and many may not catch up. *Health Aff (Millwood)* **31:** 1803–1813.

Perls T. 2003. Genetics of exceptional longevity. *Exp Gerontol* **38:** 725–730.

Perls TT, Silver MH. 2008. *Living to 100.* Basic Books, New York.

Perls TT, Wilmoth J, Levenson R, Drinkwater M, Cohen M, Bogan H, Joyce E, Brewster S, Kunkel L, Puca A. 2002. Life-long sustained mortality advantage of siblings of centenarians. *Proc Natl Acad Sci* **99:** 8442–8447.

Riley JC. 2001. *Rising life expectancy: A global history.* Cambridge University Press, New York.

Sierra F, Hadley E, Suzman R, Hodes R. 2009. Prospects for life span extension. *Annu Rev Med* **60:** 457–469.

Vaupel JW. 2010. Biodemography of human ageing. *Nature* **464:** 536–542.

Vaupel J, Lundstrom H. 1994. Longer life expectancy? Evidence from Sweden of reductions in mortality rates at advanced ages. In *Studies in the economics of aging* (ed. Wise DA), pp. 79–102. University of Chicago Press, Chicago.

World Health Organization. 2009. *State of the world's vaccines and immunization,* 3rd ed. World Health Organization, Geneva.

Index